Language, Space and Mind

The idea that spatial cognition provides the foundation of linguistic meanings, even highly abstract meanings, has been put forward by a number of linguists in recent years. This book takes this proposal into new dimensions and develops a theoretical framework based on simple geometric principles. All speakers are conceptualisers who have a point of view both in a literal and in an abstract sense, choosing their perspective in space, time and the real world. The book examines the conceptualising properties of verbs, including tense, aspect, modality and transitivity, as well as the conceptual workings of grammatical constructions associated with counterfactuality, other minds and the expression of moral force. It makes links to the cognitive sciences throughout, and concludes with a discussion of the relationships between language, brain and mind.

PAUL CHILTON is Emeritus Professor of Linguistics at Lancaster University.

Language, Space and Mind
The Conceptual Geometry of Linguistic Meaning

Paul Chilton

CAMBRIDGE
UNIVERSITY PRESS

University Printing House, Cambridge CB2 8BS, United Kingdom

Cambridge University Press is part of the University of Cambridge.

It furthers the University's mission by disseminating knowledge in the pursuit of education, learning and research at the highest international levels of excellence.

www.cambridge.org
Information on this title: www.cambridge.org/9781107010130

© Paul Chilton 2014

This publication is in copyright. Subject to statutory exception and to the provisions of relevant collective licensing agreements, no reproduction of any part may take place without the written permission of Cambridge University Press.

First published 2014

Printed in the United Kingdom by Clays, St Ives plc

A catalogue record for this publication is available from the British Library

Library of Congress Cataloguing in Publication data
Chilton, Paul A. (Paul Anthony)
Language, space and mind : the conceptual geometry of linguistic meaning / Paul Chilton.
 pages cm
ISBN 978-1-107-01013-0 (Hardback)
1. Space and time in language. 2. Geometry. 3. Mathematical linguistics.
4. Computational linguistics. I. Title.
P37.5.S65C45 2014
401'.9–dc23   2013047093

ISBN 978-1-107-01013-0 Hardback

Cambridge University Press has no responsibility for the persistence or accuracy of URLs for external or third-party internet websites referred to in this publication, and does not guarantee that any content on such websites is, or will remain, accurate or appropriate.

To my children, Jonathan and Emily

Contents

List of figures	*page* ix
List of tables	xiii
Preface	xv
Acknowledgements	xviii

1	Introduction: space, geometry, mind	1
	1.1 Language and mind	2
	1.2 Formalisation	4
	1.3 Using geometry	7
	1.4 Space, situation and deixis	9
2	Viewpoint, reference frames and transformations	15
	2.1 Physical space: prepositions, deixis and reference frames	16
	2.2 The abstract deictic space	29
	2.3 Further characteristics of the deictic space	42
3	Distance, direction and verbs	50
	3.1 Vectors, discourse entities and reference frames	52
	3.2 Displacement vectors and verbs of motion	60
	3.3 Force vectors and transitivity	71
4	Event types and cognitive operators	106
	4.1 Temporal aspects of happenings: event types	108
	4.2 Tense forms as cognitive operators: instancing and presencing	116
	4.3 Instancing and presencing in the past	131
5	Times, tenses and reference frames	133
	5.1 A present of present things	135
	5.2 A present of past things	140
	5.3 A present of future things	143
	5.4 The putative future: a reference frame solution	151
6	Counterfactual reflections	157
	6.1 Counterfactuality	158
	6.2 *If*-sentences and counterfactual conceptions	159
	6.3 Tense in the modal mirror	163
	6.4 The geometry of *if*-sentences	167

| | 6.5 Through the looking glass: counterfactual *if*-sentences | 173 |
| | 6.6 Concluding reflections | 176 |

7 Reference frames and other minds — 178
  7.1 Epistemic reference frames — 179
  7.2 *That*-ness and other-ness — 180
  7.3 Other minds as reference frames — 183
  7.4 Connections and disconnections across parallel worlds — 200

8 Mental distance and complement clauses — 210
  8.1 Verb meanings and clausal complements — 210
  8.2 The meaning of *that*, *to*, *ing* and zero — 215
  8.3 Constructions with the verb *seem* — 221
  8.4 Further notes on seeming — 227

9 Verbs, complements and their conceptual effects — 229
  9.1 *to* constructions and grammatical subjects — 229
  9.2 Modelling *ing* constructions — 239
  9.3 Modelling zero constructions — 243
  9.4 Overview of alternations and restrictions — 248

10 The deontic dimension — 256
  10.1 Deontic meanings presuppose epistemic meanings — 256
  10.2 Deontic reflections — 260
  10.3 The deontic source — 274
  10.4 Thoughts on *ought* — 277

11 Concluding perspectives — 281
  11.1 Questions — 282
  11.2 Space, the brain and language — 284
  11.3 Deictic Space Theory and the brain — 299
  11.4 Deictic Space Theory and the mind — 305
  11.5 In conclusion: Deictic Space Theory and metaphor — 311

Appendix — 313
References — 315
Index — 330

Figures

Figure 2.1	*in front of*: analysis (i)	page 21
Figure 2.2	*in front of*: analysis (ii)	22
Figure 2.3a	*in front of* as translation	24
Figure 2.3b	*in front of* as translation plus rotation	25
Figure 2.3c	*in front of* as reflection	26
Figure 2.4	Prepositions as vectors (after O'Keefe 2003: 79)	27
Figure 2.5	The fundamental coordinate configuration	30
Figure 2.6	Relative distance from S on d-axis	31
Figure 2.7	Attentional distance metaphorically projects onto temporal distance	34
Figure 2.8	The fundamental deictic space	41
Figure 2.9a	Example (6) John does not own the car	46
Figure 2.9b	Example (7) John does not own a car	47
Figure 3.1a	Example (1) John is in front of the tree	53
Figure 3.1b	Example (2) The tree is in front of John	54
Figure 3.2	Example (7) The linguist is in a good humour	57
Figure 3.3	Possession as position	58
Figure 3.4	Property and entity relation as position vector	59
Figure 3.5	Example (11) Li travelled	62
Figure 3.6a	Example (12) from Beijing to Guangzhou	64
Figure 3.6b	Example (13) to Guangzhou from Beijing	64
Figure 3.7a	Example (14) Li travelled from Beijing to Guangzhou	66
Figure 3.7b	Example (15) Li travelled to Guangzhou from Beijing	67
Figure 3.8	Example (16) From Beijing Li travelled to Guangzhou	68
Figure 3.9	Example (27) The sky reddened	71
Figure 3.10a	Example (33a) Jake gave the code to Bert	77
Figure 3.10b	Example (34a) *inactive* Bert received/got the code from Jake	78
Figure 3.10c	Example (34a) *active* Bert got the code from Jake	79
Figure 3.11	Example (37c) James wiped the counter clean	81
Figure 3.12a	Example (38a) John broke the vase	84
Figure 3.12b	Example (38b) The vase broke	84

List of figures

Figure 3.13a	Conceptual structure of event in (47b)	95
Figure 3.13b	Combining the vectors in Figure 3.13a	96
Figure 3.14a	Example (48a) The lads loaded logs onto the lorry	97
Figure 3.14b	Example (48b) The lads loaded the lorry with logs	98
Figure 3.14c	Combining vectors in Figures 3.14a and 3.14b	98
Figure 3.15a	Passive construction as refocusing of entities	101
Figure 3.15b	Passive construction after vector combination	101
Figure 3.15c	State ascription reading of (49c) The vase was broken	102
Figure 3.16	Role of the m-axis in modelling *build*-verbs	104
Figure 4.1	Geometric schema for states	109
Figure 4.2	Semelfactive event	112
Figure 4.3	Homogeneous activity	113
Figure 4.4	Process type: accomplishment	114
Figure 4.5	Process type: achievement, sentence (5) Hillary reached the summit	116
Figure 4.6	Instancing (simple present operator) operating on a state schema	120
Figure 4.7	Instancing (simple present operator) operating on (i) a stative schema and (ii) a process (accomplishment) schema, resulting in (iii) an 'instance' or 'instant' corresponding to the simple present tense form	122
Figure 4.8	Presencing in the (d, t) plane, applied to a process event	125
Figure 4.9	Effect of *ing* presencing operator on a state event schema	128
Figure 4.10	Insertion of process schema and application of presencing operator	130
Figure 4.11	Progressive (presencing) in the past relative to S	132
Figure 5.1	Timeless simple present for (1a) and (1b)	138
Figure 5.2	Transforming of reference frame for historical present: (5b) In June 1520 Henry sails to Calais	141
Figure 5.3	Present progressive in the past: (6) In 1520 Henry is sailing to Calais	143
Figure 5.4	Example (7a) Henry visits Calais this Thursday	147
Figure 5.5	Frame shift for (8a) Henry is visiting Calais this Thursday	149
Figure 5.6	Example (11) Henry is going to/gonna visit Calais this Thursday	151
Figure 5.7	Example (7) Henry will be visiting Calais this Thursday [non-putative]	153
Figure 5.8	Example (14) Henry will be visiting Calais (now) [putative]	153
Figure 5.9	Example (15) Henry will have visited Calais [putative reading]	155
Figure 6.1	Reflection of time onto modality	164

List of figures

Figure 6.2a	Conditional sentence (1) present tense	169
Figure 6.2b	Conditional sentence (2) past tense	170
Figure 6.3	Modalised apodosis: sentence (7)	171
Figure 6.4	Counterfactual sentence (3): first approximation	174
Figure 6.5	Examples (3) and (12)	175
Figure 7.1	Possibility within probability: (1) John probably has children and it's possible his children are bald	180
Figure 7.2	Example (2a) John knows that Mary wrote the report	185
Figure 7.3	Example (2b) John knows that Mary might have written the report	188
Figure 7.4	Example (3a) John might know that Mary wrote the report	190
Figure 7.5	Example (3b) John might know that Mary might have written the report	191
Figure 7.6	Example (4) John does not know that Mary wrote the report (= it is not the case that John knows that Mary wrote the report)	192
Figure 7.7	Example (5a) John believes that Mary wrote the report	194
Figure 7.8	Example (5b) John believes that Mary might have written the report	196
Figure 7.9	Example (6a) John might believe that Mary wrote the report	197
Figure 7.10	Example (7b) John disbelieves that Mary wrote the report	198
Figure 7.11a	Example (8) Hob believes that a witch has blighted Bob's mare	202
Figure 7.11b	Part of sentence (8) Nob believes she has killed Cob's cow	203
Figure 7.11c	Coordinated spaces: (8) Hob believes that a witch has blighted Bob's mare, and Nob believes she has killed Cob's cow	204
Figure 7.12a	Normal representation of other mind communication	207
Figure 7.12b	Possible autistic representation of other mind communication	208
Figure 8.1	*It seems that* construction: (5) It seems that Mary wrote the report	224
Figure 8.2	*Seem to* construction: (6) Mary seems to have written the report	226
Figure 9.1	Subject-control structure (equi NP, subject): (1) John expects to write the report	231
Figure 9.2	Object-control structure (equi NP, object): (2) John urged Mary to write the report	234

xii   List of figures

Figure 9.3	Example (3) John persuaded Mary to write the report	236
Figure 9.4	Raising to object: (4) John expects Mary to write the report	238
Figure 9.5	Example (5) John imagined writing the report	241
Figure 9.6	Example (8) John imagined Mary writing the report	244
Figure 9.7	Example (10a) John saw Mary write the report	247
Figure 10.1	Base axis system and reflected copy	262
Figure 10.2	Obligation expressions	264
Figure 10.3	Permission and exemption: example (8)	270
Figure 10.4	Conceptual structure of *may*-prohibition: removal of prohibition	274
Figure 10.5	Example (1a) Mary must write the report	276
Appendix Figure	Two degrees of central embedding	314

Tables

Table 5.1 Correspondences between present-tense forms and deictic time reference *page* 134
Table 8.1 Syntactic description crossed with zero, *to*, *ing* and *that* constructions (sample) 212

Preface

This book explores a theoretical format I call deictic space. The term *deictic* comes from the Greek word that means 'to point'. Humans are probably unique among primates in their ability to point. They point in order to establish joint attention with other humans. It is impossible to point meaningfully unless one is in a certain position, and one's interlocutor is aware of that position. One's pointing is relative to a spatial frame of reference. Deixis, shifting points of view, frames of reference, are fundamental to human communication.

Arrows conventionally stand for pointing in a certain direction. Linguists are always using arrows in their diagrams but often in highly abstract ways (as in the re-write rules of Phrase Structure Grammar). Mathematicians use arrows too, standing for vectors, which have distance and direction, within coordinate systems. Frames of reference are needed in order to navigate our environment. They are also needed for directing our actions on that environment using our limbs, primarily arms and hands. Reaching, grasping, pointing and the visual attention that guides them depend on frames of reference.

Abstractly, we can think of navigating, reaching, pointing and attention in terms of geometrical vectors in frames of reference. This is essentially what the book sets out to explore. The starting point for this exploration is in the conceptualisation of space as organised by language. The most obvious spatial expressions in language are prepositions but from there we can proceed to far more abstract conceptual spaces, speculating as to how far elementary structures and operations that geometry has developed can assist us. Basic Euclidean geometry can be regarded as embodied, related to human experience in relation to the space, earth, direction and motion.

The book is full of diagrams of coordinate systems that are meant to evoke the abstract three-dimensional space that I call deictic space and which I think may be the most fundamental part of our language ability. I hope these will not cause the reader too much double vision. I use them not only because the visual is sometimes clearer than the verbal but also because visual cognition (and its cross-modal versions) is so much a part of our spatial experience. The diagrams are based on very elementary geometrical ideas and these have some standard logical implications that make it possible to explore the ways in which spatial conceptualisation might – or might not – be part and parcel of our language-based conceptualisation. But since I argue that simple spatial representations can lead to abstract and complex meanings in language, some of these diagrams do end up complicated. I can only beg the reader's patience. The difficulty of 'reading' some of the diagrams is a reminder that after all these are mere attempts to model complex operations that our mind–brains handle with unconscious fluidity.

Just how far we can go in this exploration of deictic space remains an open question. But I suggest in this book that we can go a considerable distance in relating some of our most abstract language-based conceptualisations to an origin in physical space. Here is a rough route map for the book, and some reasons why I take certain paths.

Linguistics has developed various formal metalanguages. Since the one I develop in this book is unusual in some respects and is heavily dependent on diagrams, Chapter 1 gives some initial motivations for adopting and adapting the key geometrical notions of coordinates, transformations and simple vectors. The most obvious application of geometrical description to language concerns literal spatial expressions, primarily prepositions. Chapter 2 develops in more detail the basic geometrical ingredients of the book. It opens with a survey of the geometrical element in spatial prepositions, though geometry is certainly not all there is to their meaning. The chapter crosses an important threshold – moving to a geometrical space that does not refer to physical space but to three dimensions of the mind that are woven into language – the three dimensions of attentional focus, time and reality assessment. The subsequent chapters explore this space, moving into increasingly abstract conceptual spaces that are linked with grammatical constructions.

Chapter 3 is at one level about the phenomenon of attention and the ways in which linguistic constructions act to direct it. At another it is also about arrows and axes, that is, vectors and coordinates, and the ways in which some of their routine properties can be used to capture the schematic conceptual meanings of predicates. The chapter again begins with the modelling of spatial expressions and moves into progressively more abstract meanings of verbs and their grammatical frames – a line of enquiry that returns in Chapters 8 and 9. First, however, Chapters 4 and 5 look into two closely related characteristics of verbal meaning – the conceptualisation of types of event over time and the placement of events in a temporal frame of reference. The purpose in these two chapters is to see if we can bring this area of linguistic enquiry into a unifying geometrical approach, hoping that along the way this approach itself will shed light on the linguistic phenomena themselves. Chapter 6 pursues this latter goal by applying a key geometrical idea that is already found in the description of some prepositional meanings – the mirror transformation of axes.

What happens if we look for such transformations in grammatical structure? Unexpectedly, it turns out that counterfactual conditionals can be so described, though controversially. Chapters 8 and 9 continue to explore transformations of axes – embedded translated axes – as a way of modelling the idiosyncratic behaviour of verbs in relation to types of complement clause. This line of argument broadly follows one line of cognitive linguistics that sees complementisers and complement clauses as conceptually motivated. Chapter 10 returns to the modelling of counterfactuals that was laid out in Chapter 6, entering into what is the furthest limit of abstraction I have chanced addressing in this book, deontic meaning. This is not the first time in cognitive linguistics that the abstractions of deontic meaning have been found to be rooted in the concrete, but I have attempted here to unify the account with the abstract-geometrical approach, with potentially controversial implications.

All this is based on the spatial hypothesis and some admittedly risky theoretical speculation. What is the theory doing? Is there any evidence that the linguistic mind–brain actually works this way? In Chapter 11, I briefly address some philosophical issues, or perhaps merely raise them for linguists to consider. I also take a snapshot of rapidly developing areas of neuroscience and neurolinguistics that may corroborate some of the theoretical speculations of the book, or at least provide further food for thought.

Acknowledgements

This book has been written over a number of years. I thank my family for their forbearance during absences both physical and mental during that time. Because the preparation and writing have been protracted, the final result reflects many sources, influences and encounters. In some cases a pointed question at a conference, a passing word of curiosity or an email enquiry has encouraged me to continue. I have also been fortunate to have colleagues and students who have engaged with me in what must have seemed peculiar linguistic ideas. I am especially grateful to colleagues at the University of East Anglia, the University of Lancaster, the University of Neuchâtel and the University of Łódź, and in particular to Bill Downes, Clive Matthews and Gabriella Rundblad for their ideas and critique. At Lancaster I benefitted from the knowledge, understanding and kindness of the late Anna Siewierska as well as from the interest shown by a number of other supportive colleagues and graduate students. I am also grateful to Louis de Saussure at Geneva and Neuchâtel, and to Piotr Cap at Łódź for ideas, insight and generosity. Vyv Evans has been supportive at various points in my explorations of spatial meanings. Thanks are due also to Bertie Kaal, Monika Kopytowska and Christopher Hart, who have discussed my ideas with me in detail and taken them further, and in different directions, in their own work. John O'Keefe, of UCL's Institute of Cognitive Neuroscience, has been a source of inspiration through his pioneering work on spatial cognition. He proposed vector grammars some time before I embarked on DST and was kind enough to discuss my initial ideas with me.

Many other colleagues, too numerous to name, who work in various branches of cognitive linguistics, have given me time, ideas, encouraging comments and critical insights: I am grateful to them all for conference questions, passing comments, emails and conversations. A very early stage of the theory presented in this book focused on social discourse rather than linguistic structure and received interest from colleagues and good friends around the world in the field of discourse analysis. Although DST may not have turned out as they expected, I thank them for many helpful ideas and much personal support. I also wish to thank two anonymous reviewers of my

Acknowledgements

manuscript for their sharp eyes and advice, as well as my understanding copy-editor and the team of editors at Cambridge University Press. All remaining errors and blunders are mine.

In the preparation of the present book I revised portions of previous peer-reviewed publications and used them in certain chapters. I am grateful to the following publishers for granting permission to use these materials.

John Benjamins for permission to reuse of parts of 'Vectors, viewpoint and viewpoint shift: toward a discourse space theory', *Annual Review of Cognitive Linguistics*, **3**, 2005: 78–116, in Chapter 1 (parts of Sections 1.2.1 and 1.4) in Chapter 2 (parts of Sections 2.1 and 2.2) and Chapter 3 (Section 3.1), and for the reuse of part of 'Geometrical concepts at the interface of formal and cognitive models: *Aktionsart*, aspect and the English progressive', *Pragmatics and Cognition*, **15**(1), 2007: 91–114, in the first part of Chapter 4.

De Gruyter for permission to reuse parts of 'The conceptual structure of deontic meaning: a model based on geometrical principles', *Language and Cognition*, **2**(2), 2010: 191–220, in Chapter 10.

Oxford University Press for permission to reuse 'Frames of reference and the linguistic conceptualization of time: present and future', in *Time: Language, Cognition and Reality*, edited by K. Jaszczolt and L. de Saussure, 2013, pp. 236–58. Revised portions of this material appear by permission of the publisher in Chapter 2 (Section 2.3.1), Chapter 4 (Section 4.2) and Chapter 5 (Section 5.3).

I am also grateful to the following rights holders for granting permission to quote short extracts from various works as epigraphs to certain chapters.

John Wiley & Sons for permission to quote from Wittgenstein's *Philosophical Investigations*, translated by G. E. M. Anscombe, 1953.

Taylor & Francis Books for permission to quote from Merleau-Ponty, *The Phenomenology of Perception*, translated by Colin Smith, 1962.

Penguin Books for the quotation from St Augustine's *Confessions*, translated by R. S. Pine-Coffin, 1961.

Professor Catherine Elgin for permission to use an extract from Nelson Goodman, 1947, 'The problem of counterfactual conditionals', *Journal of Philosophy*, **44**(5), 1947: 113–28.

Penguin Random House for permission to quote from Proust's *In Search of Lost Time: The Guermantes Way*, translated by C. K. Scott Moncrieff and Terence Kilmartin, revised by D. J. Enright, 2001.

Princeton University Press, who hold the copyright of the translation by Edwin Curley, 1996, of Benedict de Spinoza's *Ethics*.

Mercure de France for permission to use an extract from Pascal, *Pensées*, edited by Philippe Sellier, 1976.

1 Introduction: space, geometry, mind

This book is about the geometry of linguistic meaning. It outlines a theory the full sense of which depends on its geometrical formalism. This is a simple but hopefully consistent formalism that most people will recognise from high-school geometry. It is perhaps an approach that may strike readers as odd. Yet I think it is a perfectly natural way to approach the concerns that linguists, and especially cognitive linguists, have had for decades, in particular the relationship between space, cognition and language. If spatial concepts play a crucial role in linguistic meaning, then geometry ought to be a useful way to describe these meanings. What is more, the linguistic literature is full of hints that geometrical modelling is a natural way to go. Our textbooks and monographs are replete with arrows, axes, and words like *perspective*, *location*, *direction* and *distance*. Why multiply vague English expressions when we have a well-defined tried and tested mathematical formalism? Even more can be said in its favour. For geometry itself, and its standard notation, is motivated in a way that cognitive linguists are familiar with. It is, in its Euclidean form, based on human bodily experience. And it is Euclidean three-dimensional geometry that we shall use. True, multidimensional vector spaces are needed if we are trying to model meaning in an information-science or a connectionist framework (see for example Widdows 2004) but we are trying here to model meaning in a cognitive embodied framework, and our fundamental spatio-cognitive understanding is in fact three-dimensional, thus natural even when spatial cognitions are projected to non-spatial realms.

This is not a book that attempts mathematical proofs of the claims expressed in the diagrams, though it is not ruled out that a more rigorous mathematical demonstration would be worth attempting. Nor does the book use algebraic formulations, although in some cases the complex abstract relations expressed in the diagrams could be more simply displayed in that way. The guiding principle has been to rely on our visual intuitions about geometrical figures, always seeking to be as consistent and clear as possible. Of course, the concern is not just to have a neater, more encompassing, better motivated notation or diagramming system. The more one explores three-dimensional figures, the more it seems that this kind of diagram is able to capture some deep facts about words and constructions and perhaps also to resolve some puzzles or pseudo-puzzles.

1.1 Language and mind

We do not study language, or a language, only for the sake of knowing more about language or languages, but because language, any language, tells us something about some of the workings of the human mind, a view that is now common among cognitive linguists. This is not quite as straightforward as it might seem and needs a little clarification. The term 'mind' is deliberately chosen: studying language does not tell us about the brain, at least only in an extremely indirect manner. Investigating language tells us something about the sorts of things the human mind does or can do. I certainly do not mean that mind is independent of brain, but investigating language does not tell us directly about neurons and neuronal networks. And by investigating language, I do not mean investigating languages or investigating examples of the use of language in attested contexts.

Language is a tool for stimulating silent conceptualisation, and I do not mean only conscious conceptualisation. What sort of conceptualisations is language capable of stimulating? If we look into this question we are going to find, if we are lucky, only a small part of the conceptual activity that takes place in our minds. The term 'conceptualisation' is being used here in preference to 'meaning', since this term is often connected to the view that language elements (lexical items, constructions) have meaning in some independent language system in the mind. However, along with other cognitive linguists, I am adopting the view that there is no independent linguistic–semantic inventory, but rather that there is conceptualisation that can happen independently of language, but which language accesses. I also take the view that the particular structure of a language – its lexis and grammar – does not privilege or make more accessible or delimit some particular kinds of conceptualisation. That is, I think there is no clear evidence for the strong Whorf hypothesis – that, for example, a classifier language will make speakers more sensitive to, or restrict speakers to, certain kinds of conceptualisation. This is not to say that some empirical work does not give convincing evidence of certain limited effects (Lucy 1992, Levinson 1996, 2003), but these effects appear not to be extensive or not sufficient to make us think that linguistic meaning is something separate from a universal conceptualising ability.[1]

---

[1] It may be different if one's focus is on discourse. It is simpler and faster to code those conceptual distinctions that one's lexical or grammatical structure encodes but the generative properties of language mean that one is not limited to those structures in communicating conceptualisations of all sorts. If anything constrains them, it is cultural practice, which is also not an absolute constraint.

This view is of course different from the cognitive view inspired by Chomsky and his followers, according to which language is autonomous, modular, interfaced with a semantic component. I am not, however, throwing out the idea that the language system may be modular in some sense – but it is not sealed off or encapsulated (in the sense of Fodor 1983): rather, it is linked inextricably to non-linguistic cognitive abilities.

It is also different from denotational semantics, which is why I use the term 'semantics' throughout this book sparingly, usually to refer to the natural concepts associated with lexical items, that is, their 'semantic frame' (Fillmore 1982b). Since the 1980s cognitive scientists and linguists have realised that the meanings of words and constructions in languages are in some way built up on the basis of our embodied experience of physical space (Lakoff and Johnson 1980, 1999, Johnson 1987, Pinker 1997). Pinker sums up some of the conceptual elements that are important for the geometrical approach:

Space and force pervade language. Many cognitive scientists (including me) have concluded from their research on language that a handful of concepts about places, paths, motions, agency, and causation underlie the literal or figurative meanings of tens of thousands of words and constructions, not only in English but in every other language that has been studied ... These concepts and relations appear to be the vocabulary and syntax of mentalese, the language of thought ... And the discovery that the elements of mentalese[2] are based on places and projectiles has implications for both where the language of thought came from and how we put it to use in modern times. (Pinker 1997: 355)

'Space' and 'force' are crucial to the approach developed in the present book, because they can be, and commonly are, formalised in geometric terms, as vectors. The nature of the linkage between spatial cognition and the structures of language is similar to metaphorical linkage: spatial, motor and visual systems have some sort of counterparts in the linguistic–communicative system. For example, humans can attend to detail or take in a gross gestalt, and linguistic constructions and lexical items also permit this kind of alternating focus. Depth perception enables us to judge one object as more distant from the self than another object opposed to another: similarly, linguistic structures can place information in the foreground or background that is, topicalise some information relative to other information (cf. Talmy 2000 [1983], Langacker 1995, Croft and Cruse 2004). Spatial expressions, and the spatial perceptual–cognitive systems, provide a source domain for linguistic expressions of time, in many, perhaps all, languages (Lakoff and Johnson 1980, 1999, Haspelmath 1997, Boroditsky 2000, Evans 2004, Evans and Chilton 2010).

---

[2] Or 'language of thought', Jerry Fodor's hypothesis that thinking is executed in a mental symbols system. It is not relevant to address the plausibility of that hypothesis here.

Such views as these are not uncommon amongst linguists. However, the apparatus that I am proposing is an unusual one in linguistics, though it is hinted at in many places, as will become clear by way of the references that I shall give as we move along. Moreover, the formal apparatus itself, as will be seen, is one that springs from natural foundations – namely, the fact (I am assuming it is a fact) that spatial location and orientation are experiences that organisms must cognise for their survival. We shall see, using the apparatus that I will outline, that these concepts of physical space can be used for other kinds of fundamental concept that languages use as their bedrock and about which they enable humans to communicate among one another. This idea is not new in cognitive linguistics (for example, Lakoff and Johnson 1980, 1999, Langacker 1987, 1991, Talmy 2000; cf. also 'localists', for example, Anderson 1971, Lyons 1977) or even outside it (cf. Jackendoff, for example, 1976, 2002a, Frawley 1992). What I'm suggesting is that there is a derived conceptual foundation that uses spatial cognition of various kinds, and that language in turn uses it as the most fundamental requirement for communication. Actually, this means that language is not simply a 'window' on the mind, but that investigating language and conceptualisation together, using the apparatus I describe below, will tell us something about both. It is perhaps not surprising that we need to view language through different kinds of apparatus and that indeed language may, after all, have components and modules that that require different instruments of investigation.

1.2   Formalisation

Linguistics is bedevilled by a proliferation of formalised notations and diagrams, the most systematic of which are bound up with propositional and predicate calculus and other elements of mathematical logic and reasoning. Others are ad hoc, intuitive and iconic. A few have origins in Euclidean geometry.[3]

1.2.1   *Discourse Representation Theory*

Possibly the most developed of discourse theories is the Discourse Representation Theory (DRT) of Kamp and Reyle (1993), a theory that has had some influence on the present approach. However, DRT does not claim to be cognitively motivated, although Kamp does occasionally maintain the cognitive *relevance* of DRT (see also Asher and Lascarides 2003: 376ff.). It is true

---

[3] This section is based on part of Chilton (2005).

that Kamp's project does resolve major problems, specifically anaphora, indefinite NP reference and definite NP reference, much debated in the logic-oriented tradition of twentieth-century philosophy. It is also true that in the more recent work, such as the SDRT (Segmented Discourse Representation Theory) of Asher and Lascarides (2003) and 'dynamic' versions of the theory (see for example Kamp *et al.* 2011), DRT goes still further in explaining various phenomena of discourse coherence and context dependence. However, the DRT apparatus itself and its newer incarnations may be constrained by their predicate calculus foundations: in any event, they do not appear to incorporate systematically the concepts of situatedness or embodiment. Discourse Representation Theory does not, for example, handle deixis in a naturalistic fashion, limiting deixis to objects in the non-linguistic context, similar to the way anaphoric referents are treated. For the essentially deictic concepts coded by linguistic tense, DRT also treats 'times' as referents more or less on a par with other discourse referents ('yesterday' has equivalent status to 'Peter' or 'the donkey'), which seems counter-intuitive. 'Yesterday' is the 'location', relativised to the speaker, at which, for instance, Peter beat his donkey, not a participant in that event. In the present theory, by contrast, it is assumed that temporal and spatial – and also modal – deixis are fundamental; consequently, the model integrates them into the representation of all utterances in discourse. It is the relativisation of 'yesterday' to another time point, that of the time of speaking, that is left out in DRT, yet this fact is precisely an intrinsic feature of discourse.

1.2.2 Mappings

Set theory and functions, alongside predicate calculus, have been fundamental to formal semantics and pragmatics. Within cognitive linguistics itself, set theory and functions have been used in a somewhat different way. Mappings across cognitively defined sets of one kind and another are systematically used in mental space theory (Fauconnier 1994, 1997), in blending theory (Fauconnier and Turner 2002), and more informally in conceptual metaphor theory (Lakoff and Johnson 1980, 1999). The strength of the concept of mapping lies in its potential mathematical clarity, which enables it to model the phenomena of mental spaces that were first precisely identified by the authors just mentioned. As in the case of predicate logic, however, there is no inherent connection between sets-and-mappings and cognitive or linguistic processes that are situated and embodied. Specifically, they do not incorporate deictic phenomena. The geometrical approach I am putting forward in this book owes much to the mental spaces approach, but recasts mappings across spaces as coordinate correspondences on three fundamental dimensions – three dimensions that we will spend some time explaining in the chapters to follow. This means that any point in the three-dimensional deictic space has three coordinates with respect to the three axes defining the space in which such a point occurs. Axis systems can be nested inside the base axis system, as will be seen, and points can be defined by their coordinates within a nested axis system. It is a logical consequence of the geometry that correspondences can be tracked across systems, relatively to the base system – or that what we might call in less abstract terms, the 'base world' or 'base reality' of the deictic centre, the speaker.

1.2.3 Diagrams

Diagrams are ubiquitous in linguistics, but in cognitive linguistics they arguably have a distinctive place in that particular approach to language. Talmy's detailed accounts, for instance, have this geometric quality (for example the account of path concepts in Talmy 2000 [1985]), as does Langacker's consistent use of pictorial diagrams (Langacker 1987, 1991, etc.). Langacker-diagrams, which are designed to capture a range of intuitions about linguistically encoded meanings (e.g. 'foregrounding', 'prominence', 'trajector', 'distance'), can probably be transposed into standard geometrical concepts. For instance, Langacker-diagrams generally encapsulate topological relations, directionality, relative distance and scalar magnitude. It is important that there is also a claim that the iconic diagrams capture *non*-linguistic perceptual or conceptual phenomena, with a particular emphasis on vision.

The most recent advocate of a more explicitly geometrical approach in cognitive grammar is Croft (2012), who develops 'geometric-cum-graph structures' diagrams for the purpose of describing lexical aspect (*Aktionsart*), the time-related event structure that is part and parcel of verb semantics, and also for the purpose of describing causal event structure of verbs such as *break*. These diagrams resemble familiar two-dimensional Cartesian coordinate diagrams. They use two clearly defined axes, a time axis and a q-axis, the latter providing a place for separate qualitative states in event structures (Croft 2012: 53–7). This system has the advantage over earlier theoretical models of making the time course of verbal aspect explicit and easily visualised, yielding a richer and more coherent description of the conceptual frames associated with the verbs of a language.

In his account of complex structure of causative verbs, Croft (2012: 212–17) says that he is using three-dimensional diagrams, the third dimension being the causal relation between two phases of a causal chain. However, the diagrams themselves do not show three dimensions defined by three axes, as one might expect. And a further detail appears: the placing not only of qualitative states q on the second axis but also participant entities associated with each q. In fact, the diagrams that model the causal structures of the sentence *Jack broke the vase* consist of two two-dimensional diagrams for two subevents: one for Jack making an impact, one for the vase's state of being broken. In addition 'the causal chain linking the individual participant subevents is represented in a third dimension'. It is not clear how this holds together geometrically – how the third dimension relates geometrically to the first two. It is also not clear how this third dimension, one reserved specially for causal relations, is to be defined. In Chapter 3, I attempt a different kind of geometrical modelling for causative verbs, bringing in another concept that comes for free with geometry, the concept of force vectors.

There are some similarities between Croft's geometric diagrams and the ones I develop in this book.[4] Both Croft's and mine use a time axis and an orthogonal axis relating to the structure of events. But there are also some significant differences. Whereas Croft's q-axis is specifically concerned with the stages in the unrolling of events (e.g. the successive states in an event such as the one denoted by the verb *melt*), my second axis provides coordinates for discourse entities, i.e. the participants in event structure. The most significant difference is that the abstract geometric space I work with is explicitly three-dimensional. As will be explained in Chapter 2, my geometric model has an essential third axis, which is epistemic.

1.3 Using geometry

Whereas Croft, and to a lesser extent Langacker, borrow elements of geometry to model meanings of linguistic expressions, the present book starts at a more fundamental level. To begin with, I attempt to apply geometry to the sorts of meaning that one would expect geometry to be well equipped to describe – spatial meanings. Then, building on the insight of many linguists that spatial meanings are somehow crucial for many kinds of more abstract meaning, I look into the potential of geometry for describing apparently non-spatial meanings, which may nonetheless be derived from, or linked to, spatial conceptualisation. This in turn means taking seriously certain elements of Euclidean (and Cartesian) geometry and developing out of them a descriptive model of language-based conceptual space. The elementary geometry of coordinate systems, vectors and transformations then becomes a heuristic for exploring linguistic–conceptual space as well as a descriptive model-building project.

The theoretical ideas that I develop in the present book draw on insights and methods found in the broad approaches outlined in the preceding sections,

---

[4] They were developed independently.

as well as many other meaning-based approaches to language. What I am putting forward is a research framework that investigates the applicability of coordinate vector geometry for the description and perhaps explanation of certain kinds of linguistic meaning. I am not suggesting that *all* linguistic meaning can be described or explained in this way, only that the range of lexical–constructional meanings that can be described in this way is wider than linguists may have thought.

There is nothing new about the use of coordinate geometry to account for spatial meanings, for example in prepositions and deictic expressions. Bühler, in writing of the 'deictic field', clearly has Cartesian coordinate geometry in mind. His description of what he calls 'the here-now-I system of subjective orientation' begins starkly:

> Let two perpendicularly intersecting lines on the paper suggest a coordinate system to us, 0 for the origin, the coordinate source ... My claim is that if this arrangement is to represent the deictic field of human language, three deictic words must be placed where the 0 is, namely the deictic words *here*, *now* and *I*. (Bühler 1990 [1934])

Bühler's idea that deixis is one fundamental aspect of human language and has something to do with self, space and time leads naturally to geometric modelling. It is in some ways also the starting point for the Deictic Space Theory that I shall outline in this book. However, the space in question will be three-dimensional and the three axes will be defined conceptually in ways that are distinct from Bühler's ideas.[5] Bühler also spoke of 'displacement'[6] of the deictic centre away from the here and now of utterance (Bühler 1990 [1934]: 136–57). As will be seen in later chapters such deictic displacements can be modelled geometrically as transformations of coordinate systems. In more recent linguistic research, spatial expressions that relate objects to reference points other than self, particularly in spatial prepositions, have also been treated in explicitly geometric terms.

Highly abstract geometry and topology have been used to model language phenomena by Thom (1970) and Petitot (e.g. 1995). Gärdenfors (2000) argues for the geometrical representation of conceptualisation. Gallistel (1990) argues that low-dimensioned geometries are fundamental in the vertebrate nervous system. Of particular relevance is the empirical work of Paillard and Jeannerod amongst other neuroscientists, which shows that visuo-spatial

---

[5] The geometric approach may be in the background to the terminology used by Appolonius Dyscolus, as suggested by Dalimier (2001). Apollonius and the Stoics drew a broad distinction between content expressions and deictic expressions, as Bühler mentions.

[6] In Bühler (1990 [1934]) the translation is 'transposition', the translation used by Levinson (2003: 51). Bühler used *Versetzung* or *Verschiebung*, the latter term being associated with contemporary psychology. Bühler (1990) is an abridged and translated extract from his 1934 work.

processing and actions such as reaching and grasping depend on the brain's ability to work with egocentric coordinates systems whose origin is located at different parts of the body (shoulder, hand, retina, for example) and in different sensory modalities (Paillard 1991, Jeannerod 1997).[7] This work builds on other important empirical work (Ungerleider and Mishkin 1982, Goodale and Milner 1992, Milner and Goodale 1995) that indicates two pathways in visual processing, one relating to the identification of objects and the other locating objects in body-centred frames of reference. Such finding are highly suggestive both for investigating the linguistic phenomenon of deixis and for using a geometrical approach to linguistic description.

While the work mentioned so far focuses on egocentric coordinate systems, experimental findings of O'Keefe and Nadel (1978) point to the brain's ability to represent allocentric and absolute maps of an organism's spatial environment, as noted by Levinson (2003: 9–10). Levinson argues strongly for the need to recognise non-egocentric systems in the spatial semantics of certain languages. However, there need be no conflict between the two research emphases. It would appear likely that both egocentric (deictic) coordinate systems and non-egocentric systems are instantiated in neural structures and mental representations, including linguistic ones.

I shall not attempt to explore this body of scientific work further, nor attempt to apply any but the simplest geometrical concepts. Sufficient has, I hope, been said to justify the scientific appropriateness of a geometric approach in cognitive linguistics.

1.4 Space, situation and deixis

In the study of spatial semantics the most precise use of geometry, in particular of coordinate systems, has been Levinson (2003). As will be clear in Chapter 2, geometrical descriptions of linguistic expressions referring to physical space provide a foundation for the present book, though the aim is to move into an abstract conceptual geometry that can handle linguistic expressions and constructions that are not spatial in any obvious sense.[8]

This book is not concerned with spatial expressions as such – and it is important to emphasise this at the outset. But there is now a rich body of research into the precise nature of spatial meaning in language and the precise nature of the relationship between language and spatial cognition (see Evans and Chilton 2010 for a sample of recent theoretical and experimental work). In this work, while geometric aspects (or geometric descriptions) of spatial cognition are well established, geometry is not the whole story. In the case of

---

[7] Bühler was already aware of the shifting origo on the body: Bühler (1990 [1934]: 146–7).
[8] Parts of this section are based on Chilton (2005).

spatial prepositions, which have the most obvious links to spatial cognition, it is now widely recognised that for spatial prepositions a coordinate geometry alone is not adequate, though it is needed and is in some sense fundamental. Many linguists and psycholinguists have shown the crucial role played also by the shape, presence or absence of surface contact and the function of reference objects. What is loosely called 'topology' is also relevant (e.g. for prepositions like *in* which involve a concept of bounded space), although topology is indeed a kind of geometry (Herskovits 1986, Vandeloise 1991, Carlson-Radvansky *et al.* 1999, Coventry *et al.* 2001, Tyler and Evans 2003). It is also the case that when prepositions (e.g. *in front of*) locate objects relative to a reference point, it is not actually a point that is at issue, but a spatial region around the reference landmark. Such conceptualised regions are affected by the function and shape of the landmark itself (see for example Carlson 2010).

But coordinate systems (or reference frames) remain the foundation of the cognitive system underlying spatial orientation. The crucial point is that a rather simple geometrical formalism is an economical way, and indeed a natural way, of describing fundamental spatial meanings. On a more abstract level, what coordinate geometry enables us to do is to analyse 'point of view' or 'perspective', both in a physical–spatial and in various abstract sense. As is now recognised (see for example Tomasello 1999), one of the reasons why language is so complex and apparently redundant if we look at it from a truth-conditional position may well be that language 'is designed' to permit viewpoint alternation. Euclidean geometry is a well-understood way to analyse space the way humans experience it. This book is proposing that it is also a natural way to describe non-spatial linguistic meanings that are nonetheless in some way rooted in human spatial experience.

Once we introduce the geometry of coordinate systems it is a natural step to introduce vectors – mathematical objects conventionally visualised as arrows that have (i) direction and (ii) magnitude. In a coordinate system the position of a point can be given by the length and direction of a vector from the origin to the point. This approach can be used for explicating the denotation of certain prepositions, including *in front of*, by specifying a vector space in which all vectors have the same origin in some coordinate system (O'Keefe 1996, 2003, O'Keefe and Burgess 1996, Zwarts 1997). This space will be the 'search domain' within which an object can be said to be, for example, 'in front of John'.

Cognitive linguists often assert that linguistic meaning is situated and/or embodied (e.g. Lakoff and Johnson 1999, Croft and Cruse 2004, and, from a philosopher's perspective, Johnson 1987). To be situated means, for human language users, that our use of language will always assume and/or refer to the place and time of speaking, and will take a perspective on the surrounding

1.4 Space, situation and deixis

physical environment. The most striking way in which situatedness is built into the 'design' of language is the phenomenon of deixis. The etymological connection with the human gesture of pointing is revealing in the cognitive perspective, suggesting as it does the embodied nature of linguistic expression. Pointing itself is a way of establishing joint attention (as Tomasello emphasises: 1999, 2008) and can only work because the direction of the extended arm has an origin in the body of the pointer, who in turn has a spatial location relative to the interlocutor. In linguistic expression, the origin is the speaker, and a pointing gesture may or may not be present. Crucially, deixis is not solely a spatial matter. The definition offered by Lyons gives the essence of the concept but mentions explicitly only the further aspect of time:

By deixis is meant the location and identification of persons, objects, events, processes and activities being talked about, or referred to, in relation to the spatiotemporal context created and sustained by the act of utterance and the participation in it, typically, of a single speaker and at least one addressee. (Lyons 1977: 637)

Spatial and temporal deixis, as reflected in words like *here*, *there*, *now* and *then* and demonstratives such as English *this* and *that*, may indeed be the most frequent and salient deictic expressions in languages. Spatial deixis, however, may be more fundamental, because it is capable of metaphorical projection into abstract domains. Linguists have also added distinct, but related, deictic categories. Person deixis, implicit in Lyons's definition, is expressed by pronouns, particularly the first and second pronouns, whose referential meaning switches according to the speaker. Social 'distance' and status have also been held to be forms of deixis.[9]

But there is another dimension of deixis that may be considered fundamental. In the prototypical situation of utterance the persons, objects, events, processes and activities being talked about are not only positioned relative to the speaker, by the speaker. Situated speakers also take an epistemic stance. They assess the likelihood of the existence of persons and objects, the likelihood of the occurrence of events, the plausibility of persons and objects having certain properties, having certain thoughts and intentions, and so forth. And speakers do this by reference to what they take to be real, i.e. known to them individually. Pragmatically no utterance is modally innocent,[10] whether modal expressions are present or not in the utterance. An utterance without modal morphemes is taken to be an assertion of a known (to the speaker) reality. Expressions of degrees of certainty and possibility (expressions such

---

[9] On accounts of deixis after Bühler, see in particular Lyons (1977), Jarvella and Klein (1982), Fillmore (1997 [1975], 1982a), Levinson (2004).
[10] See Recanati (2007: 71) on the idea of 'simple' sentences that are neither grammatically nor conceptually modalised.

as *might*, *may*, *must*, etc.) are thus relative to a fixed reality point in the speaker's mind and can be regarded as deictic. A deictic definition of modality is outlined in Frawley (1992). The epistemic dimension is the third fundamental dimension in the three-dimensional space I shall develop in Chapter 2.

These three dimensions are always 'anchored' at the speaking self or in the mind of the speaking self. It is only when we use language in context – that is, in discourse – that we produce deixis. For this reason, I have sometimes called the theory outlined in this book Discourse Space Theory (DST), but because deixis is the overarching concept of the theory it makes better sense now to call it Deictic Space Theory. Throughout the book I shall simply refer to it as DST.

The three-dimensional space of DST is an abstract space. However, the theory is grounded in the embodied cognition of physical space, Euclidean three-dimensional space as it is, or may be, represented by the human nervous system (cf. O'Keefe and Nadel 1978, Gallistel 1990, Paillard 1991, O'Keefe 1996, 2003, O'Keefe and Burgess 1996, Jeannerod 1997, Gärdenfors 2000). This approach has two important implications. The first is that in order to model and explore the way language is grounded in the experiential situatedness of human individuals, we need a notation with which to work. The second implication is that any such notation will need to be well motivated, by which is meant here that it will need to be more or less directly linked with the way we model the behaviour of human cognition and perception in general.

The main aim of this book is to propose that rather simple vector geometry in coordinate systems allows us to model some fundamental properties of human language – or, to be more precise, the kinds of conceptualisations that human languages prompt their speakers and hearers to use in communication. The aim is not to account for all properties of human language, but to explore those aspects that the geometrical approach illuminates, and these properties we will expect to be those that are adapted to the situated use of language.

One way of seeking to justify the proposal is to investigate the extent of its efficiency for descriptively capturing the linguistic data – the usual procedure for any proposed linguistic theory (or any other theory, for that matter). However, it has long been argued that this is not enough, and in cognitive linguistics, we should want our descriptive apparatus to go some way towards interfacing with the descriptive approaches of neighbouring sciences, such as cognitive science and neuroscience.

This does not mean that cognitive linguists need to be specialists in two disciplines, but it does mean that we should try to propose descriptive means that can be assessed by the appropriate specialists. One way to do this is to use

1.4 Space, situation and deixis

the common language of mathematics, in the present instance the mathematics of space, namely, geometry. The chapters that follow try to demonstrate that there are advantages in thinking in terms of coordinate systems, one of the key advantages being that this enables us see the individual cogniser and speaker as situated at the intersection of different dimensions. What the DST model itself consists of is an *abstract* three-dimensional space in which we use geometric (also known as Euclidean) vectors[11] to represent (i) spatial locations (position vectors), (ii) movements (translations) and (iii) forces. The use of vectors in linguistics is also not entirely new. It is position vectors that are used by some authors to model (at least partially) the spatial regions picked out by spatial prepositions (Zwarts 1997, 2003, 2005, Zwarts and Winter 2000). O'Keefe, a cognitive neuroscientist, uses vectors for describing spatial prepositions, which he links with the neural mechanisms of location and navigation (O'Keefe and Nadel 1978, O'Keefe 1996, 2003, O'Keefe and Burgess 1996; also Burgess 2002, Burgess *et al.* 2003). Zwarts (2010) uses force vectors to model certain types of spatial preposition that involve more than relative position and Wolff uses force vectors to model causative verbs, making more mathematically explicit Talmy's (2000 [1988]) iconic drawings illustrating 'force dynamics' (Wolff and Zettergren 2002, Wolff 2007).

It is important to make clear from the outset that I am talking here not only of lexical expressions that conventionally encode meanings having to do literally with physical positions, movements and forces. The meanings of the words and constructions that we shall investigate are related metaphorically to physical spatial meanings and many are usually thought of as 'abstract', if not as meaningless or purely functional. In fact, the three-dimensional conceptual space in which we shall operate is itself derived metaphorically from humanly cognised physical space. The idea that abstract meanings are metaphorically related to basic physical spatial ones is scarcely a new idea in cognitive linguistics, but the point here will be to use elementary geometry to describe the spatial and trans-spatial meanings at issue. This exploration will bring together and offer a unified account of, an unexpectedly diverse range of lexical and grammatical phenomena.

A property of coordinate systems, which we shall make use of, is that they can be transformed: that is, the origin can be shifted relative to the base set of coordinates. There may be limitations on what we can do in this way. In general, however, adopting the type of geometric formalism just outlined does

---

[11] I stress *geometric* vectors. High-dimensional abstract vectors are essential to connectionist models of language (Neural Theory of Language, Latent Semantic Analysis). But the present approach stresses elementary geometric vectors in three dimensions because of their rootedness in the human cognition of physical space. Indeed, one can speculate that the origin of abstract linear algebra lies in the visual and kinaesthetic experience of space, an example of the embodied origin of mathematics not noted by Lakoff and Núñez (2000).

manage to integrate the notions of situatedness, embodiment and speaker–hearer orientation, by taking, as its starting point, the speaker in a physical orientation to physical space as perceived and conceived by human beings.

As already stated, the theory is built on the idea that not only any and all utterances and their associated conceptualisation are relativised to a minimal communication pair (speaker and hearer) but also that they integrate three conceptual dimensions. These dimensions are scalar: attentional distancing from S, temporal distancing from S and epistemic-modal distancing from S. Here and throughout the book S stands for speaker, self and the experiencing subject. The first dimension, referential distancing, is based on the perceptual and cognitive fact that in binocular vision, we see objects in perspective, as closer or more distant, foregrounded or backgrounded. Moreover, we can apply attentional cognition to these objects. Similarly, in conceptualising entities we are talking about, whether they are physical and visually perceptible or not, we can conceptualise such entities as conceptually closer and/or of greater attentional interest. The second dimension is temporal distancing from S. I will not attempt to resume the entire history of thinking about time, in philosophy, theology, physics and linguistics. More will be said in Chapter 2, but to anticipate now, the temporal dimension is framed in DST as relative distance from the *now* of S. There is an important area of discussion that will be touched on in Chapter 2 concerning the relationship between linguistically mediated conceptualisations of time, non-linguistic conceptualisation of time and 'real' time, the time of the space-time continuum studied by physicists (see for example the important work of Jaszczolt 2009). Suffice it here to note that DST claims that all utterances are not only events that occur in real time, but also always represent states, processes and events that are time-anchored relative to the time of utterance by S. The third dimension is epistemic distancing relative to S. The claim here is that epistemic cognition is as fundamental as the other two dimensions and is also necessarily involved in the conceptualisations associated with any and all utterances that represent states, processes and events. The epistemically modal dimension is modelled in DST in a particular manner that will be explained in more detail Chapter 2. Essentially, the idea is that what is closer to the cognising subject is more evident, more certain, while what is more distant is less certain, even counterfactual.[12] The claim is, again, that in modal cognition is always present in the conceptualisations associated with any and all utterances that represent happenings and states of affairs. Indeed, the three dimensions are integrated in such conceptualisations.

---

[12] The modal axis is thus linked with evidentiality, despite the common view that epistemic and evidential modalities are distinct. See, however, Jaszczolt's (2009: 39–40) arguments for linking them.

2 Viewpoint, reference frames and transformations

> How does it come about that this arrow ➔ *points*? Doesn't it seem to carry in it something beside itself? – "No, not the dead line on paper; only the psychical thing, the meaning, can do that." – That is both true and false. The arrow points only in the application that a living being makes of it.
> Ludwig Wittgenstein, *Philosophical Investigations*, p. 454

It is impossible to have no viewpoint, at least in the physical sense. As we move around the physical world the angle between our line of sight and various objects to which we attend varies continually. In practice, it is not only the line of sight that is at issue, but the coordinate systems located, so to speak, at various points on our bodies – for example, our limbs, particularly our hands, which are designed to grasp and tweak. But we can have a 'hearpoint' also, since we detect direction of sound sources too, if less finely with our ears than with our eyes. Viewpoint is thus, to a certain extent, a misnomer but we shall keep it because line of sight for sighted people is the probably the dominant perceptual modality.

From this premise we can go in various directions, which are not spatial or physical at all. One direction is temporal. For example, it is possible to adopt a 'view' of one's present situation by imagining oneself expecting the present this morning, or last year. English enables us to talk about this by means of its tense system, as when we might say, 'This time last year, I was expecting to be in New York.' We can do the same with the future, as we see it at some particular time point, as in 'On 2 June next year John will have been given his degree certificate.'

But more abstractly, and in fact more commonly, when we say we adopt a viewpoint, we mean something that is neither spatial nor temporal. Idiomatically, in English and other languages, when we speak of our point of view we mean something roughly like 'opinion'. What is the connection between such an abstract notion as 'opinion' and physical viewpoint? The vision-is-cognition metaphor is well known: *I see what you mean* is metaphorical (Lakoff and Johnson 1980, Sweetser 1990). There are indefinitely many physical viewpoints. Opinions in this sense are complex cognitions dependent on all kinds of cultural factors. But a fundamental ingredient would seem to be the degree to which an individual judges some proposition or set of propositions to be true (or real) or false (or unreal). This is the sense in which the present chapter relates epistemic judgements to spatial and temporal viewpoint. From the next chapter on shall we shall leave the purely spatial almost entirely behind. In the present chapter, however, we review some of the theoretical and empirical work in linguistics that has investigated the language–space interface. This will lay the groundwork on which is built the DST framework.

2.1 Physical space: prepositions, deixis and reference frames[1]

The geometrical notion of 'reference frames' (or 'frames of reference') has been important in the study of spatial prepositions, in particular in the precise form used by Levinson (2003). Experimental work has shown the relevance of this approach (Carlson-Radvansky *et al.* 2002, Carlson 2010). We shall suggest ways in which frames of reference are not only constitutive of spatial meanings in language but can be seen to constitute an abstract space essential for explaining many kinds of non-spatial conceptualisations that language gives us.

For modelling the meanings of spatial prepositions geometry may well not be enough. Coventry and Garrod (2004) review geometric models of spatial prepositions and convincingly argue that these accounts are insufficient alone. On the basis of experimental evidence they propose that two further kinds of cognitive input to the appropriate understanding and production of spatial prepositions are required: (a) the spatial properties of reference objects themselves and (b) the motor schemas for human interaction with such objects. However, despite the need for supplementation, the geometric component (whether axial or topological), if not sufficient, seems to remain necessary.

Research into spatial prepositions in general indicates that geometry is essential for their description and, furthermore, that humans use geometries in some form in applying prepositional expressions, filled out, however, by other types of cognitive schemata. As far as DST is concerned, the implication is that geometry is not merely an arbitrary theoretical means of description. If humans use geometrical schemas for spatial concepts, it is worth speculating that geometry is involved in abstract representations. Indeed, with regard to all kinds of spatial concepts, this is a key assumption behind cognitive metaphor theory. There is a particular class of prepositions known as 'projective' prepositions and it is this class that is of special significant for

---

[1] This section uses parts of Chilton (2005).

the DST because all analytical accounts make use of 'reference' frames in one form or another. Reference frames are simply three-dimensional axis systems with the origin located either at ego or some other referent. They are thus intrinsically deictic and by extension intrinsically express 'point of view'.

For this kind of preposition, we need not only three orthogonal axes but geometric transformations (translation, rotation and reflection) of axes, as we shall see for the spatial expression *in front of*. It is this class of prepositions that is especially pertinent to the fundamental conception of DST. Such prepositions are essentially deictic: they depend on the orientation of the speaker and of reference objects in relation to them – and vice versa. Another way of saying this is to say it depends on 'point of view', the point of view of the self or of some other object (or person). The reflection transformation can be thought of as representing an 'opposing' point of view. Taking terms like 'point of view' and 'opposing' in a metaphorically abstract sense, we have grounds for speculating that a kind of 'metaphorical geometry' may not be irrelevant to understanding a certain range of conceptualising phenomena that arise with many linguistic expressions.

The literature on spatial expressions in language, especially prepositions (and adpositions in general), is abundant and growing and it is not to the point to review it all here (for an overview see Evans and Chilton 2010). While geometrical modelling of spatial semantics is certainly not the whole story, certain geometrical approaches are well established. It is these that ground the abstract spaces of DST and it is these we focus on in this section.

2.1.1 *Projective prepositions: egocentric and allocentric*

What we have said so far assumes an egocentric and unitary set of axes, but it is well known that the human representations of space operate with multiple axes and can shift from one to the other.[2] In fact different brain areas appear to be responsible for the different spatial coordinate systems and their transformations (see for example Gallistel 1990, 1999, 2002, Andersen 1995, Colby and Goldberg 1999, Burgess 2002).

In cognitive linguistics, shifting viewpoints are frequently mentioned and they are fundamental. It is also well known that language enables us to encode a shift from an egocentric speaker's viewpoint, to an allocentric viewpoint, i.e. there is a shift of deictic centre or geometric origin. Langacker (1995) and van Hoek (2003) go a long way in demonstrating the extent of such phenomena in language, using Langacker-style pictorial diagrams. However,

---

[2] Levinson (1996, 2003) considers the phenomenon cross-linguistically and has demonstrated also the existence of absolute (geophysical) frames of reference (e.g. in Tzeltal and Guugu Yimithirr).

geometrical ideas and formalisms provide much of what we need in this regard, and more. If we adopt coordinate systems as our basis, then translation of axes, a standard operation, is implicit in the geometric framework, and gives us a natural way to describe viewpoint shift and related phenomena. We might indeed hazard the guess that the viewing organism is computing an analogous operation, and the linguistic coding of such shifts can be represented quite simply as vector spaces in alternate coordinate systems. Familiar cases such as (1) can be handled in this way:

(1)   John is in front of the church.

Since some buildings, in some cultures, have an orientation calqued on human orientation, (1) is ambiguous.

The conceptual alternation one may experience in reading (1) corresponds to the switching from a coordinate system centred on the human speaker to one whose origin is positioned on the landmark object (the church) and whose axes are directed relative to it. What is the relation between the two coordinate systems? In one of the interpretations of (1), the interpreter is not adopting a viewpoint located at the church; the base coordinate system (based on speaker's and interpreter's real-space location) is not actually shifted. This is the egocentric viewpoint, where ego is the speaker, S, who is oriented toward (i.e. is facing) the church and *John*'s position is between the oriented speaker and the church. In an alternative conceptualisation of (1), *John* is located relative to the church's orientation – specifically, relative to the church's 'front'. This interpretation is an allocentric representation: the centre of the axis system is located at the distal church, not at the speaker S. But the church still remains in the space viewed by the speaker. In the straightforward case, the church itself is located by a vector whose tail is at the origin, S, of the base system, that is, the world as viewed by the speaker.[3]

However, in many real situations, no such vector can be computed. For instance, one may say 'Go into the piazza; the statue is in front of the church', without either speaker or interpreter needing to calculate the orientation and distance of the church's coordinate system with respect to their own. A shift to an allocentric representation is clearly involved in such an example. Each person has their own reference frame, grounded in their bodily orientation, which moves and turns with them. The point remains that egocentric and allocentric spatial coordinate systems are a natural part of the human cognitive equipment and that linguistic expressions cue one or the other depending on contextual factors in the real world of discourse processing. In any event, we should not be limited by the single example of (1). In fact, when

---

[3] Axis systems are in fact anchored at various origins of the human body in spatial cognition: e.g. eyes, effector limbs.

2.1 Physical space

allocentric spatial viewpoint shifts are constructed in discourse, the object to which axes are shifted are not always in view of the interlocutors and their position relative to the interlocutors may not be known and is not interpreted:

(2)   Go to the end of the bridge: the church is in front of you.

We might imagine that something like the following cognitive operations occur during processing of (2). (i) The hearer, H, can locate the bridge in his own mental representation of physical space (egocentric representation). (ii) To understand *to the end of the bridge*, he then has to initiate an allocentric representation in which *the end of* is defined relative to the bridge, such that the bridge is aligned with a path that has a beginning and end, i.e. a vector whose tail is at the start of the bridge. Note that this arrangement will not be entirely allocentric, since which end of the bridge is selected as the starting point depends on the way the bridge is positioned in the egocentric space represented in (i). Then, (iii), H has to decide the interpretation of *the church is in front of you*. The hearer has now mentally shifted his own (egocentric) axes to the end of the bridge, indeed has aligned his sagittal axis with the structural axis of the bridge itself and the location of the church is given by *in front of* interpreted relative to the egocentric axis system of H. Presumably, then, understanding (2) requires several representations involving axis shift, and these representations are not only sequential but nested.

The point of these examples is to show how discourse seems to trigger axis shifts from egocentric to allocentric representations of physical space. But shifts of 'viewpoint' occur also in the domain of abstract discourse processing, as Fillmore, Langacker, Talmy and others have amply demonstrated. What later chapters of this book aim to do is to show how a projection of these mechanisms may underlie key aspects of semantic processing at the more abstract level that linguistic forms appear to prompt.

2.1.2   Axes within axes and transformation of axes

The notion of transformation will be of importance throughout this book. It should not of course be confused with the term 'transformation' as it was used in the early transformational-generative grammar of Chomsky. The sense in which we use it here is taken entirely from geometry, and the claim in DST is that geometrical transformations correspond to cognitively real operations. We can see this very clearly in the case of spatial terms, such as prepositions, which is why we are spending time on prepositions in this chapter. Spatial conceptualisation is the foundation of more abstract conceptualisations.

As Levinson (2003: 25–34) shows, the terminology can be confusing in that terms such as egocentric, allocentric, relative and deictic are used in the different disciplines (philosophy, psychology and brain sciences, for example, as well as linguistics) that take an interest in spatial representation in language. In his own classification of reference frames Levinson discards the terms 'egocentric' and 'allocentric' but I shall retain them in the sense that will emerge in what follows. One reason for this is that in the study of linguistic constructions in general, many cognitive linguists and developmental psychologists believe that language (and language use) reflects and facilitates perspective-taking, that is, the conceptual adopting of the viewpoint of another speaking subject. This kind of alternation can be understood, as indeed it is also by Levinson (2003: 44–7, 84–92), in terms of geometric transformation of coordinate systems. Levinson's important and influential account of linguistic expressions for horizontal spatial directions proposes three main reference frames. These are 'intrinsic, 'relative' and 'absolute' reference frames. I shall not be concerned with absolute reference frames, which are coordinate systems whose axes are defined relative to the geographical environment. I shall, however, discuss 'intrinsic' and, more extensively, 'relative' coordinate systems, combined with the egocentric–allocentric alternation, while emphasising the role of transformations between different systems.

Levinson uses the term 'frames of reference' in his main account of spatial expressions (Levinson 2003: 34–56). The term is also used by experimental psychologists. In this book I use also the terms 'coordinate system' or 'axis system', to make clear the connection with geometry. It is important to note that psychological experiments indicate that geometrical axes and transformations are alone insufficient to account for the range of semantic judgements prompted by the projective prepositions *in front of*, *to the left/right of*, *over/under* and *above/below* (Coventry and Garrod 2004: 91–112). Investigations show that such judgements also involve the function of located objects, knowledge about both located object and reference object and functional properties of the situation in which such judgements are made. But these findings do not mean that the axis systems are not operative – in fact it appears to be the case that they are a necessary component. Axis systems and their transformations are needed for modelling spatial cognition and spatial language. I will show in the main part of this book that the geometry needed for projective spatial prepositions is very similar to the geometry that is needed for an insightful modelling of much more abstract conceptualisations related to point of view and non-actual 'worlds'. We may speculate about the connection between the physical–spatial and the highly abstract mental representations we shall look at – and there are plenty of claims about the metaphorical projection from the spatial to the abstract in cognitive linguistics. We need not pursue such speculation at this point.

Let us turn to the details of the geometrical component of the conceptualisations associated with the projective spatial prepositions. As we have noted

2.1 Physical space

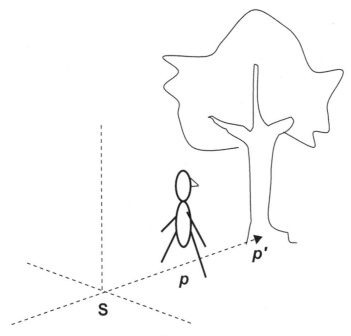

Figure 2.1 *in front of*: analysis (i)

in the previous section, people (and many living creatures) are represented as having intrinsic axes, oriented in the direction of their typical motion. Vehicles and some kinds of building (though they do not move) are also treated as having intrinsically oriented axes. But not all objects are so treated and in such cases people and their languages exhibit some interesting conceptual phenomena that have not always proved straightforward to describe (Levinson 2003: 1–23). To see this, let us consider a slightly different version of (1) in (3) below:

(3)   John is in front of the tree.

This example will illustrate different conceptual types of what Levinson terms the 'relative frame of reference' (Levinson 2003: 44–7, 85–9). The prepositional phrase in sentence (3) can be analysed either as Figure 2.1 or as Figure 2.2.

Figure 2.1 shows the conceptualisation in which the coordinates are centred on the speaker S; the locandum (the object to be located, here *John*) is between S and the landmark location. Geometrically, there are two points *John* and *tree*, call them p and p' respectively, and these points are such that the distance from S to p is smaller than the distance from S to p'. In this

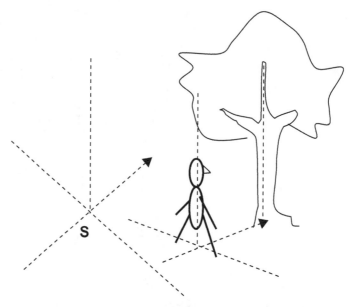

Figure 2.2 *in front of*: analyis (ii)

understanding of (3) the point *p* is positioned in S's axis system, i.e. *John* is positioned relative to S in terms of distance; the point *p* has to be in this position, to satisfy the meaning of *in front of*. Thus the coordinates within which this reading of (3) emerges are egocentric coordinates, taking S as ego. As for John's coordinate system and orientation, these do not make any difference, though the diagram indicates *John* in a specific orientation ('facing' the tree).[4]

Turning to the speaker's orientation, in real life the speaker S of (3) may be facing in any direction (including toward the tree), as suggested in Figure 2.2. In such a situation, the sentence may be acceptable if John is bodily oriented 'facing' the tree. Putting things geometrically, we can truthfully assert (3), if his sagittal axis is a vector with its head located on the vertical axis of the tree and has a length that is variable within relevant limits, but we will not be communicating his precise position in a region around the tree, since trees have no intrinsic 'front' (or 'face') of their own. What seems to be happening is that John's axes are 'projected' onto the tree, whatever his position around the tree. Unless the hearer H of (3) can see the situation, H's mental representation based on the linguistic utterance alone, will be indeterminate – H will not know

---

[4] Geometrically, the orientation of John's axes relative to S's axes is definable in terms of the angles between the intersection of axes in the two systems.

2.1 Physical space

exactly where in the region John is positioned, only that he is 'facing' the tree. This is the case however H and S themselves are actually oriented.

What I have described so far is alternative mental representations of the spatial expression *in front of* and the point is that there are alternative reference frames, some of which are dependent on real-world spatial relativities and some of which are mentally represented or 'virtual' by way of language. This is, however, not the end of the story. I have spoken of John's axes being 'projected' onto the tree. What does this mean cognitively and geometrically?

It might be said that, denotationally speaking, it is sufficient to say that the meaning of *in front of* is given by requiring that John's sagittal plane includes the central vertical axis of the tree (and similarly for any locandum and location). However, in the English expression the tree has a *front* and this needs to be explained – at least if we want a cognitively motivated explanation rather than one that assumes all linguistic expressions to be arbitrary.

The problem (for the expression *in front of*) is that trees and similar objects, unlike vehicles and certain kinds of building, have no intrinsically directed axes that give them 'fronts' or 'faces'. How, then, does the tree appear to get directed axes in the expression *in front of the tree*? Levinson's explanation (2003: 85–9) is that S's axis system is copied under three possible kinds of transformations onto the tree. Once the tree gets some directed axes, they are like intrinsic axes so far as John's relation to the tree is concerned, oriented in such a fashion that the tree has a 'front'. This requires a certain kind of geometrical description, which is discussed below. First, however, the question arises as to whose coordinate system is projected. It is seems to be equally possible for it to be that of S or that of the locandum (here John). If this is the case, then it may be that two kinds of conceptualisation involving projected axes are possible in practice, one egocentric (that of the speaker) and one allocentric (the one attributed by the speaker to the locandum). Here, I shall discuss only the former possibility, since it seems to give rise to additional geometrical and conceptual possibilities.[5]

In Figure 2.3a we have a translation transformation: S's sagittal plane runs through the tree's vertical axis and a copy of this axis system is translated shifted onto the landmark, giving a new virtual origin S' at the tree. Since the set of translated axes is directed, it gives the tree its own sagittal axis and a 'front' with respect to which John can be positioned.[6] John's orientation may

---

[5] The following account broadly follows Levinson's account (2003: 85–9), which is based on transformation of the speaker's axes.
[6] Strictly speaking, his position has to be a position defined by a set of position vectors whose tails are defined by an angle subtended at the origin (the tree). The same point is relevant to the cases discussed below.

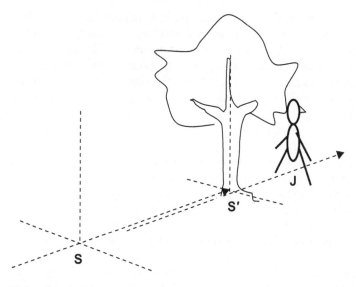

Figure 2.3a *in front of* as translation

be in any direction. In fact, this kind of conceptualisation, though possible, is not typical for English *in front of*: the situation in Figure 2.3a is more likely to be expressed as *John is behind the tree*.[7] In this type of conceptualisation it is a copy of S's coordinate system that is translated, not that of the locandum (John). S's base system remains. In this sense, we seem to have an egocentric rather than allocentric conceptualisation, though once the translation is effected the result 'feels' conceptually allocentric.

In Figure 2.3b a copy of S's axes is first translated so that their origin S' is at the tree and then rotated through 180°, with the result that the tree has a virtual orientation that gives it human handedness as well as a front. This virtual arrangement of axes requires that S's sagittal plane runs through the tree. John's position is given in relation to the virtual axes mapped onto the tree, whichever way he is oriented. He can be said to be *in front of the tree* if he is positioned in the appropriate region between S and S', as indicated in the diagram. Again, we can analyse this conceptualisation as involving an egocentric transformation, though the result may 'feel' allocentric.

However, English speakers often are unsure what they mean when they say something like 'John is to the right of the tree'. Is it the tree's right or the speaker's right? This is evidence that speakers naturally use also another type

---

[7] Languages such as Hausa do, however, seem to involve conceptualised translation transformations such as that shown in Figure 2.3b for the equivalents of *in front of* (Hill 1982).

2.1 Physical space   25

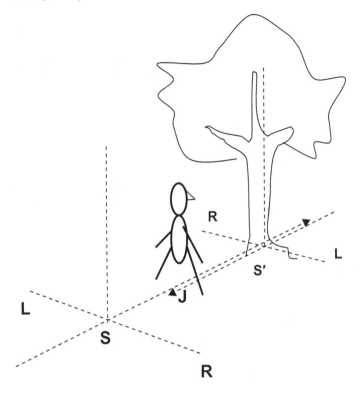

Figure 2.3b *in front of* as translation plus rotation

of transformation of axes, a reflection transformation. A reflection transformation is a mapping that changes the initial set of axes into its mirror image. All points on the axes are mapped perpendicularly onto a mirror plane. This results in a reversed image of S's directed sagittal axis, while from S's point of view left and right stay as they are.

Again, as for the translation-plus-rotation transformation, the initial coordinate system can be either at S, provided his sagittal axis runs through the tree, or at John, providing his sagittal axis runs through the tree. But for John to be in *any* orientation and still 'in front of the tree' we have to assume that it is S's axes that are reflected, as shown in Figure 2.3c. The mirror plane has to be assumed to lie halfway between S and the tree, in order that the reflected origin S' is sited on the tree. As before, a conceptualisation of this type can be analysed as egocentric but may be experienced as allocentric because the virtual axes are centred at the distal location, the tree.

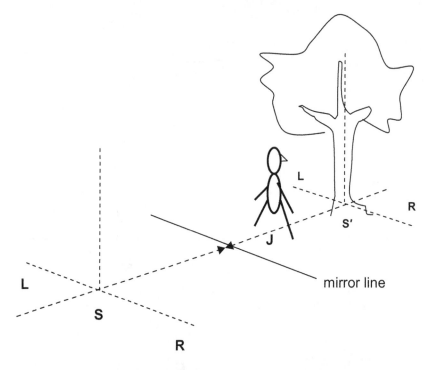

Figure 2.3c *in front of* as reflection

The rotation and reflection conceptual models are reminiscent of the experience of meeting another individual face-to-face sometimes called a 'canonical encounter' (Clark 1973), an embodied and situated experience that may in part account for why humans have conceptualisations of the kind described here. The main point, however, is that in order to describe the basic meanings of spatial expressions geometrical analyses are crucial, and are probably part of the neural architecture need for organism's self-positioning and navigation in physical space. I have entered into some detail, because the idea of coordinate transformations will play an important part in later chapters of this book – under a highly abstract conceptual definition, it must be emphasised, rather than with reference to three-dimensional physical space.

2.1.3 Vectors

The use of vectors to describe the workings of spatial prepositions is less widespread, although vectors are a natural complement to coordinate systems. It is possible to use vectors to define a truth conditional semantics for

2.1 Physical space

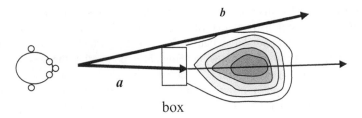

Figure 2.4 Prepositions as vectors (after O'Keefe 2003: 79)

prepositions (as Zwarts 1997, 2003 and Zwarts and Winter 2000 have done), but O'Keefe's work is of particular interest because it is grounded in the modelling of the neurological correlates of spatial cognition in the mammalian hippocampus. According to O'Keefe, the firing of hippocampal 'place neurons' represents a probabilistic model of 'place regions', such that places are represented as concentric tear-shaped regions whose centre is located at the head of a vector whose tail has coordinates at the point of perceiving organism.

For example, O'Keefe defines the English preposition *behind* as the set of vectors V such that any member of V has its tail at the deictic origin, is greater in magnitude than a reference vector drawn from the origin to the object, and has an angle smaller than that between the reference vector a and the vector b tangential to the reference object (see Figure 2.4). In O'Keefe's account, the region is structured in such a way that there is a greater probability of finding the object to be located (corresponding to the NP complement of the preposition) at the centre of the region. Experimentally, if people are asked which points in space best satisfy the subspace denoted by *behind the box*, the acceptability rating is highest in the peaked central region.

Similar vector/region representations can be established for most of the other spatial prepositions, using a three-dimension coordinate system. Prepositions pick out regions in all three dimensions of Euclidean space as perceived by humans. The geometric intersection (origin) is equivalent to the deictic centre, i.e. the speaker of the current utterance in a discourse – and it is this that gives one aspect of speaker orientation, embodiment and situatedness. Human embodied cognition imposes orientation on the three axes, by coordinating the gravitational signal, the sagittal back–front axis and its orthogonal complements in the horizontal plane, i.e. the left–right axis. First-order predicate logic and propositional calculus cannot easily accommodate such things (as argued, for instance, by Landau 2003).

Carlson (2010) also approaches conceptualisations of *in front of* experimentally and demonstrates that the region to which *in front of* corresponds is three-dimensional and somewhat more internally variable than the theoretical vector space implies. It does not logically contradict the Zwarts or the O'Keefe vector approach but adds detail and complexity. It should be noted that Carlson's approach also assumes the existence of coordinate systems ('frames of reference'). Carlson bases her experimental investigation on Herskovits's (1986) suggestion that there are 'ideal' or 'underlying' meanings of spatial prepositions that are geometrical, these being modified by contextual features including, for instance, typical functional interaction with the landmark object; Carlson also notes there are similar ideas in Vandeloise (1991), Tyler and Evans (2001, 2003) and Coventry and Garrod (2004). An important development of the vector approach is the 'attention vector sum model' of Regier and Carlson (2001). The experiments based on this model draw on findings that perception of overall direction is represented neurally as a vector sum and show that judgements about the applicability of the preposition *over* to different spatial configurations of two objects also depends on whether the focus of the attentional beam is wide or narrow.

Two important and well-known riders need to be added. First, not all prepositions utilise all three dimensions: English *at*, for instance, uses one, denoting a point, *across* uses two. Second, as argued by Herskovits (1986), for example, additional image-schematic concepts may be required for a complete description of prepositional concepts. Similar points are made by Vandeloise (1991 [1986]) and by Tyler and Evans (2003). However, it is possible that the meanings expressed by prepositions all have a conceptual component that can be geometrically described, even if for certain cases geometry is not sufficient for complete description. Again, in DST it is important to realise that we are not modelling spatial expressions as such but applying the most general properties (frames of reference or coordinate systems), drawing on some of their mathematical properties and exploring their applicability in understanding some of the abstract conceptualisations that languages afford their users. The vector space approach does not claim that the entire semantics of prepositions can be so specified, since a full account seems to require a 'functional' semantics as well (Vandeloise 1991 [1986]; Carlson-Radvansky *et al.* 1999, Sinha and Jensen de Lopez 2000, Coventry *et al.* 2001, Tyler and Evans 2001, 2003, Zwarts 2003: 43, Coventry and Garrod 2004 Carlson 2010).[8]

---

[8] Prepositions like *in(side)* and *out(side)* are often said to involve a concept of 'containment' or to be loosely 'topological' (Herskovits 1986, Vandeloise 1991 [1986], Levinson 2003); the prepositions *on* seems to require contact and support (Herskovits 1986, Tyler and Evans 2003). Topology is of course a kind of geometry. Further, Zwarts (2010) shows how *in(side)*, *out(side)*, Dutch *aan* and *op* and English *on* can be modelled by force vectors, again a

2.2 The abstract deictic space

It is extremely important to emphasise that the three-dimensional diagrams we shall use in subsequent chapters of this book (which we will call Deictic Space Models or DSMs) are not the three-dimensional diagrams we have used above in discussing the preposition *in front of*. The axis systems that appear in Figures 2.1, 2.2, 2.3a, 2.3b and 2.3c represent the three dimensions of physical (Euclidean) space; the axes that constitute DSMs represent an *abstract* space, with three dimensions that will be defined below. To be sure, DST regards its three dimensions as corresponding to cognitive dimensions *derived* from spatial cognition, but they are not representations of physical space as such.

What we have shown in this section is that geometry provides coordinates, coordinate transformations and vectors for the description of physical spatial conceptualisation in language. The same resources can be applied to construct an abstract deictic space.

2.2   The abstract deictic space

In this section we explain in what sense each of the three dimensions of DST abstracts away from the material spatial domain. Figure 2.5 represents a perspective sketch of the abstract three-dimensional space with which we shall work.[9]

This configuration is idiosyncratic to linguistic conceptualisation and is not the familiar three-dimensional depiction of Euclidean physical space with axes x, y and z. For instance, two of the three axes are half-lines and the m-axis is finite. We now say a little more about each of these abstract axes in turn. The geometric origin 0 is also understood as the S (Speaker, Self and Subject).

2.2.1   The d-axis

The d-axis gives mental 'locations' for referential entities (discourse referents) presupposed by the speaker S. It corresponds to a scale of relative conceptual (not necessarily physical) distance from the deictic origin (see Figure 2.6). Conceptually, it is closest of the three axes to what we understand by spatial in a physical sense. However, it does not represent the three Euclidean dimensions discussed above in connection with spatial prepositions. Rather, the d-axis is a representational abstraction that reduces the three Euclidean dimensions to different spatial concepts, namely, direction and distance, which are part of human embodied conceptualisation. The

---

geometrical tool. On topological spatial prepositions, see Levinson and Meira (2003) and for a summary Kemmerer (2010).
[9] Parts of this section are developments of Chilton (2005).

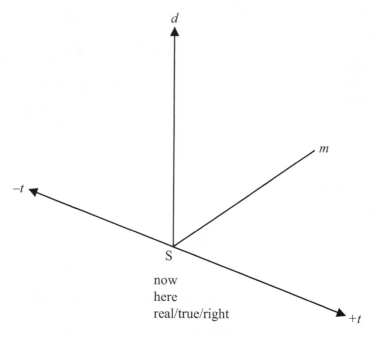

Figure 2.5 The fundamental coordinate configuration

d-axis is not meant to give an objective metric for any point location on this continuum. The relevant metric is scalar and relative to a reference frame.[10]

The d-axis includes space-related expressions such as *this* vs. *that*. Such expressions do not indicate precise measurements, but the entities referred to are distinguishable in terms of their *relative* distance from the speaker. We do not confine the relative positioning of discourse referents to relatively positioned spatial objects. In general, discourse referents are located on the d-axis at points varying in relative distance from S or from some reference point that is also relatively close to or remote from the S. Further, events (represented by a vector) are directed towards or away from the speaker at the origin or towards or away from a discourse entity. The distances and the directionality are in the conceptualisation not necessarily in the objective world. In linguistic expression, entities that are 'closer' to the speaker in this sense, are indicated by grammatical means, including, for English, word order,

---

[10] Remember that DST is trying to model an abstract conceptual space. It is only part of the overall processing of discourse, which involves very many dimensions. Physical relations in space, of the kind discussed in Chapter 2, in connection with prepositions, are of course a subset of those dimensions of global discourse processing.

2.2 The abstract deictic space

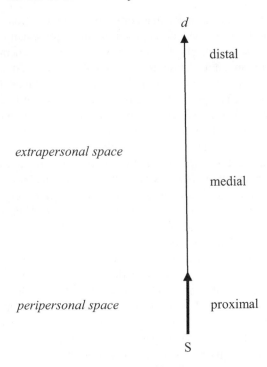

Figure 2.6 Relative distance from S on *d*-axis

grammatical subjecthood and objecthood and various topicalisation devices. However they are marked, the conceptual effect is of relative closeness and distance and will be shown in the same way on the *d*-axis, whatever the way in which a particular language expresses them. In cognitive terms, the relation between entities on the *d*-axis can be thought of as a figure–ground relation or in terms of attentional focus. The axis is labelled *d* as a mnemonic for (attentional) distance, direction and discourse referent. We are thinking of distance here not in metric terms but in terms of relative conceptual distance along the *d*-scale, grounded in psychological and linguistic considerations.

Proximal distance can be defined in terms of peripersonal space. This is a psychologically real and neurally instantiated region that is part of human consciousness and behaviour (on peripersonal space, see Rizzolatti *et al.* 1997, Weiss *et al.* 2000, Berti and Rizzolatti 2002, and on its relevance for spatial language see Kemmerer 1999, 2006a, 2010, Coventry and Garrod 2004, Coventry *et al.* 2008). The extent of peripersonal space is also the distance that we can take to be relevant for the spatial component of transitive verbs in English such as *touch, hit, grasp, hold, knock, kick, break.*

And it may also be peripersonal space that corresponds to the region denoted by proximal demonstratives (*this* as opposed to *that*, for example). All languages have at least two deictic demonstratives: a proximal demonstrative for referring to an entity close to the deictic centre, and a distal demonstrative for referring to entities located at some indeterminate distance from S, in the extrapersonal space. In the case of languages that have three terms, Diessel (1999: 39–41) distinguishes between 'distance-oriented' and 'person-oriented' systems. The third demonstrative in three-term systems is for referring to an indeterminate medial location relative to S (distance orientation) or to a location near S's addressee (person orientation). An example of a distance-oriented system is Spanish (*este* proximal, *ese* medial, *aquel* distal, relative to S). Latin on the other hand can be considered person-oriented, having *hic* (near S), *iste* (near addressee) and *ille* (remote from both S and H), which makes H a potential secondary deictic centre. Although it is claimed that more than three demonstratives are found in some languages, it appears that in the distance-oriented systems there are usually not more than three demonstratives strictly definable solely in terms of distance and direction (Fillmore 1982a: 48–9, Diessel 1999: 40).[11]

Using these observations, DST treats the d-axis at having three fundamental points. It should be emphasised that these are not determinate geometric points, nor do they refer to physical space. This abstract approach to the dimensions of the deictic space in DST is compatible with the way deictic demonstratives work in practice. Despite the distinction between distance orientation and person orientation, there are grounds for postulating a universal cognitive attentional scale common to both. Demonstratives can in fact be seen as devices for whose fundamental role is to establish joint attention in communication in combination with selectively directing attention to entities in the foreground and background of a deictically centred space. Precise location points cannot possibly encoded in word forms: relative degrees of distance are established by contextual cues and background knowledge, often at a very abstract level (Tomasello 1999, Enfield 2003, Diessel 2006). It is this abstract level of attentional distance relative to a deictic centre that motivates the deictic space as defined in DST.

We shall propose that the proximal–medial–distal distinctions are projected onto the two remaining axes. But it is also important to repeat that the relative points on the d-axis are not spatial either. They can represent at least two kinds of attentional distancing that are expressed in linguistic form and are

---

[11] Languages vary in the way they encode distance and attention. Further, although not all languages have demonstratives that express, in themselves, distance of an object relative to the speaker, they have the capability to combine demonstratives with adverbials to this end (Diessel 2006). There are no known languages lacking demonstratives.

incorporated into the DST approach. One of these is remoteness in the sense of perceptual–conceptual accessibility. Thus 'far away' referents will be relatively distal on the d-axis. A second has to do with 'importance' (Piwek et al. 2007), that is importance for the communicative interaction initiated by S. Thus even a 'far away' or 'inaccessible' entity can be 'brought into focus' or 'foregrounded' by giving it a proximal position on d.[12] Although I am using spatial and visual metaphors here, and such metaphors are also in the cognitive psychology literature, the cognitive phenomenon I have in mind is that of selective mental attention in general, going beyond vision and the other sensory modalities. The d-axis is the *distancing* (and also proximising) axis in the general sense of cognitive attention to cognised entities and their foreground–background relationship.

2.2.2 The t-axis

As shown in Figure 2.7, the d-axis projects metaphorically onto the temporal axis (t) – which gives us relative temporal 'distance' from the origin in two directions, past ($-t$) and future ($+t$), both on scales of 'distance' relative to time of utterance. The suggestion here is that the space-time plane in human deictic space involves a temporal dimension that is conceptualised by analogy to spatial distance and direction, as indicated by linguistic expressions: the time-as-space metaphor is well known. Thus there are temporal points (or better, zones) on the t-axis that correspond to proximal, medial and distal. The conceptualisation of times, however, is heavily dependent on context and languages also have means by which one time is used a reference point for another; more will be said about times in Chapter 4.

The t-axis is not 'time's arrow' as represented conventionally (e.g. in space-time graphs), but directed according to the speaker's (S's) viewpoint: events can be relatively 'close' or relatively 'distant' in the past, and similarly for the future. The two opposed directions are conventionally labelled $-t$ and $+t$. In English these directions correspond to a systematic front–back orientation found in metaphorical expressions such as *look back to the past* vs. *look forward to the future*, which are derived from natural human orientation. However, the precise axis of orientation in human three-dimensional space is not relevant. Some languages, for instance, appear to have up–down rather than front–back metaphors for where time goes to and comes from (e.g. Haspelmath 1997, Yu 1998, Radden 2004). What is relevant in all conceptual spatialisations of time is bidirectionality and distance. One can conceptually orient to the future or to the past and an event can be 'closer' or more 'remote' in

---

[12] In another terminology, thematised or topicalised.

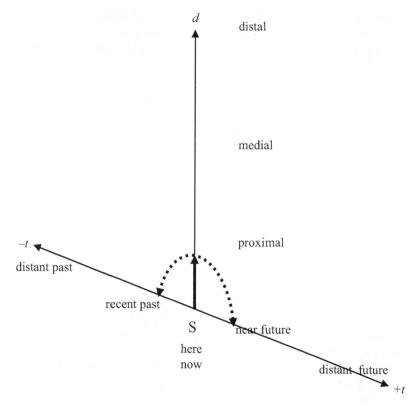

Figure 2.7 Attentional distance metaphorically projects onto temporal distance

either direction. Jointly orientating to a temporal 'location' is what tense (however it is morphologically manifest in the world) enables communicators to do.

The DST framework also suggests that proximal temporal space corresponds in some way to spatial peripersonal space. Like other 'locations' in the deictic space, events can be viewed egocentrically or allocentrically, that is, taking 0 as reference point or some other time relatively distal to 0 as reference point. The conceptual structure of time vectors thus abstracts from some of the conceptual properties of spatial vectors.

But what kind of time do the bidirectional arrows in the DST represent? The time axis in the fundamental DST model clearly does not represent real time, in the sense of time as understood by physicists. The t-axis here does not represent time as it is manifest to physics in the second law of thermodynamics or the expansion of the cosmos since the Big Bang. Nor, self-evidently, do

2.2 The abstract deictic space

the diagrams of DST represent four-dimensional space-time. Rather DST is concerned with conceptual time, and more specifically conceptual time as communicated via linguistic expressions. This is not to say, however – and the case has been carefully discussed by Jaszczolt – that there is no relationship between what the mind holds concerning time and real external time. It may be helpful here to adopt the supervenience relation, as Jaszczolt does (Jaszczolt 2009: 2, 19–21, 31). Necessarily simplifying, the mind is supervenient on the biological brain and the brain is supervenient on the physical universe and thus on real time, which is an element of space-time as understood by physicists. The task that DST sets itself is to model the interaction of human time conceptualisations with the other two conceptual dimensions of DST, modality and attentional distance. But we still need to consider the grounds for representing the t-axis as bidirectional – which, of course, real time is not.

In psychological accounts there is also often reference to the one-directional arrow of time. The past is experienced because of memory, working memory, short-term memory and episodic memory in particular, and the constant expansion of memory is experienced as qualitatively different from the experience of the cognitive processes of intention, anticipation, planning, hoping, and so forth. A unidirectional time arrow emerges in consciousness that must be the bedrock of linguistic conceptualisation. Philosophers trying to make further sense of how the human mind relates to real time (and whether real time is real at all) frequently refer to McTaggart's 'A-series' and 'B-series' conceptualisations. The former adopts the temporal perspective of a person speaking of time or times, and is influenced by the tense conceptions of (or about) certain languages: future events become the present and in turn become the past one after another. The B-series view sees events as strung one after the other in before/after relation to one another, with no flow. The DST representation of the time axis is different from both A- and B-series to the extent that it is bidirectional. It resembles the B-series to the extent it does not seek to represent flow (which is not to deny that time metaphors, as is well known, represent time as motion). But it resembles the B-series to the extent that the present is focal as a temporal deictic centre. The DST seeks to model the ordering effects of the use of tense forms and temporal adverbs, which speakers use with reference both to times preceding to time succeeding their experienced *now*. When a speaker refers to time, these are treated as deictic, and in fact vectorial, since they have both direction (toward the past or toward the future) and distance (near past, remote future, etc.) However, as later chapters will show, tense forms are certainly not unequivocal temporal markers and their effects have to be accounted for by making use of secondary reference frames located within the fundamental deictic space. Other philosophers have been concerned with the phenomenology of time, notably Bergson, Husserl and Heidegger, and something of their discursive explorations of time will surface in Chapter 5 where we probe the conceptualisation of 'simple present' and 'progressive' forms in English. Such aspects of experienced time are communicated by language and to some extent can be accommodated in the geometrical principles of DST. The conceptualisations of time communicated by language are not only positions on a line; the t-axis needs also to be thought of as allowing for the communication of the experience of durational continuity.

2.2.3 The m-*axis*

Humans are capable of detached representations; their mental representations of the external environment are not dependent on stimuli in the environment. They are capable of representing processes and states that took place in the past or that are expected to take place in the future relative to the time of the mental activity of representing. In DST this ability is accounted for in the t-axis. Humans are also able to represent processes and states that have not happened at all, do not obtain in the present, and that they do not expect to happen in the future. And they have the ability to organise environmental information in terms of reliability of the source and degrees of epistemic certainty (see Premack and Woodruff 1978, Cosmides and Tooby 2000, and accompanying papers in Sperber 2000).[13]

On a priori grounds one might assume that the ability to think of events that have not happened or are not happening is linked to the ability to think of future events, since future events themselves have not happened. Perhaps the human mind acquired the ability to assign some sort of probability to the events of futuredom. I shall not pursue this idea further here or attempt to build it into the basic model. What is clear is that humans have the ability to mentally represent events that might happen, might not happen, might not have happened, or that even could never happen. One might think of the epistemic cognitive ability I am evoking here as metarepresentation, since one might propose that a representation of an event is formed first and then assigned a modality, making a representation of a representation, i.e. a metarepresentation. However, while this epistemic ability can be deployed consciously and calculatedly, I want to claim that it is also spontaneously and unconsciously active and that it is *integrated in* acts of linguistic communication too (see above Chapter 1). Pragmatically, an epistemic evaluation is communicated whether an uttered sentence is morphologically modalised or

---

[13] Note also that Hurford, on the basis of primate and other animal studies, is inclined to think that what he terms 'metacognition' and what other researchers call 'uncertainty-monitoring' is found to some degree even in non-humans (Hurford 2007: 29–35).

2.2 The abstract deictic space

not. This does not mean that modally unmarked sentences are not, in some sense, the default. Perhaps, as Cosmides and Tooby (2000: 61) suggest, a 'true-is-unmarked' cognitive architecture is the natural starting point ontogenetically and phylogenetically. However, in the evolved cognitive architecture of modern humans, and in their languages, unmodalised and modalised expressions exist in the same system and contrast with one another, which implies that modally unmarked utterances will also communicate a modal value. This can be taken to further imply that separating a purely propositional ('modally innocent') formula from one that is modalised is not going to yield a naturalistic model of pragmatic communication.[14] On the assumption that this is the case I propose the third dimension of the basic deictic space.

This third axis is called here the modal axis (m-axis), and its role is to reflect what seems to be the case for all utterances, namely that speakers give them a valuation in terms of their subjective truth evaluation, or to put it slightly differently, in terms of the extent to which speakers decide to epistemically 'detach' or 'distance' the contents of an utterance from absolutely true assertions. States of affairs are always 'positioned' along a scale such as the following: 'certainly true' relative to the speaker's knowledge state, 'certainly not true' relative to the speaker. It is natural to model modality in this sense by projecting spatial concepts. Langacker, for instance, talks of modality in spatial terms, using the metaphor of 'epistemic distance'. Interestingly, he does so in terms that almost formally define a position vector:

> ... the modals can be described as contrasting with one another because they situate the process at varying *distances* from the speaker's *position* at immediate known reality. (Langacker 1991: 246; emphasis added)

Similarly Frawley (1992: 384–436), drawing in part on Chung and Timberlake (1985), outlines a deictic theory of modality that also has distance and direction as intrinsic features in direct correspondence with spatial deixis and the spatialised deixis of time. In fact modal expressions can be ordered in relation to a 'position of known reality' in a way that bears out these intuitions. As a working hypothesis, the modal axis refers to a scale based on an intuitive grading of the English modal adverbs, adjectives and auxiliaries (Werth 1999: 314–15) as outlined below:

is, necessary, must be,	probable, should be,	might,	possible, uncertain, may	might not,	improbable, may not,	impossible, can't be	is not

realis ⎯⎯⎯⎯⎯⎯⎯⎯⎯⎯⎯⎯⎯⎯⎯⎯⎯⎯⎯⎯⎯⎯⎯⎯⎯⎯⎯⎯⎯⎯→ *irrealis*

epistemic distance

S

---

[14] Modelling situated sentences in logical form involves modal and temporal operators acting on a proposition. The advantage of logical modelling is that axiomatisation permits valid inferencing within the logical system. What is lost is the potential for a cognitively natural account.

The terms *realis* and *irrealis* are used to characterise the limiting points of this epistemic scale. They are not used here to denote grammatical categories as they are in the description of some languages but to denote cognitive states, with which linguistic expressions may be associated. A realis representation is a cognitive state in which S, the speaking subject, takes some cognised entity, state of affairs or happening to be real, or 'there' or 'in' the world (phenomenologically immediate or present) or consistent with S's encyclopaedic knowledge, i.e. mediate experience taken to be true by S. An irrealis representation is a cognitive state in which S has a mental representation that S understands as being removed from realis cognition to some degree. The fact that modal expressions can be graded indicates that there are degrees of non-linguistic assessment of subjective certainty. The scale can be thought of as metaphorically spatial, in the sense that realis representations are positioned at S's known reality; what S considers irrealis is 'distant' from S, or 'remote', 'located' at degrees of distance, with a limiting point that is counter to fact, i.e. 'opposite' to S's known reality. It is the conceptual elements of direction and distance that justify epistemic modality being characterised as deictic (as argued by Frawley 1992: 387-9, and assumed in the notion of 'epistemic distance' used by Langacker 1991: 240-9 and also by Dancygier and Sweetser 2005: 56-65).

Some cognitive linguists (notably Talmy 2000 [1988] and Sweetser 1990) claim that epistemic modal expressions derive by metaphorical extension from deontic (or 'root') modals, and the implication seems to be that deontic modal mental states and acts themselves (e.g. wanting, requesting, ordering) are somehow more basic.[15] The semantic change in many languages of deontic modals into epistemic modals appears to support this idea (Heine 1993, Hopper and Traugott 2003), as do studies showing that children learn deontic modals earlier than epistemic ones (see Papafragou and Ozturk 2006 for empirical findings that point in the opposite direction). Whatever the significance of this evidence, it remains unclear what it would mean to say that the deontic is more basic than the epistemic in adult cognition and language. I return to these questions and propose an alternative model of deontic meaning in Chapter 10.

Givón's perspective on the question is interesting (Givón 2005: 114-16). First he notes several reasons why, though appealing, the deonticity-first claim may not be right. For example volitional expressions (thus 'deontic' in some accounts) already presuppose the speaker's knowledge state: *I want to leave the room* presupposes that the speaker knows she is in the room and that she has in mind a future non-real condition where she is outside the room. This is an important consideration that may point to the

---

[15] Givón (2005: 115) notes that this idea is hinted at in the classic paper on theory of mind by Premack and Woodruff (1978).

2.2 The abstract deictic space

way a language-using organism functions. But since the idea that deontic states are more basic remains intuitively appealing. Givón's solution is that deontic cognition and action could be evolutionarily early but would have been dependent upon epistemic input. Deontic and epistemic modalities would both be early in human development. At some stage, Givón proposes, the epistemic ability separates itself from the deontic. If Givón is right, then humans (or human adults using language) have an independent epistemic ability (perhaps module) that assesses the realness of incoming environmental information. Be all this as it may, we should note the following. First, the precise sense in which deontic (or 'root') modals are prior to epistemic ones has not been stated. Second, even if deontic cognition precedes epistemic assessment either phylogenetically or ontogenetically or both, the structure of complex linguistic meaning is another matter and may be entirely distinct. Scalar epistemic conceptualisation may be an integral part of a great deal of linguistic structure and, moreover, be inherent in every situated utterance. This is the conjecture that is built into DST. Demonstrating how a variety of linguistic expressions can be coherently and economically modelled on this basis, is, in part a test of the epistemic priority conjecture for linguistic conceptualisation. In Chapter 10 I return to the question of the relation between epistemic and deontic meaning.

The m-axis is scalar but also has structure that will be important in the investigation of grammatical constructions in later chapters. What is close corresponds to what is most real for S and what is maximally distal modal corresponds to what is counterfactual, negated or unreal for S. DST posits a midpoint on the m-axis on the basis of two kinds of linguistic evidence. The first is that sentences such as those in (4) are not self-contradictory, whereas those in (5) are:

(4)   a   John might go to the party and/but he might not go
           b   Possibly John will go to the party and possibly he won't

(5)   a   *John must have gone to the party and John must not have gone to the party
           b   *Probably John went to the party and probably he didn't.

The second kind of evidence, the existence of modal 'Horn-scales' (see Horn 1989, Israel 2004) also suggests symmetry. While the scale considered above is a unidirectional conceptual gradient from *realis* to *irrealis*, the semantic implication relations in each half of the scale are mirror images of one another. For example:

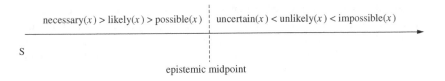

Epistemically 'stronger' terms implicate 'weaker' ones: if x is necessarily true, it is also likely and possible; if x is impossible (necessarily untrue), it is also at the least unlikely and uncertain. This scale has a reflection point at a conceptual zone between possible and uncertain. Nonetheless there is a scalar opposition ranging from what S conceives as real or true to what S conceives as irreal or counterfactual. The conceptual area around the reflection point is an area of cognitive indeterminacy that is important: it is the point that stands for the cognitive state in which S is not epistemically committed.

To summarise, the DST postulates a conceptual space consisting of three dimensions that are taken to be specific to the kinds of conceptualisation drawn on and induced by linguistic forms. While the morphological and syntactic means of languages vary, the hypothesis is that all languages will use the same conceptual space. Coordinate geometry is a system for specifying points in a space. This means that discourse entities and the relations between them (which we characterise as vectors) can, in theory, be positioned anywhere in the space. Figure 2.8 fills out the basic diagram representing the conceptual space we shall explore.

To repeat the point, the three dimensions of this figure are not to be confused with the three spatial dimensions of perceived physical space. Rather, the idea is to represent a conceptual space representing an integration of discourse referents (the d-axis), conceived time (t-axis) and epistemic modality (m-axis). The basic conceptual components are relative distance and direction. The three axes are not the usual x, y, z axes of spatial geometry but abstract dimensions: d is attentional distance of discourse referents from S (foregrounding, middle ground, background), t is deictic time and m is epistemic distance from S. 'Distance' here is relative cognitive distance, not a metric. Linguistic constructions give rise to conceptual representations, the fundamental structure of which is the deictic space. Diagrams like Figure 2.8 are intended to sketch a conceptual reality space, R, whose origin or zero point is the conscious now-here-self, designated S (self, subject, speaker). The reality space is not absolute or a 'possible world', but a conceptual space relative to some conceptualiser S. All points on the three axes are relative distances from the origin S. But the axes are not to be understood as metrical; they are scalar and points lying on them are merely positioned relative to one another and to the origin. The intersection of the axes, the origin, defines the viewpoint of the subject S. The coordinate system described in this way corresponds to the speaker's self, in the sense that it corresponds to the speaker's cognisance of what is *here* (the graspable in primary peripersonal space defined physically), what is *now* (what is temporally within reach, that is, peripersonal space projected onto time) and what is *real* (what can be 'grasped' cognitively).

2.2 The abstract deictic space

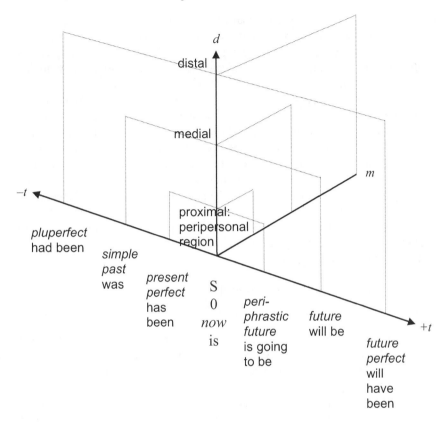

Figure 2.8 The fundamental deictic space

As we have seen, the relative distances on each axis are motivated by the kind of relative conceptual 'distancing' communicated in demonstrative expressions. There are three relative positions: *proximal*, *medial* and *distal*, in addition to the zero point (i.e. deictic and geometric origin of the three dimensions). The linguistic and psychological evidence suggests that we need to think of the intersection of the axes not only as a point but as a peripersonal conceptual region. This is the psychological space that is defined in part by the reach of an organism's limbs and which includes personal space, the space occupied by the body. We shall speculate, for the purpose of investigating tense-conceptualisations, that peripersonal space is transposed onto the time axis: *peripersonal time*. It is important to note that psychological peripersonal space can be extended depending on situations and activities and I also speculate that in linguistic communication this effect may also arise by way of linguistic and pragmatic cues. If it is reasonable to speak of peripersonal *time*, I suggest an analogous effect may help in understanding some of the conceptual effects of tense semantics. What would a peripersonal temporal space be like? A reasonable guess is that it consists of memory of the recent past and anticipation and planning for the immediate future: memories of actions just performed and experienced simultaneously with intentions for one's next actions. Analogously, the DST framework suggests a 'peripersonal' region projected onto the epistemic modal axis, giving a proximal epistemic region of the real perhaps extending to high certainty – what is within the 'grasp' of the subject S. Both the *d*-axis and the *t*-axis thus can be seen as having epistemic projections onto *m*. What is within our peripersonal space is not only spatially close, it is close in terms of the attention we give, it is epistemically 'close' – it is what we can 'grasp' in both senses of the word, and what we can grasp is epistemically most certain. The spatial foundation of the three dimensions of DST thus gives us an integrated correlation of the attentional, the temporal and the epistemic.

2.3 Further characteristics of the deictic space

2.3.1 A note on time and modality

Are time and modality the same? Since the future is unknown, and the more remote it is the more unknown it seems to be, it has often been pointed out that thoughts about the future have an essentially modal character. Jaszczolt (2009) argues that tense forms in languages are in effect modal. It seems intuitively plausible to think that future events are, to the human mind, inherently less certain than present ones. However, DST does not attempt to build this idea directly into its architecture. If we were to revise the DST geometry in the light of the claim that time equals modality, one axis would disappear. Another possibility would be to design a geometry that would make the *t*-axis not orthogonal but a curve such that +*t* curved asymptotically toward distal counterfactuality on *m*. And since it might well be argued that also the past becomes epistemically less certain the more distant it is, the same could be the case for −*t*. Such a geometry might capture the conceptual structure of the human mind but for present purposes it would immensely complicate the theory. I shall simply assume that the relative distances on +*t* and −*t* may indeed correspond to possible degrees of epistemic certainty. Another reason for not revising the fundamental geometry in this way is the following. We are concerned with conceptual representations brought on by linguistic constructions and the fact remains that languages generally combine the indication of time location relative to *now* with modal representation of degrees of certainty. Thus we can say, referring to a future event 'Mary will write the report next week', 'Mary might write the report next week', 'it is

2.3 Further characteristics of the deictic space   43

probable Mary will write the report next week', etc. We need to be able to model such combinations and this can be done in the three-dimensional deictic space of DST, even if we were to make the t-axis curved. In Jaszczolt's model (Jaszczolt 2009: 140–54) there is a scale of epistemic certainty from most to least certain tense forms: tenseless future (Peter goes to London tomorrow), futurative progressive (Peter is going to London tomorrow), periphrastic future (Peter is going to go to London tomorrow), regular future (Peter will go to London tomorrow), epistemic necessity future (Peter must be going to London tomorrow), epistemic possibility future (Peter might be going to London tomorrow). Apart from the fact that the intuitive semantic judgements involved here are debatable, it seems to me that placing all these expressions on a single scale loses distinctions with regard to the futurative use of epistemic modals.

2.3.2   Discourse entities

What we are calling *discourse entities* (or referents) are abstract conceptual objects which semantically may be arguments of predicates, but in DST are treated as end and tail points of vectors. They are understood in DST not as denotative language-to-real-world relations but as selected and variably focused objects of the attentional beam. Reference is understood cognitively as attention. Objects of attention do not have to be physical objects or events in physical space-time. They do not even have to be thought of as real or existing at all. In the DSM diagrams they are 'labelled' by the English words denoting them in the sentence that is being modelled. Any discourse entity has a point on the d-axis, at some scalar distance relative to 0. A particular discourse entity will always have the same d-coordinate, but might appear at different points on the t- and the m-axes. This is useful because it enables us to track anaphoric reference graphically. In particular, it enables us to track anaphoric reference across temporal and modal planes.[16]

What is meant by this is that we can identify – as we do in real life and in real discourse – individuals at different times and in different kinds of epistemic representation. What is meant by the latter point is the following. Consider the sentence *Mary would like John to slay the monster*. From S's cognitive viewpoint Mary is real, and so is John. From Mary's viewpoint also John is real. Both John and Mary have coordinates on d, and a coordinate on m in the realis plane (i.e. $m = 0$). But the complement clause *John to slay the monster* represents the world of Mary's desires, and in this world John initiates a

---

[16] This is a complex matter that I do not treat directly or in detail in this book.

particular action. It is not a real action, either from Mary's point of view or from that of S. Whether the monster is real too depends on S's conceptualisation. Possibly S's understanding of Mary's mind and the world in general is that monsters do not exist, or that the particular monster of Mary's imagination does not exist: in this case, the monster will have a coordinate at the irrealis end of the m-axis. In short, the deictic space as we have defined it makes it possible speak of real and unreal referents, and to track anaphora across belief spaces. Exploring the details of this potential is left to Chapter 7 but here we need to say a little more about the reference and discourse entities.

Labelled discourse referents are positioned in deictic space by an ordered pair of coordinate points (d, m). They are generally treated as having temporal permanence, so they actually are positioned by three points $(d, m, 0)$, i.e. with $t = 0$ and understanding the temporal zero point not only as S's *now* and also a-temporality (cf. below Chapter 4). For the sake of illustration, let us say that the relative position points on axes that we have discussed are numbered 1 (proximal), 2 (medial) and 3 (distal). A discourse referent might have, for example, a position in deictic space specified by (1, 1, 0). This would give the deictic space location for *Fido* in the sentence *Fido ate the bone*.

It is part of DST that entities are represented schematically and are thus merely 'labelled'. There is no semantic detail. This is because DST aims to investigate the fundamental scaffolding language related to spatial cognition. In full model of human language, the 'labels' can be thought of as providing access to cognitive frames that combine many representations derived from experience and other stored knowledge bases (see Evans 2009). However, each schematic discourse entity d can in principle be the origin of its own coordinate system, providing the scaffolding for conceptualisations associated with grammatically embedded clause structure such as relative clauses.[17]

The coordinate (axis) system of which S is the deictic origin – that is, S's reference frame – might attribute different kinds of ontological value to discourse entities. It might be thought by S that some discourse entity d_i really exists, potentially exists, possibly exists, possibly does not exist, definitely does not exist. DSMs can reflect this kind of epistemic variability by varying the m coordinate. Another conceptual possibility is that S does not label a particular discourse entity at all, though some such entity that plays a role in some vent denoted by is assumed to exist, as for example in agentless passive constructions. The DSM can thus have a coordinate d_i but no label for that d_i. This is not to say does not have a specific referent in mind, or even a specific linguistic label, but S does not spell it out in the communication of which the particular DSM in question is a model.

---

[17] See Appendix for a sketch of how this might be handled.

2.3 Further characteristics of the deictic space

What might referents that are not realis be? This is a question that could take us too far afield; but a few speculative points are in order. The modal dimension of the deictic space makes it possible to model the different status of referents in de dicto as opposed to de re sentences. Consider a sentence like: *Bill wants to marry a doctor*. As has been said many times, *a doctor* may or may not exist. In DST we are saying, *a doctor* may not exist in S's cognitive frame of reference. In other words, *a doctor*, might be real for S and thus have an *m*-coordinate at $m = 0$ (at the realis end of the *m*-axis). Alternatively, S might have no idea whether *a doctor* exists or not, or have no interest in the matter, in which case *a doctor*'s *m*-coordinate is at the epistemic midpoint on *m* (see above, Section 2.2.3).

Positioning conceptualised entities and relations between them at the irrealis end of the spectrum raises a number of issues that can only be addressed briefly. Linguistic negation, for example, is a complex matter and cannot be given adequate treatment in this book. However, the deictic space clearly does provide potential for handling several related negation phenomena. For example, a negated sentence such as (6) implies that both *John* and *the car* actually exist at the time of speaking: in the DSM, therefore, each will be labelled on the *d*-axis at $m = 0$.

(6)   John does not own the car

(7)   John does not own a car.

Reflecting word order *John* will be relative proximal, the car relatively distal on the *d*-axis itself. The verb *own*, however, is represented as a vector lying in the irrealis plane, i.e. with tail and tip coordinates at $m = 0$, while for the *d*-coordinates of the *own* vector are those for *John* and *the car*, as shown in Figure 2.9a. The difference between (6) and (7) is that the negation also affects the ontological status, within S's epistemic frame of reference, of the entity *car*. The DSM handles this by the specification of the coordinates for *car* – in other words, its position in the modalised deictic space. For (6) the label for *car* is positioned on the irrealis plane in Figure 2.9a.

Figure 2.9a also shows a dashed vector for *own* lying in the realis plane. This is partly to illustrate the general form of declarative sentences that will be used throughout the book. But it also addresses another point concerning negation. It is frequently noted that, pragmatically speaking at least, negated sentences in some sense presuppose an affirmative. If this is the case, DST can represent such a conceptual configuration in the model when appropriate. In general presuppositions can be modelled in this fashion.

Figure 2.9b is a model for sentence (7), the intuitive understanding of which seems to be that neither the relation *own* nor *car* is conceptualised as

46 Viewpoint, reference frames and transformations

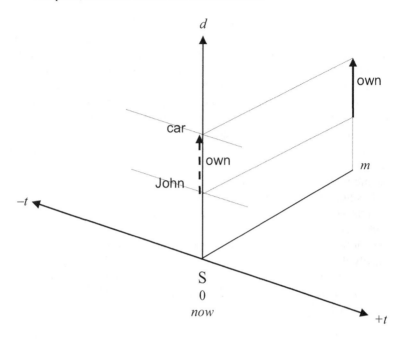

Figure 2.9a Example (6) John does not own the car

realis; both seem in fact to be irrealis. The DSM of Figure 2.9b shows this explicitly by the positioning of the discourse referent label *car* and by the position of the *own* vector.

If negated sentences may presuppose affirmative ones, the reverse can also happen. In a sentence like *I wish I had a good theory*, for instance, there is a presupposition corresponding to *I do not have a good theory*. It can be claimed that two mental representations are entertained simultaneously. DST can clearly model such a phenomenon along the lines of Figure 2.9a. This potential will become important in looking at counterfactuals in general in Chapter 6.

There is a further potential for the irrealis plane that I will only touch on here. Sentences such as *the mayor's limousine keeps getting longer*, as distinct from, say, *the mayor's limousine is very long*, refer to entities that are not physical instances but categories that are known to the interlocutors to be instantiated in various forms at various times. Fauconnier (1994: 39–81) calls this kind of expression 'roles'; Langacker speaks of 'fictive' or 'virtual' entities (Langacker 1999, 2001). While they seem to be cognitively different, such cases do resemble the entities evoked in negated sentences such as (6)

2.3 Further characteristics of the deictic space

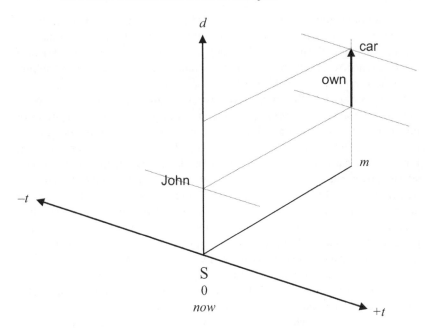

Figure 2.9b Example (7) John does not own a car

and (7) in requiring some kind of schematic mental representation that includes the awareness that the represented entity is not being represented as currently realis. The irrealis plane could perhaps be understood in a way that includes such conceptualisations.

2.3.3 Vectors again

In Section 2.1.3 we saw that geometric vectors provide a natural tool in the analysis and description of spatial prepositions. In his description of the properties of the nervous system Gallistel (1990: 475) notes that 'the aspects of reality – space, time, probability – ... lend themselves to vector representation'. Interpreting vectors in the geometric sense, DST follows this idea in setting up non-spatial axis systems defined in conceptual terms for attentional distance, deictic time and epistemic distance. Vectors are located in axis systems, that is, in reference frames, and, viewed in their basic spatial form, are mathematical objects that have (i) magnitude (i.e. length) and (ii) direction. Both negative and zero vectors are defined.[18] Not all the mathematical

---

[18] A useful overview of vector algebra is Anton and Rorres (1991).

properties of vectors will be relevant to us: the point is to begin an exploration of how many of them are helpful to a precise understanding of some key aspects of linguistic meaning. In many cases, the length (magnitude) of a vector is not relevant and we shall assume a unit vector for modelling predicates (prepositions and verbs).

It is important for DST that, as we have noted already, vectors are conventionally interpreted in various applied sciences as not only (i) specifications of position relative to the origin of the basic reality space or to some other point in the space, but also as (ii) movement (displacement, translation) and (iii) forces. They can, in other words, relate entities to one another in the deictic space, movements of such entities (relative to one another and to S), and can be combined such that an entity can exert force (maybe in a metaphorical or abstract sense) causing another entity to move (physically or in some other, maybe metaphorical, sense). These standard properties of vectors are, it is postulated, natural and appropriate for modelling certain conceptual structures that language forms are associated with. The conjecture is that they are also appropriate not only for modelling prepositions, which are prototypically to do with spatial relationships, but also for many verbs, many of which also concern the conceptualisation of spatial relations, spatial movement and direction of force, whether in a physical or metaphorical sense. This proposal is explored in Chapters 3 and 4.

2.3.4 Transformations and reference frames

The reality space R, modelled as a three-dimensional space with axes defined as in the way we have in this chapter, is S's frame of reference. It is a deictic space, a three-dimensional cognitive version of Bühler's 'deictic field'. What Bühler called the 'displacement' (or 'transposition') of the deictic origo of the deictic field, the shifted deictic centre as later linguists have called it, is extremely important in the structure and use of language; it is what constitutes perspective taking and alternating points of view. To probe this phenomenon further, and to render it more precise, DST borrows standard geometric tools. Geometric transformations turn one kind of reference frame into another.

As I shall use it in the context of DST, a 'transformation' is a function that establishes a copy R' of a base system R and shifts R' to a specified position relative to the base system R. Various kinds of transformation are recognised in geometry – translation, rotation and reflection have already been mentioned.[19] The transformations that seem to be of particular use in the abstract

---

[19] There are also dilation and contraction transformations, which expand and reduce an axis system by a given factor. These may have applications in linguistic–conceptual modelling in DST but I have not explored this possibility.

2.3 Further characteristics of the deictic space

deictic space are indeed the ones that have been used to model spatial prepositions. In the case of the preposition *in front of* we have an initial – or 'base'– axis system within which a second set of axes is embedded, and whose relation to the base system can be described as a copy that is located at some other point within the base axis system, by being moved in parallel with the 'horizontal' axis, turned through an arc (say of 180°) or reflected as a mirror image. When we look more closely in later chapters into how such configurations can be applied, beyond spatial expressions, to more abstract kinds of meaning, it turns out that it is translations and – maybe surprisingly – reflections that are of particular relevance.

Transformation of coordinate systems also gives DST a way of nesting further coordinate systems in a way that keeps all axes parallel. We may think of 'parallel worlds'. One useful implication is that we automatically get 'counterparts' and anaphoric relations across the coordinate systems (or 'spaces', to use Fauconnier's term). This, as will be seen, makes it possible without additional apparatus to model Fauconnier-type mappings between spaces in an economical fashion. But this is far from being all, since embedded and transformed axis systems provide a way of modelling relative reference frames that are required to model the conceptual effects of a range of grammatical constructions. These are akin to 'mental spaces' in Fauconnier's sense but bring with them the enrichment of the three abstract cognitive dimensions represented by the three axes d, t and m.

3 Distance, direction and verbs

> To pay attention is not merely further to elucidate pre-existing data, it is to bring about a new articulation of them by taking them as *figures*.
> Maurice Merleau-Ponty, *Phenomenology of Perception*

Deictic Space Theory abstracts away from the three-dimensional spatial reference frames, vectors and transformations discussed in Chapter 2. It is important to repeat: the d-axis is not intended to represent physical space. But it is derived from our spatial cognition. What is intended by the d-axis is that it stands for what I have called 'attentional distance'. When we represent discourse referents – the things to which we intend to draw the attention of our interlocutors in communication – we present them sometimes as physically closer or more distant, sometimes as metaphorically closer or more distant. In the latter case, syntax and discourse structure enable speakers to present particular referents as having conceptual priority, more salient, the theme at issue, and so forth. A corollary of this is that we can view the same referent or referents in alternate ways, as conceptually positioned 'nearer' or more 'distantly'.

Syntax enables us to give more or less thematic prominence to a referent, or indeed to an event or action, depending on the perspective we wish to take for the communicative purposes we wish to engage in. The simple geometric approach we are developing enables us to depict this possibility, because we are using the axes to represent relative 'distance' from the speaker S, and because we are also using vectors. This means we can have the referents in a different 'focus', so to speak, relative to S, while maintaining the same directed relationship between two referents, since we can have the arrow pointing in the relevant direction irrespective of which point is in focus. Thus, for example, if the agent is in focus (i.e. 'nearer' to S) the vector points towards the more 'distant' patient; if the patient is in focus, as in passive constructions, the vector points from the more 'distant' agent to the 'nearer' patient.

Though I have spoken of the deictic space as abstract, it is intended to be understood as grounded in embodied cognition. The d-axis can be understood in at least three ways, drawing on well-known ideas in the psychology of perception – visual focus, figure–ground perception and attention. The eye and visual pathways naturally focus selectively on entities in relation to their surroundings over varying distances. Figure–ground relations are studied in Gestalt psychology and include the ability to separate whole forms from their surroundings. Attention is the complex cognitive system that enables the directing of consciousness onto particular areas of ongoing experience. At a very general level, we may speak of foreground and background, perhaps of varying degrees, common to all perceptual modes. Cognition of foreground and background cognition is grounded in the embodied needs of humans and other organisms to relate to the physical world. It is based on distance and direction: perception of what is closer to and further away from self. I make the assumption that it is these systems that are available in the conceptualisation that linguistic forms conventionally code. Relations between points (discourse referents) on the d-axis provide a way of formalising, and of course crudely oversimplifying, such relations between foreground and background. We will see how the geometrical scaffolding unifies several different phenomena related to figure–ground cognition and linguistic construction.

Distance and direction are grounding concepts in DST. The three axes, described in the previous chapter, are vectors in the sense that they 'point' and also have relative distances. On the d-axis, discourse entities are located at relative mental 'distances' from S (speaker, subject, self). But there are also relations between these entities and in DST these are also specified by simple geometric vectors. In very general terms, vectors stand for predicates, including prepositions, and, more prominently, verbs. Because different verb meanings implicate different ways of conceptualising discourse entities that are typically associated with each verb, this chapter will be discussing how, and to what limited extent, vectors can be useful in elucidating this part of the fundamental architecture of language. So discourse entities and vectors will be frequently discussed together. Moreover, discourse referents and vectors make sense only in relation to reference frames, so our account will also include some discussion and speculation about how this most fundamental concept can assist in modelling linguistically communicated concepts of various kinds.

It is important to stress that geometrical ideas and diagrams are not meant to model the whole of language syntax and semantics. Nor is it the intention in this chapter to give an exhaustive account of verbs, for which one may turn to Croft (2012). The point is to explore the extent to which geometrical models, rooted as I think they are in embodied cognition, can tell us something about some basic characteristics of language.

3.1 Vectors, discourse entities and reference frames[1]

Vectors, in their basic form, are a formal device for the analysis of physical space, motion and forces. This is why they are of interest if one entertains the possibility that much of language-related conceptualisation is built upon spatial representation. But one should not expect linguistic conceptualisations to match space exactly. Even more importantly, we need to think of vectors as being incorporated in linguistic conceptualisation in a way that is humanly relevant. Assuming that it is relevant to start from the three conventionally recognised types of vector – position, translation and force vectors – we need to speculate as to how each might be related to what has independently been observed about language-based conceptualisation.

3.1.1 Perspectives on spatial relations: position vectors

Geometrically, position vectors in three-dimensional space anchor one entity at another, the tail at an anchor point in a coordinate system (at the origin or some other point) and the tip of the vector is given by coordinates for the positioned object. In DST we are concerned with linguistic conceptualisations and one important way in which we depart from a strict adherence to geometrical formalism is to work with a *relative* metric only – i.e. no attempt is made to give a real scaled measure of the length of a vector. In fact, as already said in Chapter 2, we use an arbitrary unit vector for all vectors standing for relations between discourse entities.[2]

Taking a closer look at the way in which DST abstracts away from three-dimensional spatial models for spatial expressions gives a clearer picture of the more general conceptual structures it is aiming to identify and describe. DST does not aim, as has been said, to model three-dimensional space, indeed it cannot do so. For spatial expressions it models just two rather abstract cognitive directional relations. One of these relations concerns what is cognitively foregrounded – and what is back-grounded. The other concerns positioning – the conceptual process whereby some entity is cognised as located at some position relative to some reference point. These are purely conceptual relations, for, given two objects neither one nor the other has an inherent objective priority as a reference anchor for positioning the other; such prioritising comes from human conceptualisation.

---

[1] The following section derives from but supersedes Chilton (2005).
[2] If there is a measurement expression, e.g. *John is ten metres in front of the tree*, this expression is absorbed into the spatial relation expression as part of the vector 'label', thus: *ten metres in front of*.

3.1 Vectors, discourse entities and reference frames

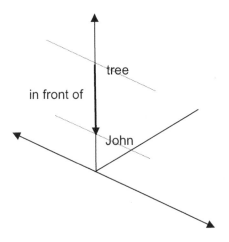

Figure 3.1a Example (1) John is in front of the tree

Consider first how the two sentences below appear in the abstract deictic space:

(1) John is in front of the tree

(2) The tree is in front of John.

For both these sentences a purely spatial three-dimensional reference frame – an 'absolute' reference frame, in Levinson's terminology – would be identical. Whether we say (1) or (2) the physical situation – the grid reference on a map – we are communicating about is the same. In conceptual terms, however, as we saw in Chapter 2, the two sentences involve alternate *conceptual* reference frames, all concerned with the three physical dimensions. DSMs are concerned with something slightly different. Between sentences (1) and (2) above there is also a difference of attentional focus, communicated by the different word order. In (1) John is conceptually 'closer' than the tree for S – he is in the attentional foreground. In (2) he is in the conceptual background. The word-order alternation enables a speaker to communicate the same physical situation from different conceptual viewpoints. These viewpoints may, in the natural flow of speech, be equivalent to what is referred to by 'discourse topic' – the current focus of communicative intention, what is in the forefront, so to speak, of ongoing talk. Figures 3.1a and 3.1b show how this kind of language-based conceptualisation can be modelled using the geometrical ideas introduced so far. The d-axis gives us relative attentional distance relative to S – which we can also think of cognitively related to visual focus and depth, foreground–background, or figure–ground.

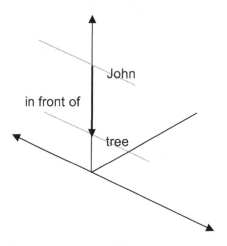

Figure 3.1b Example (2) The tree is in front of John

Another way of making the point is to note that in objective reality, and truth conditionally, all physical objects referred to (here John and the church) have the same absolute spatial positions, but in the conceived situation, which is what the DSM represents, the relationships are represented differently. This is what deictically anchored coordinate systems and position vectors enable us to model explicitly. The DSM does not of course contain all the information needed for a full spatial conceptualisation, since it does not represent the three spatial dimensions. But as already noted, the conceptualisation that arises in the course of processing discourse involves many parallel systems, one of which is the three-dimensional spatial system of representation. What the DSM does is model the fundamental deictic scaffolding.

Consider another spatial preposition, *opposite*. Once again, this preposition requires its own spatial model, but such a model, while it will tell us about the schematic semantics of this word in English, will not tell us about distancing and directional effects that emerge in discourse. It is just the latter that DST tells us about. Specifically, the d-axis is used for the mental distance and direction that arises in discourse rather than in objective physical space. The sentences (3) and (4) are denotationally equivalent, but have different conceptual significance.

(3)   The pub is opposite the post office

(4)   The post office is opposite the pub.

Loosely, we can say that in (3) *the pub* is foregrounded (the sentence is 'about' the pub), while in (4) *the post office* is foregrounded (the sentence

3.1 Vectors, discourse entities and reference frames

is 'about' the post office). Because the d-axis is a scale giving us relative closeness to S, and because vectors give us directionality, we can represent discourse foregrounding phenomena by positioning one referent closer to 0 on the d-axis than another referent. In visual experience the ground is what appears relatively more distant (backgrounded) in space. In the abstract DSM these phenomena are of course *quasi*-spatial (or *metaphorically* spatial) and constitute what might be called 'conceived' distance. The DSM diagrams handle this very simply by changing the relative positions of the referents: in (3), for example, *the pub* is 'closer' (more foregrounded) than the post office and the reference object is the *post office* more 'distant' (more backgrounded). The tail of the position vector is at *the post office* and its head at *the pub*. The case is reversed for (4).

This section has outlined how the DSM uses the d-axis to model both conceptualisations of (reduced) physical space and also indicates how the model gives rise to the notion of attentional distance. The most obvious use of such a notion is to integrate foreground–background construals into the geometric modelling of discourse. It is assumed also that, for English at least, the linear ordering of NPs in clauses will be reflected in the ordering of referents on the d-axis, barring grammatical constructions for marked topicalisation. In the DSMs, the subject NP of a declarative clause will be treated as 'closer' to S than an object NP, for example. We have seen that vectors can be used to represent the relation of objects to landmarks in three-dimensional space. In DST they are also treated as representing the direction of activity from a source to a goal. In terms of thematic roles, this means that the tail of the vector might, for example, be located at the coordinates of an agent, the head at the coordinates of a patient. Of course, since the vector is directional, the orientation can be varied in combination with the relatively proximal or relative distal location of the referents with respect to S. This in turn means that it is possible to model the conceptualisations associated with passive–active constructions, raising constructions and similar constructions in a unified and natural way. We look at these constructions in these terms in later sections and chapters.

Some linguists find problematic transitive verbs that express spatial relations. Langacker (1991: 311) examines sentences such as:

(5)   a fence surrounds his property

(6)   the parcel contained an explosive device.

Langacker notes that they are 'static' and have no 'active' element in their semantics. This may be the case if one reinterprets such sentences in the light of our metaknowledge of spatial experience. But the semantics of *surround* and *contain* is open to alternate DSMs, i.e. alternate ways of using and interpreting vectors in the coordinate system, a fact that may reflect the somewhat variable meanings associated with these and similar verbs. We can consider using position vectors: for (5) tail at *property* and head at *fence*, i.e. the position of the fence with respect to property. Similarly, for (6) the tail of the vector is at *explosive device* and the head at *parcel*. But neither of these solutions seems intuitively satisfactory or complete. The other modelling possibility is to take literally the grammatical transitivity relation (despite Langacker's apparent unease about there being any implied 'active' element). Why not follow through the principles of cognitive semantics and accept that *surround* and *contain* are transitive because experience of these spatial relationships is indeed associated with force in some sense? In this case *surround* and *contain* can be modelled as force vectors. There is justification for this in the semantics of the spatial preposition *in*, to which the verbs *surround* and *contain* are conceptually related. Surrounding and containing may indeed involve force in various types of sentence and context that easily spring to mind. It has been argued by several linguists that *in* cannot be modelled simply by a geometrical schema, by which they mean a purely spatial configuration (cf. Herskovits 1986, Vandeloise 1991[1986], Tyler and Evans 2003, Evans and Tyler 2004b). In fact Zwarts (2010: 194–9) surmounts this problem by using force vectors.

Further, some such spatial verbs denote relations that are geometrically symmetrical, although linguistic reversal of word order gives them the feel of being not symmetrical conceptually, as in (3) and (4). These kinds of symmetrical relation verbs appear to have a simpler conceptual representation, since they can be adequately modelled by means of position vectors and relative positioning of the entities labelled on the d-axis.

Examples (1) to (6) show how position vectors can be used in the obvious way – to model physical spatial location, abstracting away from the three physical dimensions and using the d-axis in the base frame of reference to model relative 'distance' of attentional focus. But in conceptual space, the positioning of two entities in a reference frame need not be confined to physical entities.

3.1.2 *Position vectors, abstract locations and property predications*

Position vectors can also be used to model certain kinds of non-spatial predicates, including state-of-mind predicates. As work in conceptual metaphor theory has shown, spatial concepts, in particular the container image schema, is a source domain for changeable emotional and other mental states (Lakoff and Johnson 1980, Kövecses 2000). In (7) the spatial positioning is metaphorical:

(7)   The linguist is in a good humour.

3.1 Vectors, discourse entities and reference frames

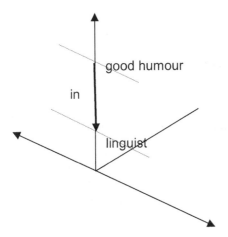

Figure 3.2 Example (7) The linguist is in a good humour

The discourse entity closer to S on the *d*-axis (i.e. *linguist*) is located at the more distal entity *good humour*. The position vector, labelled *in*, has its tip at *the linguist* and its tail at *good humour* – the vector locates the linguist in a particular state conceived as a container. The DSM configuration, shown in Figure 3.2, is the same as in Figures 3.1a and 3.1b, and rather than denoting a physical location denotes a temporary mental state – a temporary property of the discourse entity *linguist*.

Possession predicates also lend themselves to treatment as position vectors. In some languages possession is expressed by spatial meanings: French has *le livre est à Philippe* (literally 'the book is *at* Philippe'), and Russian *u menya kniga* (literally 'the book is *at* me'). An English sentence such as (8):

(8)   John has/owns/possesses the book

can be represented by a DSM configured in such a way that the vector for *have, own, possess*, etc. locates *book* at *John* – i.e. the vector points from proximal *John* to distal *book*. That is, *John* is at the forefront of attention, the book is the secondary focus of attention, and the positioning relation is given by the direction of the vector (Figure 3.3).

Thus Figure 3.3 is the geometrical model for 'the book is at John', the spatial origin of the quasi-spatial sentence. The verb *have* has the same conceptual function as a spatial preposition, say *at*: it locates on entity at another more focal entity. A position vector gives the position of point relative to a reference point. Such a point is often the origin of a coordinate system. Note, however that position vectors in DST generally relate one point

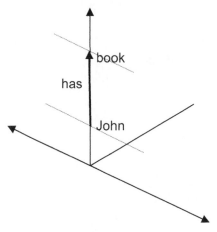

Figure 3.3 Possession as position

to another point that is not the origin of the base reference frame. Potentially all reference entities are origins of embedded reference frames, especially in the modelling of constructions that represent other minds: individuals other than oneself have their own reference frames, i.e. coordinate systems.

Taking account of the fact that languages use quasi-spatial locations to communicate the conceptualisation of possession relations leads us to a motivated way of modelling property predications in DST in terms of spatial locations (cf. Anderson 1971, Lyons 1977: 718–24, Frawley 1992: 229–32). There are a number of considerations and linguistic facts that further justify modelling of property predications as quasi-spatial position vectors. Sentences such as (5) can be regarded as predicating a property – the temporary property of being good-humoured – and the means of expression is clearly spatial.[3] The sentences (9) and (10) illustrate two ways English has for predicating a property of an entity:

(9)   Bill has wisdom

(10)   Bill is wise.

The proposal is to model properties, whether expressed by a nominal as in (9) or as an adjective as in (10), in the configuration shown in Figure 3.4. The

---

[3] There is a difference between being in a good humour and being good humoured in that the latter can be construed as a permanent property. DST does not discriminate such semantic differences between properties, nor does it discriminate classes of property concept as, for example, Dixon (1982) does. This does not affect the central claim being made here: that vector modelling of property predication in general is justifiable because property predication involves spatial conceptualisation.

3.1 Vectors, discourse entities and reference frames

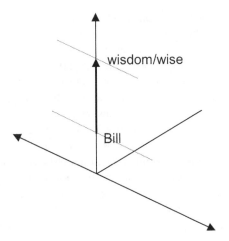

Figure 3.4 Property and entity relation as position vector

vector is unlabelled; it simply positions the separately conceived property at the entity *Bill*. (All entities potentially have their own axis systems, in which the entity itself is the origin.)

Abstracting from the grammatical difference between noun and adjective and between *have* and *is*, there is a common conceptual core that puts the property wise/wisdom and entity (Bill) together. This is not to say that there is no conceptual difference between these two constructions but there are grounds for treating them as having a common cognitive basis. Francez and Koontz-Garboden (2012) point to the occurrence of possessive strategies for property predication across a range of languages including Hausa (a Chadic language), Huitoto (Huitotoan, spoken in Columbia), Bisa (Mande) and Ulwa (Misumalpan, Nicaragua), as well as Germanic and romance languages. Possessive strategies may apply to only a small lexical subset. Spanish makes use of the possessive relation in *Juan tiene sueño* analogous to *Juan tiene un carro*. In French, the English *John is afraid* has to be expressed as *Jean a peur*. The point is that possessive constructions surface across a range unconnected languages; it is a natural means of expressing inherence of a property in an entity.

Francez and Koontz-Garboden (2012) argue that grammatical possessive strategies are semantically linked to property predication. Their argument is that properties such as *strength/strong* are 'abstract mass substances' just as *water*. In formal semantics, strength would then denote the set of all portions of strength, just as water denotes the set of all portions of water. But to predicate a portion of strength to an entity does not yield the required meaning but something like 'entity *e* is (a portion of) strength'. They note that it is common to speak of 'having' properties. One might add that the etymology of 'property' is also suggestive. They further note that properties are integral parts of the entities of which they are predicated and that accordingly languages use the possessive in order to express 'integral parthood'.

The proposal that properties are 'abstract mass substances' is certainly relevant to the present account, since it justifies the modelling of properties as reference points in the same way as mass nouns such as water would be. The notion that properties 'belong' to an entity is also crucial. But DST emphasises that this relation is essentially spatial: what one owns or is an integral part of one is conceived as a spatial co-location. Now DST does not seek to model integral parthood as such. In fact the position vector approach reflects the curious patterns that language can impose on conceptualisation. In human experience properties are surely not perceived separately from entities, but linguistic expressions make it possible to conceptualise emotional states and abstractions as independent, most clearly when expressed as nouns but also when expressed as adjectives. Because of cognitive comparison abilities the human mind apparently can abstract states such as *happiness*, *wisdom*, *red*, *tall*, *independent* conceptually from their integral manifestation in physical entities. Language makes separation of entity concepts and autonomous property concepts available.

One final point should be made about the kind of vector used to model possession and property constructions. For consistency and simplicity I have used a unit vector, as for other transitivity relations and as for spatial relations. However, an interesting alternative is to use zero vectors. A zero vector has no magnitude (length) and has arbitrary direction. Since we can plausibly think of possessions and properties as located at the same point as the entities, using a zero vector would make sense. The direction could be stipulated as pointing from the entity (in focus as the grammatical subject in the above examples, thus 'closer' on the d-axis) to the possession or property. This becomes relevant to the way we treat change of state in Section 3.2.3 below.

3.2   Displacement vectors and verbs of motion

A displacement (translation) vector[4] gives the shortest distance between two points – whatever the route taken. They are imaginary directed straight lines. Mathematically, and most abstractly, it is a function that maps the position of one point in a reference frame to another point in that reference frame. Conventionally, displacements are diagrammed as arrows and in practical applications may represent actual straight-line motions. This need not concern

---

[4] Throughout I use the terms 'displacement vector', 'translation vector' and 'movement vector' in the same sense.

3.2 Displacement vectors and verbs of motion

us too much. In DST we are using relevant aspects of vectors to model relations between entities; in particular, we are not concerned in most cases with the actual measure of distance. The crucial point to remember is that, abstract though vectors are, they are likely to derive cognitively from embodied concepts of bodily motion as experienced by humans (cf. above, Chapter 2, Sections 2.1.3 and 2.4.3 and Lakoff and Núñez 2000). Vectors, and their arrow images, are thus well motivated for modelling linguistic expressions – verbs – that denote physical motion. As we shall note, however, not all motion meanings relate to physical space.

3.2.1 Spatial motion

The vector notation straightforwardly represents start points as the coordinate of the tail and the end point of the motion as the coordinates at the tip. However, the human experience of locomotion and the conceptualisations communicated by human language go well beyond such notation. Verbs denoting spatial displacement from one point to another – as distinct from verbs that denote motion without such displacement, such as *shiver*, *tremble*, etc. – are inherently directional, whether volitional or not, whether the manner of motion is incorporated in the verb or not (e.g. *crawl* vs. *move*). For count nouns the assumptions is for single linear direction; for mass substances (mud, slime, toothpaste, blobs ...) the directions may be multiple. Since the *d*-axis is not directly spatial it is not concerned with the number or length of directions concerned.[5] Verbs such as *fall*, *drop*, *rise*, *climb* have particular directions in three-dimensional frames of physical space, but since DST does not directly model three-dimensional physical space only direction is represented. Verbs that do not have directional specificity, such as *move*, *travel*, *translocate* and the like, nonetheless assume inherent direction. For human conceptualisation directionality is not simply physical displacement in a reference frame but includes what is expressed and experienced as 'forward' motion, that is, willed self-locomotion related to the positioning of sense organs. One would expect intentional human movement to be goal-directed, or *telic*, especially if one is thinking of the biological evolution of self-propulsion – movement towards a goal can be claimed to have evolved purely for the goal of obtaining food, for example. Human language can nonetheless express aimless, non-goal-directed, i.e. *atelic* movement. It seems intuitively, however, that goals are often pragmatically assumed even though languages make it possible not to express them overtly. In English spatial goals are

---

[5] Motion verbs denoting motion but not displacement, such as *shiver*, *shake*, *tremble*, *wriggle* ... simply do not occur in constructions involving starting and ending points. DST still treats them as vectors; cf. the treatment of atelic verbs.

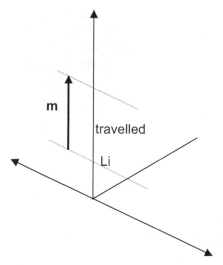

Figure 3.5 Example (11) Li travelled

explicitly expressed by way of the preposition *to*, which is already contained in the conceptual content we are attributing to the directionality of vectors. This point will become significant again in Chapter 8, when we consider more abstract conceptualisations associated with the preposition *to*.

Start points and end points are inseparable from conceptualisations of human movement, just as they are in the definition of displacement vectors. If they are not overtly expressed in linguistic expressions they are highly likely to be pragmatically presupposed (except for certain atelic expressions) and also related to the location of the speaking subject S. Further, in human language expressions the start point and end points can be foregrounded or backgrounded. One can focus on the source from which someone is travelling or on their destination. Using displacement vectors in the deictic space as we have defined it makes it possible to formalise some of these features.

Sentence (11), which leaves start and finish points unexpressed and is open to an atelic interpretation, can be modelled simply as in Figure 3.5.

(11)   Li travelled.

The discourse entity L_i has a label on the *d*-axis: it exists for S and is in focus, that is, relatively 'close' to S. The vector is of arbitrary unit length. It has direction, as is implied by the semantics of *travel*. But this is not the same as having a specific goal: the DSM thus has no label for the tip of the vector. It is perfectly possible to travel without intending a specific goal, though the default assumption is that travelling will end at some point in space (and

3.2 Displacement vectors and verbs of motion

time). It is of course not possible to travel without a determinate start point – but it is not necessary to mention it overtly in an utterance: accordingly Figure 3.5 does not have a label for the start tail of the vector either. The vector in Figure 3.5 is to be interpreted as a displacement vector, displacing the point labelled *Li* to the unlabelled point at the tip of the vector. As already noted, we have to understand this bare geometrical definition as standing for the complex human understanding of directed movement.

Languages make it possible to name either start or finish points. If the finish point is referred to (or is known from context), the construction is conventionally called telic, otherwise atelic. Start points can also be named (or known from context) or not named. Beyond what language expresses, conceptual schemas for motion (human knowledge of motion) always have starts and ends. If I say 'John walked for an hour', 'John travelled for five years' – expressions that are atelic – the background assumption is, even if there is no time expression of the duration of the motion ('John walked', 'John travelled'), that there were two points in physical space that count as starts and ends. Though directional the motion is always an integration of from and to. The moving entity ('theme') is, conceptually, integral with motion.

Motions have inherently both starts and ends, and this is reflected in the composite and conventionally iconic notation of the arrow. The concept of movement is a gestalt summarising where you are being displaced both from and to simultaneously; if you are moving you are necessarily travelling to and from, from and to. However, human conceptualisation can focus either on the start or the finish of a movement and this is precisely what words such as the English prepositions *to* and *from* do (cf. Tyler and Evans 2003, Evans and Tyler 2004a, Zwarts 2005). It is important to note, however, that both these prepositional meanings depend on the overall motion gestalt. Without making detailed proposals here, we can say that the conceptual schema for *from* is a displacement vector in which some portion of the vector including the initial point is in focus and the rest occluded, and conversely, for *to*, the tip is in focus and the rest occluded. In representing from and to, therefore, I will use an arrow labelled *from–to*, since we are trying to model a direct motion that the mind conceives as continuous while simultaneously choosing to focus on, attend to, some initial or final part.[6] It must be remembered that arrow vectors are merely notations, however: one would guess that the conceptual representation of starts and finishes are not visual images.

Despite the fact that *from* and *to* make sense only in relation to an overt motion concept with both start and finish, language supports the

---

[6] There is thus a composite image schema, a vector combining start, direction and finish, together with a cognitive process of attention focus. In Chilton (2009) I refer to a similar process for the image schema associated with the verb *get*.

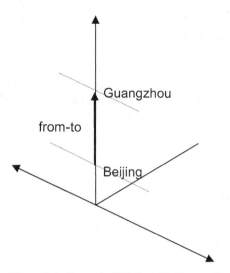

Figure 3.6a Example (12) from Beijing to Guangzhou

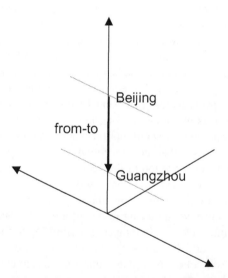

Figure 3.6b Example (13) to Guangzhou from Beijing

conceptualisation of starts and finishes separately and in alternate order. That is, start and end points, when communicated as two prepositional phrases, can be in both possible orders, though possibly the default is the natural sequence of start–finish:

(12)   from Beijing to Guangzhou

(13)   to Guangzhou from Beijing.

This alternation is easy to represent in DST (see Figures 3.6a and 3.6b above).

3.2.2   Combining motion concepts with location referents

How does a language combine reference to the motion of an entity with explicit reference to starts and finish points? In English this is done by prepositional adjuncts after the verb. It is important to note that, although movement concepts associated with verbs necessarily include conceptual starts and finishes, verbs do not generally incorporate particular kinds of beginnings and end locations in the verb semantics (Talmy 2000 [1985]), although Frawley (1992: 173–4) points out that such incorporation does occur (*deplane, exhume, imprison,* ...). This produces a situation where foregrounding–backgrounding reflected in word order is disrupted. In an unmarked example such as (14) below the moving entity and movement are expressed in the verb, the start and finish points optionally appear as prepositional phrases (either and perhaps both of which may be omitted). The start and finish are conceptually presupposed but do not both have to be overtly named. This means that the associated conceptualisation for the whole sentence must somehow have integrated as a gestalt what is presented in separate linear pieces.

I have suggested that the d-axis can account for foregrounding/backgrounding effects induced by linear word order – but while this seems a workable principle, it should not be rigidly imposed. As just noted, languages (at least on Talmy's evidence) do not on the whole incorporate specific source-goal referents – there are presumably cognitive and communicative motivations for this that we can't go into. Considering the three referring noun phrases in a sentence like (14) and (15) there are some word-order choices that are potentially significant in terms of conceptual focus: grammatical subject and verb come before the two adjunctive prepositional phrases, and the two prepositional phrases themselves can be ordered in two. I assume that what we need to model is a conceptual coalescence of (i) *Li travelled* and (ii) the selection and ordering of the *from* phrase and the *to* phrase.

This is not as straightforward as one might expect: what follows is a proposal. The task is to design a DSM that will combine the concepts of motion, direction, foregrounding and backgrounding that are intuitively involved in (14) and (15):

(14)   Li travelled from Beijing to Guangzhou

(15)   Li travelled to Guangzhou from Beijing.

In both (14) and (15) *Li* is in the focus main of attention as the traveller. The start and finish points are in secondary focus and differently focused with respect to one another in the two sentences. Sentence (14) is a coalescence of *Li travelled* with *from Beijing to Guangzhou*; sentence (15) coalesces *Li travelled* with *to Guangzhou from Beijing*. Figures 3.7a and 3.7b show this by a displacement vector for *travel* in Figures 3.6a and 3.6b.

In (14) and Figure 3.7a, it is Li's travelling from Beijing that is in the foreground for the speaker S; in (15) and Figure 3.7b, it is Li's destination of Guangzhou. The process of combining the travel schema with the naming of start and finfish means that the DSM results in double focus (double labelling of the proximal coordinate): Li and Beijing in Figure 3.7a, Li and Guangzhou in Figure 3.7b. DST has to allow for the natural cognitive operation of placing two entities at the same point in space. Notice that the direction of the arrow in Figures 3.7a and 3.7b is not to do with the spatial position of the speaker S but of direction with respect to what is foregrounded by S, namely *Li* and *Beijing* in one case and *Li* and *Guangzhou* in the other.

There is an interesting restriction on the ordering of *from* and *to* phrases when one is placed sentence-initially:

(16)   From Beijing Li travelled to Guangzhou

(17)   *To Guangzhou Li travelled from Beijing.

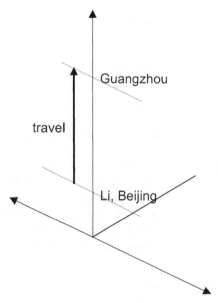

Figure 3.7a Example (14) Li travelled from Beijing to Guangzhou

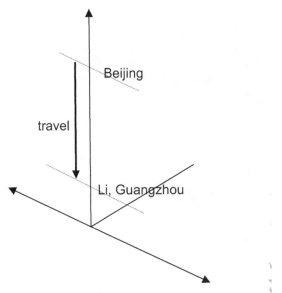

Figure 3.7b Example (15) Li travelled to Guangzhou from Beijing

While it seems to be natural to prepose a *from*-phrase, the effect of preposing a *to*-phrase seems curiously confusing. The explanation may lie in the implicit iconicity of linear sequencing and processing in language, if we assume that there is an expectation that start points are mentioned first because they come temporally first in natural experience of displacement. We need therefore attempt to model only sentence (16), in which the *from*-phrase naturalistically precedes the *to*-phrase. This is the case also for unmarked (14), so we need to seek to model the difference in cognitive effect brought about by the surface separation of the two prepositional phrases. The default assumption in cognitive-linguistic methodology is that such word-order alternations are indeed potentially meaningful. Intuitively – and in this case we seem to have 'only' intuition to go on – sentence (16) would occur in discourse when the location of Li in Beijing has already been mentioned or is somehow known to the interlocutors. In other words placing the *from*-phrase sentence-initially foregrounds Li's position at the start of his travelling to Guangzhou; it is made more explicit. Tentatively, a relevant way to model this, using the elements and conventions established so far, is to use a position vector to model sentence-initial *from Beijing*, as shown in Figure 3.8.[7]

---

[7] I have used a vector of arbitrary unit length, but it would be appropriate to use a zero vector, which would have arbitrary direction but no length and would locate Li at Beijing and at the start of the *travel* vector.

68 Distance, direction and verbs

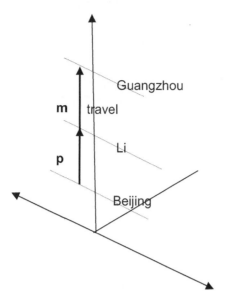

Figure 3.8 Example (16) From Beijing Li travelled to Guangzhou

The oddness of (17) is not an arbitrary syntactic matter affecting only prepositional phrases with *to* and presumably has to do with the non-naturalness of attending to arrivals before departures. One should expect DST, if its principles are plausible, also to have difficulty in designing a DSM for a sentence like (17) and it seems likely that this is the case.[8]

3.2.3 Displacement as change of state

The most abstract change-of-state verb in English is perhaps *become*. Its etymology, like that of its equivalent in Romance languages, is telling. Both the verb *become* ('about + come') and the French verb *devenir* ('from + come') recruited motion verbs in order to express passage over time from one condition to another. Etymologies do not tell us that current users of these

---

[8] In outline the argument is as follows. DST logic does not permit a consistent DSM for (17). Consistently with the treatment of (16), (17) can be said to foreground the position of Li implied in *to Guangzhou* and this could be modelled as a position vector, giving Li's position at the end of the movement. The DST logic expects Li to be at the tail of the travel vector, since Li is the moving entity. But this leads to representing Li as being in two places at once – both Guangzhou and in Beijing (iconically he's there before he leaves Beijing!). Another way of looking at this is to add the two vectors, assuming we can legitimately add a translation vector to a position vector. Since they would point in opposite directions, addition results in cancelling out: Li goes nowhere.

3.2 Displacement vectors and verbs of motion

words conceptualise change as movement, of course, but they do indicate that in order to form lexical items conventionally denoting abstract change language users initially utilised conceptualisations of physical motion. Motion takes place over time, but if the spatial dimension is removed, if that is imaginable, the passage from 'place' to 'place' happens over time but on the spot: going from, say, Manchester to London would constitute Manchester changing into London – certainly a change of state. This cognitive move motivates many English expressions, as has long been noted (e.g. Lakoff and Johnson 1980). Initial states are places from which a metaphorical motion occurs; final states are places to which there is motion. This is suggested by the following examples:

(18)   The liquid went from green to red

(19)   The economy moved from recovery to expansion

(20)   The economy went (?came) into recession

(21)   The economy came (went) out of recession

(22)   From being a hamster Fred turned into a monster

(23)   Jane went (*came) berserk

(24)   Jane came (*went) to her senses.

In (18) the properties red and green are analogous to spatial starting and finishing positions, though the liquid is not displaced. In (19) similar points can be made. It is true that there are restrictions on the verbs of motions: *move* will not work in (18), though *went* does work in (19). DST is not able and not concerned to explain such restriction, interesting as they are. The point is that motion verbs are used to denote non-spatial change of state. Initial and final states are often further conceptualised in terms of the container image schema, as shown in (20) to (24). These examples show that some changes seem to be sensitive to direction, where 'direction' must have an abstract meaning by way of metaphor. Again, the crucial point is that motion verbs are used in the first place; the present aim is not to attempt to model each of the above sentences.[9]

Examples such as (18) to (24) suggest that it is appropriate to use the geometrical notion of a displacement vector to model change-of-state meanings that are expressed as spatial movement metaphors. There also seems to

---

[9] The explanation for the *come/go* restrictions is a separate matter that is worth separate detailed treatment. For example, (23) and (24) have to do with alienation from self.

be enough motivation to do the same for verbs that have no movement-related morpheme as their root but that do leave grammatical traces, in the form of *from-* and *to-*phrases, that path conceptualisations are involved, as in (25):

(25)   The weather changed from bright to cloudy.

The next move is to extend this characterisation to verbs that equally denote change of state, even though they do not leave overt evidence of spatial movement concepts, as is the case for (26) to (28) below:

(26)   The weather became cloudy

(27)   The sky reddened

(28)   The vase broke.

The verb *become* in (26), despite its etymological origin, probably does not activate movement concepts, at least directly, in the minds of contemporary speakers of English. This is not to say that spatial movement is absent from the conceptualisation of change in general. In fact, I am proposing, for simplification, that there is reasonable motivation for modelling all change of state verbs as displacement vectors in DST. Thus (26) to (28) are all represented by displacement vectors. There are theoretical alternatives to this proposal. In extremely cursory terms, one is to use components such as BECOME in a decompositional logical formulation (cf. Dowty 1979, Jackendoff 1990). Though they do not make cognitive claims, such accounts seem to assume the reality of abstract components such as BECOME. The present account, in following through the general claims of cognitive linguistics, is maintaining the hypothesis that all change-of-state concepts may in some way be linked with the cognition of spatial motion and that in consequence displacement vectors are an appropriate and well-motivated modelling notation, whatever the morphological structure of verbs expressing change of state. We have already argued the case for viewing properties as points in deictic space linked to an entity by a position vector. Changes of state are now modelled by the movement of an entity from one property point to another, making the assumption that the entity has already been positioned with respect to one property, even though this initial property location is not explicit in the verb itself. Sentence (27), for example, would be diagrammed as in Figure 3.9. There is no label for an initial property location (contrast the labelling in Figures 3.7a and 3.7b for initial spatial location).

Figure 3.9 can be glossed as 'the sky went red', 'the sky became red', 'the sky changed to red', 'the sky reddened'. There are subtle meaning distinctions between these variants but DST is unable to distinguish them; what it does do is capture the generalised conceptualisation that they all share. Figure 3.9 is

3.3 Force vectors and transitivity

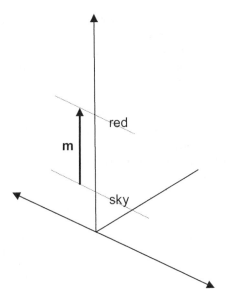

Figure 3.9 Example (27) The sky reddened

analogous to Figure 3.7a, which models the purely spatial sentence *Li travelled from Beijing to Guangzhou*, except that there is no source *from* phrase. Change-of-state verbs that incorporate a property term, such as *redden, blacken, stiffen, dry, freeze* ..., do not allow a source term (**the sky reddened from grey*), but if the change-of-state sentence were manifest as, say, *the sky went from grey to red*, then the starting property *grey* would be diagrammed in the same way as *Beijing* in Figure 3.7a.

Sentences like (28) present events as taking place without causation, though typically, in human experience, there is a causative force. The same modelling device applies – that is, the vase shifts to the property broken, and the change is diagrammed as a vector, specifically a displacement vector. In fact, many change-of-state verbs participate in alternation patterns that enable them to take on causative meanings, as in *Billy broke the vase*. It is not difficult to see how DST can model externally caused change-of-state verbs but first we need to consider the more general role of force vectors.

3.3 Force vectors and transitivity

In mathematics, physics and engineering force is an influence that causes an entity to go through some sort of change, for example movement, direction of movement or alteration of shape – usually stated as changing an object's velocity, including causing it to move from a state of rest. Though Newton gave us a mathematical formulation, force is an embodied concept: push and pull. In physics and engineering forces are represented as geometric vectors, visualised as directed line segments – i.e. arrows. Forces have sources with positions in reference frames and are directed at particular objects with their own positions. We shall ignore the convention that the length of the arrow is proportional to the force but apart from that there is good reason to exploit this already standard approach to the natural concept of 'force' as understood in cognitive semantics since Talmy (2000 [1988]). The link between the geometrical concept, human cognition and the semantic category of 'force dynamics' was noted by Deane:

Talmy's theory is a striking example of a psychologically plausible theory of causation. Its key elements are such concepts as the (amount of) force exerted by an entity, the balance between two such forces, and the force vector which results from their interaction. Such concepts have an obvious base in ordinary motor activities: the brain must be able to calculate the force vector produced by muscular exertion, and calculate the probable outcome when that force is exerted against an object in the outside world. (Deane 1996: 56)

The aim here is not to replace Talmy's theory or its principled diagram system. Rather, the aim is to show that the vector representation system in DST is compatible with it, albeit a simplification, and the purpose of this is to make it possible to use the geometric properties of vectors in frames of reference, a different project from Talmy's. DST's use of force vectors is an abstraction with respect to force dynamics. Unlike Talmy's system, in DST the use of vectors does not distinguish between stronger and weaker forces or between objects tending to rest and objects tending to act, nor does it seek to discriminate between types of causal meanings such as *let*, *keep*, *prevent*, *help*, *hinder* in addition to *cause*, *make*, etc. It should, however, be noted that verbs such as *help*, *prevent*, *hinder*, *stop* pose some particular questions of conceptual modelling that are not addressed in force dynamic semantics. The complement clause constructions are variable distributed across these verbs. For example, *let* takes a zero complement (let her do it), *help* may take zero or *to* (*help her do it* or *help her to do it*), *prevent* takes a verb with *ing* ending (*prevent her doing it*). Moreover, these verbs involve an epistemic dimension: for example, the complement clause of *prevent* represents an event that does not take place. We revisit these matters in more detail in Chapters 8 and 9. In general, however, we shall simply use force vectors to model all such causal relations: the tail of the vector is the source of some force, the tip is the entity toward which this force is directed.

A crucial feature of this approach, as in Talmy's, is that force concepts underlie not only meanings relating to physical causes and effects but also the meanings relating to psychological, logical, social, psychosocial 'forces' and other relations of a more general kind. The fact is that one and the same grammatical construction – transitive verb constructions – lumps together relations of directed force many of which do not involve what modern physics would regard as force. Language is likely to be based on an instinctive 'naïve physics' that is part of our nervous systems. This is in fact Langacker's starting point: human language presupposes a 'billiard ball model' of physical interaction (Langacker 1991: 13, 282ff.) and his 'action chain' model of events as conceptualised by way of verb semantics is presented in terms of a 'transfer of energy' or 'energetic' interaction between linguistically conceptualised entities. Croft has also adopted Talmy's force-dynamic account in his analysis of the causal structure of events (Croft 2012: 197–217). The diagrams used by Croft are geometric to a limited extent and do not make use of the potential of force vectors, which provide a natural geometric instrument for modelling force concepts in lexical and grammatical structures.

3.3.1 Physical force transitives

We have used positional vectors to stand for relations between entities in physical space, relations expressed by adpositions (prepositions and postpositions) in languages. We have also used displacement vectors for modelling spatial movement, in which one entity is involved. Syntactically languages use the prototype intransitive clause (subject noun phrase plus verb) structure. The relations among entities that natural languages express are not only spatial. In addition to prepositional expressions, two objects are related by the prototypical transitive clause structure (subject noun phrase, verb, object noun phrase) structure, and three objects by ditransitive structures (subject noun phrase, direct and indirect object noun phrases). Semantically these grammatical structures prototypically express one-place, two-place and three-place predications. While grammatical subjects can have some form of grammatical marking (inflection, word order), they also have semantic significance. The tendency is for grammatical subjects to be agents. In DST they will always be at the tail of vectors – the source of some degree of force (or in Langacker's terminology, 'energy flow'), not necessarily physical force, of course. Subjects also tend to have cognitive prominence: they tend to be foregrounded as a focus of attention, and are likely to be *topics* in ongoing discourse, and are thus modelled in DST as 'closer' to self on the d-axis. This is the prototypical pattern: as is well known, grammatical constructions of various types (e.g. passivisation) make it possible for grammatical subjects and semantic agents to be separated and to be backgrounded or excluded, in such a way that sufferers of actions (patients) can be grammatical subjects and foregrounded. Further, grammatical subjects in prototypical (non-passivised) clauses need not be agents – instruments can be subjects, as in *the knife cut the string*. Knives cannot be agentive in the same sense as human actors. But they can, as Langacker (1991: 391) argues, be agent-*like*. Merely being a grammatical subject confers this agentive element of meaning.

The problem of transitive subject addressed by Langacker (1991: 309–11) arises in part because of his postulate that 'transfer of energy' (i.e. force relations) between entities is fundamental to the way humans conceive events. The prototypical relation between entities is, in Langacker's perspective, a flow of energy from a source object to another object. Languages encode this conception in transitive clause structures, and extend this structure to non-prototypical directed relations between objects. The problem lies in explaining how the non-prototypical cases are related to the prototype. Many transitive verbs seem not to involve physical force, or even contact or influence. Langacker therefore seeks an overarching schematic concept of the agentive subject that will cover all kinds of transitive verb constructions.

The kinds of transitive verb construction that cognitive linguistics is trying to unify can be illustrated in the following sentences, and are widely known in the literature. Verbs that actually denote physical force do not necessarily denote causative action or resultant effect. Certain verbs denote a directed force but no resultant effect:

(29)   a Jake leaned on the counter
         b The counter supported Jake.

It is arguable that the appropriate conceptualisations for (29) should involve force directed by Jake and an equivalent countervailing force from the counter – a conceptualisation that can easily be represented by two equal and opposing vectors, which cancel out in stasis. But this is what we know from mechanics. Linguistic constructions do not explicitly express such a conceptualisation; rather they privilege one source of directed force or the other, irrespective of what is known, as is clear from (29). These two sentences also show that it is intentional force directed by an animate agent that is privileged; (29b) cannot even really be claimed to be the exact inverse of (29a). The DSM for (29a) would simply be proximal *Jake* at the detail of one vector whose tip is at *counter*.

Many verbs, however, require a complement that denotes the effect of an applied force:

(30)   a The burglar forced the crowbar into the gap
         b The burglar forced the householder to hand over the cashbox

3.3 Force vectors and transitivity

    c John made Mary write the report
    d The gale caused the tiles to shift.

The resultant effects of force application are expressed either in prepositional phrases as in (30a) denoting a movement or in some type of complement clause (*to* complentiser in (30b and 30d) and zero complementiser in (30c)) – about which more is said in Chapter 9. Effects, or resultant states, are not necessarily expressed but are implicit in the semantics of certain verbs.

(31)   a John pushed/pulled/shoved/dragged the cart to the gate
         b He opened/closed the window.

Such verbs inherently imply an effect – often as in (31) caused motion, though this may depend on conceptual knowledge concerning the objects at which force is directed (some objects are known to be unmovable or may be known to be unmovable in a particular context):

(32)   John kicked the bolted door in frustration.

The simplest verb types to model are those that typically cause an impacted object to move. The DSM would combine two vectors: a force vector whose tip is at the impacted referent and a displacement vector whose tail is at the same referent – making that referent simultaneously the recipient of a force and the moving entity. The tip of the displacement entity may be unspecified and unlabelled, or it may be specified and labelled as for (31a).

    There is a set of verbs whose semantics involve force applied by hand and arm leading to change of position in relation to the self. The verb *hold* is one of these and can be represented by a force vector: an animate agent uses the hand (claws, paws, etc.) to exert force and retain an object within peripersonal space. Some such verbs may also possession concepts and thus a position vector; *have* appears to involve position but not force. The verbs *take*, *put* and *get* have interesting internal semantics involving force and translation directed in specific ways. While individuals verbs have their own associated conceptual schema (on the semantics of *get*, for instance, see Chilton 2009), in DSMs they are simplified to single force vectors.

3.3.2   *Combining vectors: transfer verbs*

Examples (30) and (31) were designed to illustrate how force vectors can model quite closely English verbs denoting physical force and causation. Combinations of force vectors and displacement vectors can be used to model the semantics of verbs that denote caused change of position or possession including the deictic element that is involved in the alternating pairs exemplified in sentences (33) and (34) below. These sets of sentences also illustrate

76   Distance, direction and verbs

the proposal that it is not only physical objects that are conceptualised as transferrable from one location to another, but also information or knowledge. They also illustrate the related proposal that it is not only physical actions and forces that cause transfer by direct contact, but also social transactions such as buying and selling, and forms of communication in teaching and learning.

(33)   a Jake gave the code to Bert
        b Jake sold the code to Bert
        c Jake taught the code to Bert.

The converses of (33) are:

(34)   a Bert received/got the code from Jake
        b Bert bought the code from Jake
        c Bert learned the code from Jake.

The verbs in (33) and (34) all use the same construction of transitive verb and prepositional phrase although the denoted events are different in kind: a and b may or may not be a physical transfer – a caused movement; the semantic frame of b necessarily includes exchange of currency; c cannot be understood as a physical transfer. The alternation of *to* and *from* phrases in sentences in (33) and (34) clearly points to the relevance of spatial, in particular, motion in attempting to generalise over specific semantic differences. The grammatical construction in effect abstracts away from these differences. The transitive construction with alternating prepositional phrase can be modelled as combining two kinds of vector, a force vector and a movement (displacement) vector, labelled **f** and **m** respectively in Figures 3.10a and 3.10b. These vectors are left otherwise unlabelled, in order to bring out the common conceptual element across the sentences of (33).

In Figure 3.10a (for sentences (33)) we can take the vector with its tail at *Jake* to be a force vector, for in (33a) we can understand Jake as applying some force to *code*, causing it to move to the location labelled *Bert* – and we are taking the step of extending such a force to more abstract cases of (33b) and (33c) as well as to (34b) and (34c). Note that in such sentences the presupposition is that the transferred object is initially at some time t_i located in the peripersonal space of the grammatical subject, here *Jake*, and subsequently at time t_j positioned at the distal location labelled *Bert*. For the sake of simplification, this time progression is not modelled in Figures 3.10a and 3.10b, though it is perfectly possible to do so in the coordinate geometry we are using; nor is the initial position of the *code* (at Jake's coordinate) indicated, though one might chose to regard the *Jake* vector as a position vector at time t_i. The vector with its tail at *code* can be understood as a movement (displacement) vector: in (33a) we can understand a physical displacement and extend this to the abstract cases (33b) and (33c). The basic

3.3 Force vectors and transitivity

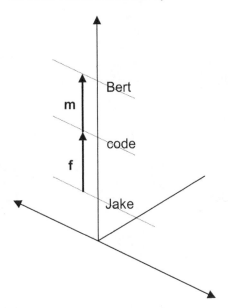

Figure 3.10a Example (33a) Jake gave the code to Bert

transfer sentences of (33) can thus plausibly be modelled by combining a force and a movement vector as in Figure 3.10a.

However, in the converse cases (34) some questions arise as to the modelling. One aspect of this is relatively straightforward. The crucial difference between the alternating converses in (33) and (34) concerns a change in 'perspective', while the direction of the transfer, i.e. the movement (physical or abstract), is preserved. The DSMs are able to capture this relationship, because they are able to represent the relative conceptual 'distance' from S of the alternating grammatical subjects and simultaneously represent the required direction of transfer. Thus in Figure 3.10b the ordering of *Jake*, the *code* and *Bert* in sentences (34) is transposed to the *d*-axis: *Bert* is in focus or 'closer' to S, *Jake* relatively more 'distant' on the *d*-axis. But – if we seek to reflect the meanings of the converses in (34) – it is not simply a matter of reversing the vectors, as one might assume. In addition to the perspective shift, the semantics of the converse verbs appears to be different. Applying a vector analysis brings this out.

Notice that all the sentences in (34) have an 'inactive' and an 'active' understanding (we are not speaking here of active and passive constructions), an ambiguity that is particularly clear in the case of *get*. That is to say, *Bert got the code from Jake* can mean either the code just came to him without action on his part, or that Bert applied some active force

78 Distance, direction and verbs

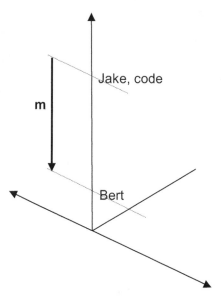

Figure 3.10b Example (34a) *inactive* Bert received/got the code from Jake

(physical or not) to obtain it from Jake, force possibly applied to the code (grabbing it, for example) or to Jake himself or both. The degree and types of force are not specified in the linguistic structure, of course. The verb *receive* may be understood 'inactively' or, depending on context, to some extent 'actively' (e.g. receiving guests); *buy* appears to involve more activity ('force' in the general sense I am using the term here), and even *learn* can be understood in terms of lesser or greater energy expenditure ('passive' or 'active' learning).[10] It appears that the converse verbs in (34) do not necessarily imply force, or at least they imply different degrees of force. Figures 3.10b and 3.10c offer a proposal for different DSMs that capture such differences. The aim is merely to indicate how this kind of vector analysis can be used to explore fundamental semantic–conceptual structures and distinctions.

Figures 3.10b and 3.10c are attempts to outline the two possible readings of verbs like *receive* and more obviously *get*. Figure 3.10b is intended to correspond with the 'inactive' reading of sentences (34): no force vector is included, merely a movement of the code from Jake to Bert. It is Bert who is foregrounded, having initial position in a sentence such as (34a).

---

[10] Compare also verbs such as *frighten*, *scare*, *embarrass*: e.g. Jo embarrassed her partner.

3.3 Force vectors and transitivity

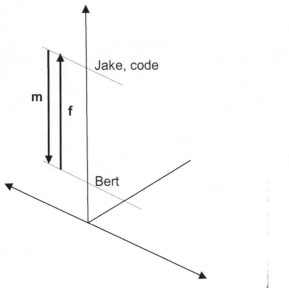

Figure 3.10c Example (34a) *active* Bert got the code from Jake

In Figure 3.10c, on the other hand, while keeping Bert in the foregrounded focal position, i.e. relatively 'close' to S, there is a force vector representing force exerted by Bert on the code, and perhaps on Jake (relevantly, the diagram does not specify this any more than does the sentence) and also a movement vector translating the code to Bert, as a caused result. There is some ambiguity here, and one simply has to stipulate that in Figure 3.10b it is the code that is moved from one coordinate to another, not Jake. As noted earlier, the schema for *get* etc. can be more fully modelled by taking the time axis into account but is simplified here.[11]

3.3.3 Force as cause: resultatives and changes of state

We move now to another semantic verb class whose fundamental conceptualisation can be understood in terms of direction, motion and force, and whose grammatical realisation in English is in the form of transitive clause constructions. The point is not to *reduce* but to reveal common conceptual scaffolding. The main principle in this approach is, as before, to recognise metaphorisation

---

[11] Figures 3.10b and 3.10c can be seen as versions of the semantic schema for *get*. In Chilton (2009) I propose that the inactive and the active readings are variants of the same schema, the 'inactive' reading having a weakly activated *f* vector.

of physical concepts as itself a fundamental feature of whatever the cognitive processes are that underlie lexicalisation of conceptual diversity in the mental representation and linguistic communication of events.

It is reasonable to assume causation can be modelled by force vectors. We have already argued that force and displacement combinations form the conceptual basis of transfer events. The following sentences resemble the directional verbs in (31) in expressing a directed force applied by an agent that causes an entity to traverse a path. The prepositional phrases denoting a type of spatial path are often sometimes called resultative paths (cf. Goldberg and Jackendoff 2004). The verb need not, unlike the verbs in (31), inherently express propulsive force, as (b) suggests (knocking something does not entail that the thing actually moves):

(35)   a   Bill propelled the stone across the ice
           b   Alf knocked Bert through the window.

Spatial prepositions appear in resultative phrases that express not a change of location but a change of form:

(36)   a   Bill cut the tree into logs
           b   break the vase into shards
           c   They laughed Sir Percy off the podium
           d   Jim walked his shoes into holes.

These verbs do not inherently express propulsion. While (36a and b) express the application of physical force they do not inherently imply propulsive force, which comes from the prepositional phrase expressing a metaphorical path (*into*) to a transformed state – from tree to logs, from vase to shards. Sentence (36c) shows that an intransitive verb that expresses no physical force at all can be interpreted as a force (here depending on background knowledge of social 'forces') resulting in a physical motion along a schematically indicated path. In the case of (36d), motion seems not to take place in the usual sense of displacement, and expresses a change of physical form. This is tantamount to a change in the properties of an entity. The parallel between the sentences in (36) is conceptually significant and is evidence that it is relevant to model changes of state as 'displacements' from one 'location' to another (compare sentences (18) to (24) above). These constructions can be modelled accordingly by a combination of a force vector – which may be already inherent in the verb's semantics (as in *break*, for example) or induced by the construction itself (as in the case of *laugh* and *walk*, which do not denote force of the relevant kind).

Sentences (37) clearly denote finite processes at the termination of which an entity has acquired properties it did not have at the start of the process. In

3.3 Force vectors and transitivity

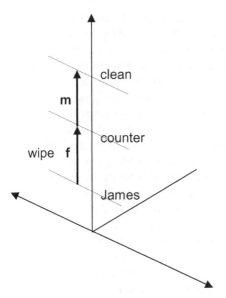

Figure 3.11 Example (37c) James wiped the counter clean

these constructions the acquired properties are expressed grammatically as adjectives (but note that there are restrictions on the combination of adjectives and resultative paths; see Goldberg and Jackendoff 2004: 558–62).

(37) a The water froze solid
      b James wiped the counter
      c James wiped the counter clean.

In Section 3.1.2 it was proposed to model property sentences by using position vectors and in Section 3.2.3 that change-of-state sentences – or acquisition of a new property by some entity – could be modelled by means of displacement vectors. In (37a) the intransitive change of state verb *freeze* has the same DSM as the verb *redden* in (27), modelled in Figure 3.9, with the difference that the displacement vector is labelled *freeze*. Sentences (37b and c) illustrate a construction that contains both a causative and a change-of-state element, modelled as a combination of a force vector and a displacement vector, as in Figure 3.11.

The **m** vector is the same representation as in models of intransitive change-of-state verbs (e.g. 27); it is also possible to interpret this kind of example as a position vector and take *clean* to be a property possessed by *counter*. As in other examples, the process of wiping takes place over time and the property *clean* is not possessed until the action is complete. In a conceptual model of the verb's semantics this would be modelled but in a sentence, tense forms also have a conceptual effect and in the present example the simple past conceptualises the action of wiping not only as past but as unitary. Notice two other features. First, the **f** vector for *wipe* does of course represent a physical action, while 'moving the counter to a property location' is an entirely abstract representation by metaphorisation. There seems to be no reason to object to this; it is just the way language and conceptualisation work. Second, the configuration of the DSM for the change-of-state verb in Figure 3.11 is the same as for the transfer verb in Figure 3.10a. DST does not model all semantic differences but aims to bring out underlying spatial and deictic scaffolding. In particular, in this chapter we are concerned to bring out the commonalities underlying transitive constructions.

A number of the above examples illustrate the well-known alternation between causative and inchoative meanings, often corresponding to transitive and intransitive uses of one and the same verb. These are verbs whose root semantics indicate the process that causes a new state, the nature of the change, and the resultant state itself. They are directional: that is, the force has a source and acts upon an entity. The most consistent way to model this conceptualisation is by two vectors. The model for the inchoative alternant follows from it.

It would be possible to argue that wiping results in the state of being wiped, indeed the state of being clean. One cannot say **James wiped the counter dirty*. However, there are apparently similar verbs that, unlike *wipe*, cannot take a resultative phrase: one does not say **John broke the vase broken*. Further resultative phrase verbs such as *wipe* do not have inchoative alternation: one can say *the vase broke* but *the table wiped* is not interpretable in that form.[12] Verbs such as *break*, which do alternate, require a slightly different account. While the basic model is similar for both kinds of verb, the *break*-type construction has a simple extension that captures the difference. In (38a) and (38b) the resultant state is the state of being broken, the result of the action of breaking.

(38)   a John broke the vase
        b the vase broke
        c the vase is broken.

Like sentence (37), sentence (38a) can be analysed as having two component vectors, one a force vector, the other a displacement (movement) vector. The force vector causes the movement. The concept of directed movement metaphorically motivates the concept of change of state (see above, Section 3.2.3).

---

[12] It can though have a 'middle' reading with adverbs, e.g. *formica-top tables wipe well*.

3.3 Force vectors and transitivity   83

The verb *break* represents a change of physical state, states being analysed as abstract locations.

As is required, Figure 3.12a for *John broke the vase* has the same form as Figure 3.11 for *James wiped the counter clean*. Turning to (38b), the inchoative alternate of (37a), in Figure 3.12b the movement vector **m** gives the conceptual structure of the 'middle voice' illustrated by (38b) the *vase broke*: the vase spontaneously 'moves to' or 'moves into' the state of being broken. This intransitive construction focuses the communication on the change of state and omits the **f** vector. Of course knowledge of the world tell us that vases do not break spontaneously, but the linguistic construction does not care about that; if there is a mentally represented vector **f** at all here, it is weakly activated. As in earlier examples (*get* is one) the vector **f** fades into the background conceptually. Further, we can derive the predicative sentence (38c) in a simple way by interpreting the change-of-state **m** vector in Figures 3.12a and 3.12b as a position vector, i.e. property vector (see above, Section 3.1.2). The same is the case for (37): 'the counter is clean' is entailed by *James wiped the counter clean* and can be modelled by a position vector. The approach outlined here makes it possible to give a unified account of sentences such as (38). It must be noted, however, that this requires including the state predicate *broken* as a label on the *d*-axis, although it is not morphologically distinct in the sentence form.

3.3.4   The transitivity of perception and mental process

In this section I return to constructions that involve a relation between two entities. The logical notion 'two-place predicate' captures only part of what grammatical transitivity in natural language is about. It is important also not to be too determinate in seeking a conceptual schema behind transitivity. Langacker (1991: 303–13) takes energy transfer in an action change as the prototypical conceptual structure of transitive constructions, with extensions for cases such as (39) below, where the verb semantics appear not to imply energy transfer. It is clear, however, from Langacker's account that the most general schema underlying transitives is unidirectionality. In other words, a vector relation, with grammatical subject at the tail, grammatical object at the head. DST models this approach in a unified fashion, incorporating a conceptual understanding of the standard types of vector. The general underlying schema of transitivity is not 'energy transfer' but rather directionality understood variously in terms of position, movement and degrees of force.

Since we know perception of an entity does not involve a force that affects the entity perceiverd, the fact that English and other languages code perception verbs in transitive constructions has often been felt to require

84   Distance, direction and verbs

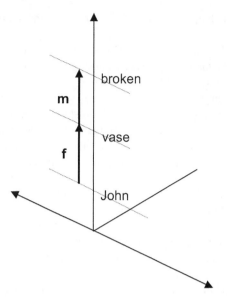

Figure 3.12a Example (38a) John broke the vase

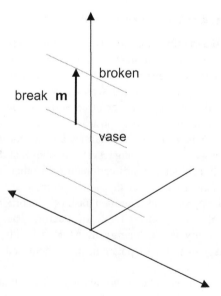

Figure 3.12b Example (38b) The vase broke

3.3 Force vectors and transitivity

explanation, or at least further description. One approach is to set up specially labelled semantic roles such as *experiencer–experienced*, with subcategories *perceiver–perceived*, *cogniser–cognised*, etc. Finer semantic role categories are possible (e.g. the list in Van Valin 2001: 29–31). However, such fine categories appear to be the consequence of a particular verb's semantics – for example, in the case of the verb *smell* one might say there is a role *smeller*, which is a subcategory of *perceiver*, which is a subcategory of *experiencer*. This kind of classification has no formal status in DST. The task now is to consider how the vector-based approach might be used to model transitivity relations between the entities involved in the semantics of verbs of perception, emotion and cognition.

English perception sentences such as *Alice saw the rabbit*, *Alice watched the rabbit* and the archaic *Alice beheld the rabbit* use transitive verbs, and thus appear to imply some kind of contact proceeding from the perceiver to the perceived object.

(39)   a   Alice saw the rabbit
        b   Alice beheld the rabbit
        c   Alice watched the rabbit
        d   Alice gazed at the rabbit
        e   Alice stared at the rabbit
        f   Alice looked at the rabbit.

In the world of physics none of the events denoted by these sentences involves an individual impacting an object. This is why Langacker feels it necessary to explain such transitive verbs as extensions of a prototype based on the tendency of the human mind to conceptualise and communicate abstract processes with reference to concrete ones. However, there is a simpler and more basic explanation. Perception, vision being the clearest case, is phenomenologically an 'energetic' or 'forceful' process anyway at the level of 'naïve theory' and even at the level of everyday thinking in modern humans. From the perspective of the experiencing self, visual attention is actively directed towards an object.

Though 'naïve' is not a term one would want to apply to Euclid and Plato, both held theories of visual emission – Euclid significantly enough stating it within his geometrical framework. Children have a spontaneous visual emission theory according to Piaget (1967) and according to a more recent study about 50% of American college students take the emission theory for granted, asserting that the eyes emit rays that impact objects, even if they have studied optics (Winer and Cottrell 2002). However, the significance of these experimental results is not that these students hold (if they do) a theory of visual emission. It is simply that, phenomenologically, visual experience involves attention and visual attention is experienced as directed at an object, the source being the experiencer. In a critique of the experimental procedure used by Winer and Cottrell, Bahr (2003) suggests that their finding may be an artefact of a 'cognitive-linguistic confound' brought about by the linguistic form of sentences using *see*. That is, Bahr suggests that the grammatical structure of *see* sentences inclines people to think in terms of visual emission. There is no evidence of any such Whorfian effect – one in which linguistic structure alone would be causing the effect observed in the experiment. What is interesting is that Bahr also suggests that English 'mirrors' the experience of directing one's attention in looking at objects. So even on Bahr's account there is a natural cognitive tendency to think of vision as the emission of visual beams impacting objects. Consequently there need be no purely linguistic confound in the experiment Bahr is concerned with. In the context of cognitive linguistics and the DST analysis of transitivity both the experiment and Bahr's comments are of interest. The point is that people appear to have a subjective experience of directing attention to a target, independently of language. It is this non-linguistic experience of directing attention that can give rise to the transitivity grammar of perception verbs like *see*.

DST provides a motivated framework for modelling these observations. The framework it puts forward is precisely a model of a cognising self diverting attention to objects in different dimensions of experience. In general, the semantics of natural language, including its grammatical constructions, is more likely to incorporate the structures of consciousness as studied in phenomenology than it is to correspond to formal logic or physics. If this statement may seem at odds with the use of coordinates and vectors, and with talk of forces and displacements, it should be recalled that vectors, in the way I am applying them here, are themselves embodied. As Wittgenstein noted, 'The arrow points only in the application that a living being makes of it.'

Even in scientific discourse the emission theory of vision appears to be natural, at least in the context of research into attention. Psychologists and cognitive scientists speak of the 'attentional spotlight' or 'attentional beam'. They are of course aware that they are using a heuristic metaphor that facilitates communication about the abstract idea of attention. Leclercq and Zimmermann (2002: 14) write:

... visual attention is comparable to a light beam, i.e. a selective lighting with a certain intensity, which would correspond to the degree of attentional investment.

These authors cite van Zomeren and Brouwer (1994), who use the same terms:

[this expression is] used in a general sense in which the spotlight is a metaphor for selection enhancement possible at different levels of perceptual, cognitive and motor representation.

3.3 Force vectors and transitivity

Not only is the need to have recourse to this metaphor evidence of the naturalness of the emitted light beam conceptualisation, but its use as a heuristic has also led to empirical findings concerning attention that are relevant to linguistic models. The metaphor entails the notion of the width and intensity of the 'beam' – the narrower the beam the narrower the selectivity (or focus), the wider and more global the attention the wider the selectivity. The narrower a beam, the more 'force' it is felt to have.[13]

Let us consider now the semantics of the vision transitives in (39), focusing on *see* and *look*. Is a vector model appropriate? Gruber (1976) argues, within the early transformational–generative paradigm, that *see* is a motion verb incorporating underlying preposition *to*, while *look* is also a motion verb but based on the preposition *toward*. Another important semantic difference is the agentivity of *look* as opposed to the stative characteristics of *see*. Gruber comments:

> ... we may consider a sentence such as 'John sees a cat' to be a metaphorical extension of 'John goes to a cat', noting that 'John's gaze goes to the cat' is 'close in meaning' to 'John sees the cat' (Gruber 1976: 941).

These observations are at least compatible with the DST approach, which is based on *directionality* from the point of view of a cognising subject. The prepositions *to* and *toward* are also clearly relevant, since both are also directional; the crucial point is that *to* more easily entails attainment of a goal, whereas *toward* is more likely not to entail attainment, whether physically spatial or abstract–intentional. Building on these points, DST may be able to give a unified conceptual account. The semantic distinction between verbs such as *see* and *look at* emerges from the standard vector concept, drawing not on motion (displacement vectors) primarily but on the force and position interpretations of vector directionality.

In (39) sentences a, b and c are syntactically transitive; sentences d, e and f require a preposition, typically *at*, although other spatial prepositions are normal. In all cases, however, it is appropriate to use a vector model. All are clearly directional. All are in addition related to force concepts, at the least to the kind of weak force indicated by 'contact', in this instance what Langacker calls 'mental contact'. The etymology of *behold* in sentence (39b) strongly suggests that a force vector is relevant to model the cognitive processes that gave rise to this verb, largely defunct though it now is.[14] The early sense in English was physical, roughly 'hold by, i.e. close to, oneself', whence observing, active looking and finally the more passive seeing. The weakening of the force component is of interest. It is also worth considering that the other vision

---

[13] The research literature actually distinguishes five kinds of attention and two 'axes', the latter being *intensity* and *selectivity*.

[14] In English. Other Germanic languages did not have this development of *behold* cognates.

transitives, *see* and *watch*, also may be conceptually linked with force, as well as direction. This would be consistent with the subjective 'beam' model of attention, which includes variation in the 'intensity' of concentration. Sentences (39a, b and c) are relatively 'weak force' compared with (39d, e and f). The semantics of *watch* contains elements in particular, temporal duration that are specific to that verb, and has relatively stronger force than *see* and shares conceptual elements with the set (39d, e and f). This is picked out by the standard test for stative meanings: *see* and *behold* cannot take the progressive (**Alice is seeing the rabbit*, **Alice is beholding the rabbit*), whereas *watch* (39c) and the verbs in (39d, e and f) can. Note, however, that though *watch* shares this 'active' sense with gaze, stare and look, (39d, e, and f) are distinct by virtue of requiring a spatial preposition.

The notion that weak and strong directed force is part of the semantics of vision verbs – conceivably also of emotion and cognising verbs – is consistent for a further reason with the intuitive accounts in the psychology and phenomenology literature. *Seeing* has lower 'force' but is more holistic, has wider scope and takes in more context, as well as being associated with a sense of lower controlling activity; *looking* is high intensity, narrow focus, active, deliberate and seeks to 'fix' an object. Subjectively, the way English *see* and *look* are used suggests the influence of phenomenologically different modes of visual cognition: it is not contradictory to say that a person is looking straight at something but does not see it.

In the light of these ideas it seems cognitively motivated to model vision verbs as force vectors with varying degrees of force. This is not the whole story, however, since the presence of prepositions with *look* and other vision verbs is also evidence of underlying conceptual structures. While both *see* and *look* can take a range of spatial prepositions – which in itself supports the vector-based approach – the prototypical preposition seems to be *at*, as in (39d, e and f). Why *at*? There are a number of considerations.

First, this preposition is associated with a geometrical schema with zero dimensions – it is point-like – in contrast with *on*, for example, which requires a two-dimensional plane as part of its conceptual schema, and in contrast with *in* and *out*, which require a three-dimensional topology. In this sense, *look at* may have conceptual connections with locations conceived in point-like ways. Again, this is consistent with the attentional spotlight model in the sense that to look at is to focus on an object with high-intensity (high force) attention, or concentration of the beam ('eyebeam'). This does not mean, however, that the entity in an *at* prepositional phrase is 'small' – it may be diffuse but is nonetheless a conceptually dimensionless portion of space on which attention is focused. The difference in meaning between *at the university* and *in the university* should make this clear. Looking into a box is not the same as looking at a box.

Second, however, the preposition construction required by *look* iconically distances the verb from the grammatical object, which, despite the active concentration, may correspond with the cognitive distancing and lower sense of contact – looking but not seeing. Metaphorically, and blending these ideas with cognising concepts, one can listen *to* but not *see* the *point*.

Third, the preposition *at* occurs with verbal constructions associated with sustained exertion of force on a point. Hitting a nail implies direct contact and successful hitting, hitting *at* it may imply failure of contact.

There is more that could be said on these issues but it seems to me that there is enough to justify treating transitive perception verbs as force vectors, with varying degrees of strength. It also seems to me that *look* involves a spatial dimension as well, which would require treating the relation between the looking entity and the entity looked *at* as a spatial relation, which would motivate the use of a position vector. Certainly, Gruber's proposal that *see* and *look* are motion verbs does not seem warranted. In fact, the DST framework outlined so far would not make sense of it, since modelling these verbs as displacement vectors would mean that the *see-er* and *look-er* themselves change location. The evidence outlined above seems to indicate that DST is 'predicting' in the right direction in this respect.[15]

It is also worth noting that *see* is not, in cognitive–linguistic terms, semantically equivalent to 'direct one's sight towards some entity'. The DSM for a sentence such as *Albert directed his gaze towards the poster* would involve a force vector with head at the coordinate on *d* for *gaze*. There would be a second vector for the spatial–directional relation *toward the poster*.

Verbs linked with the other sensory modalities are also amenable to an analysis in terms of natural geometric vectors and the psychology of attention. I shall not pursue a detailed analysis here but simply make the assumption that space, direction, position and force also underlie the semantics of verbs such as *hear* and *listen*, *feel* and *touch*.

Can emotion verbs (also known as 'psych-verbs'), such as *fear*, *terrify*, *amuse*, *admire*, etc. also be modelled by drawing on vector-based concepts? Emotion verbs are transitive, even though they are often described as 'abstract'. Their grammatical subjects are animate (saliently, human) and fall into two types (cf. Levin 1993). The first type induces a meaning in which the entity denoted by the subject noun experiences an emotion, as exemplified in (40a). In the second type it is the grammatical object that denotes a sentient

---

[15] This may be too strong. A displacement model might correspond with the morphological structure of perception sentences in Samoan (as described by Langacker 1991: 304), where there is a surface morphological parallelism between 'the boy went to the store' and 'the boy saw the ship'.

being experiencing an emotion, while the subject noun causes this state, as is the case in (40b):

(40) a The Gauls feared the Roman army
      b The Roman army frightened the Gauls.

In standard theories differences of this kind are handled by labelled semantic roles attached to the grammatical roles subject and object, such that in (40a) the subject (*the Gauls*) is labelled 'experiencer' and the object *the Roman army* is labelled 'stimulus' (or something similar), while in (40b) it is the other way about. It is possible, however, to model the semantic difference in DST without recourse to such labels.

Consider first *fear* in (40a). There is intuitively no sense in which the Gauls in (40a) are agentive. The *fear* type is also stative: **The Gauls were fearing the Roman army*.[16] There is reason, then, to treat *fear* as denoting a state or property. In Section 3.1.2, it was argued that an appropriate way to model property predications is by position vectors. One might want to say that it is not sufficient to say that (40a) is equivalent to something like 'The Gauls were in a state of fear with respect to the Roman army', and what seems to be missing in that kind of paraphrase is precisely the directedness that is part of the meaning of natural vectors. In particular, we need to allow the understanding of the fear vector that relates *Gauls* and *Roman army* to retain something of the concept of the attentional beam. For in fearing the Roman army the Gauls are, one would say, paying attention to it.

The verb *frighten* (40b) requires a different analysis, though it is one that is already familiar from the analysis of the causal verb *break*. The verb *frighten* can be treated as a causal verb: if *x* frightens *y*, then *y* is in a state of fear. Building on what was said in Section 3.1.2 to motivate the modelling of property ascription by a position vector, an appropriate DSM would be exactly as Figure 3.12a. On this analysis, *The Roman army frightened the Gauls* is analogous to *John broke the vase*, and causation of a change of physical state is analogous to causation of a change of emotional state. It is natural for a transitive construction to be associated with causative emotion verbs.

There is one difference that should be noted. Whereas the middle construction, seen in (38b) *the vase broke*, is perfectly acceptable with *break*, it is questionable with *frighten*: ?*the Gauls frightened*, though the modified form *the Gauls frightened easily* is acceptable (in an ergative reading). There is variability for physical event verbs also. A rather subtle

---

[16] *fear* is probably a less common way of expressing the emotional state in question than is the predicate *afraid of*, as is true also in other languages (e.g. French *avoir crainte/peur* vs. *craindre*). But this does not generalise to other verbs of this type.

3.3 Force vectors and transitivity

difference between emotion causation and physical event causation in this respect arises if one considers two possible readings of (40b) repeated below:

(40) b The Roman army frightened the Gauls, just the sight of them
      b′ The Roman army frightened the Gauls by banging their spears on their shields.

In (40b) the Romans are not active, in (40b′) they are. It is possible to capture this difference by the choice of semantic role label. For example, *The Romans* in (40b) would be 'stimulus' but in (40b′) they are 'agent'. However, consistently with the use of vectors I have been proposing, there are two ways the difference in question might be dealt with. One is by treating the 'weaker' meaning of (40b) as a 'weak' force vector and the meaning in (40b′) as a 'strong' force vector. Alternatively the 'weak' meaning in (40b) could be treated as a position vector: this would keep the directionality of the transitive construction, treating the *frighten* relation as a position vector relation, while the 'strong' meaning of (40b′) remains modelled by a force vector (causal) and a movement vector (change of state).

Cognising verbs such as *know, learn, understand, comprehend, grasp, conjecture, guess, suspect, anticipate, remember*, etc. may seem even more abstract than emotive words. Cognising verbs are verbs that enable speakers to make a certain kind of truth claim about the state of someone's mind (and their own) in relation to some entity or some proposition. The fact that these verbs are transitive may need explanation if transitivity is thought of as prototypically physical and energy-transmitting. However, if transitivity is thought of as primarily directional, with some kind of residual concept of weak or strong force, the transitivity of cognising verbs is natural and can be modelled naturally in terms of vectors.

It is impossible here to delve into the specific semantic frames for each verb in this category. Consider just the following:

(41)   Jake knew the code

(42) [= (33c)]   Jake taught the code to Bert

(43) [= (34c)]   Bert learned the code from Jake

(44)   Bert learned the code [on his own].

In (41) *know* is a stative verb (we cannot say **Bert is/was knowing the* code) unlike *learn* in (42), (43) and (44), which denotes mental 'action' (and we can say, as for other actions *Jake was/is teaching/learning the code*). Sentences (42) and (43) include syntactic adjuncts in the form of spatial prepositional phrases, both with directional meaning. These facts suggest we are dealing with a fundamentally spatial and vectorial conceptualisation, even for highly abstract cognising verbs. We can be more specific. The stative expression (41) can be related to the modelling of states and properties that was outlined above in Sections 3.1.2 and 3.2.3. Like adjectival expressions of properties and locative expressions for states, stative verbs can also be treated as position vectors. We return to the more detailed modelling of stative verbs in relation to the t-axis in Chapter 4.

DSMs for sentences (42), (43) and (44) by contrast use **f** and **m** vectors (force and movement), since they involve the application of mental energy that bring about a change of state, which, as argued in Section 3.2.3, involves a movement vector. In fact, the underlying conceptual structure of cognising verbs is analogous to that for transfer verbs, discussed in Section 3.3.2 and modelled in Figures 3.10a, 3.10b and 3.10c. Figure 3.10a has the structure that corresponds to (42) – *teach* is analogous to *give*, since there is 'movement' of a mental entity from one mind to another. Figure 3.10b has the structure that models (43) – where *learn* is analogous to *receive* and *get*. The fact that one can idiomatically say that 'Bert got it', in the sense of 'understood it', supports this analysis, although there is more to say about the conceptual relationship between *learn* and *understand*. Figure 3.10c models the basic transitivity structure of *Bert got the code from Jake* but also, it is proposed, of *Bert learned the code from Jake* (43). Figures 3.10b and 3.10c concern the expression of 'weak' and 'strong' force. Is there any such distinction in the meanings of abstract cognising verbs?

The verb *learn* can be understood in both an 'inactive' and an 'active' (weak force, strong force) sense, just as for example can *get*. Learning involves the attentional beam directed at some entity but may involve more or less attentional energy. In example (43), both weak and strong sense may be available, but different referents in the *from* prepositional phrase, and other contextual elements guide interpretation. For example, 'Bert learned the code from a manual' is more conducive to the 'active' or 'strong' reading than 'Bert learned the code on the job'. Either reading is available in a sentence like (44) above, which omits the source. For (44) and the relevant DSM would simply be as Figure 3.10b, with no label for the source (*Jake*).[17]

---

[17] The configuration sketched in Figures 3.10a, 3.10b and 3.10c has the drawback that it does not formally distinguish between source (Jake in the examples) and the entity (the code in the examples) that moves. This is not beyond formalisation, since the *from-* prepositional phrase can be represented by a position vector with tail at *Jake* and tip at *code*: i.e. the code is positioned at, is in the possession of, Jake, in line with the representation of property/ possession relations. For simplicity, I have not explored this more elaborate modelling here.

3.3 Force vectors and transitivity

Though the class of cognising verbs may seem particularly abstract, their appearance in transitive constructions has to do not only with the primary attentional directedness of transitivity, but also with the underlying conceptual source in embodied experience – the directing of mental attention, physical positioning, directed physical locomotion and the directed application of degrees of force. Diachronic evidence also supports the modelling of the transitivity of the 'abstract' cognising verbs by means of natural vectors. *Understand* involves a physical posture (*stand*) with a relation (*under*, meaning something like 'by' or 'close to' in Old English); *comprehend* and *grasp* involve metaphorical transfer from grasping with the hand, i.e. prehension. The idiomatic sense of *get* (get it?) is also based on obtaining by the hand (cf. Chilton 2009).

3.3.5 Problem transitives

Certain verbs in English require grammatical direct object but if transitivity is understood conceptually this requirement poses problems of modelling. Along with this goes the problem that these transitivity constructions, unlike all others, cannot be passivised. The following are the classic examples:

(45)   The package weighed two kilos

(46)   a Albert resembles Bill
        b Bill resembles Albert.

How can we explain the frequently noted fact that (45), (46a) and (46b) cannot be passivised? In the DST approach we want to know if vector modelling makes sense. It is tempting to consider that the verb in (45) might be modelled as a force vector: after all, weighing is related to gravitational force, at least in physics, and phenomenologically weighing might be attributed to the entity that is expressed as grammatical subject, *the package*. However, such an account is not plausible for *resemble* in sentences (46), where there is a symmetrical relationship: truth-conditionally, (46a) is equivalent to (46b). The conceptual difference between (46a) and (46b) has to do with attentional foregrounding, that, is with relative 'distance' from S's cognitive viewpoint. Moreover, one entity is presented as having a property or state relative to another, so there remains a deictic relation that can be appropriately modelled by a vector. There are two alternative approaches.

First, as I have just suggested, these verbs are property-ascription or state-ascription verbs, and as we saw in Section 3.1.2, the relation that such verbs represent can be reasonably modelled in DST as position vectors. The other transitive verb types we have examined have involved force vectors, whether the force is 'strong' or 'weak', and these verbs have no problem with the passive construction. It may then be the case that the passive construction is conceptually sensitive to the type of vector – force or position. Only the force vector verb types can be passivised. It may be objected that if the passive is about switching attentional focus, then one might expect the quasi-spatial position vector to allow for quasi-spatial refocusing. There is perhaps a cognitive answer to this question. Agents, from whom or through whom or by whom force is channelled, are natural candidates for close focus, as are receivers of such force. They are also discrete entities. Properties and states are not easily individuated or attributed with agentivity: they therefore appear unnatural in foregrounded position. The active–passive alternation comes out as specialised for permitting the communicating of alternate attentional focus on participants in *force-based* events – including cognition, perception and emotion verbs, some of which have semantics involving some degree of directed 'mental force', as argued above.[18]

Second, in line with what has been proposed above concerning properties as locations, the appropriate type of vector would be a position vector. We can base on this a somewhat unexpected explanation of why passivisation does not work. It was noted at the end of Section 3.1.2 that the position vector for possession and property could be treated as a zero vector: after all, properties and possessions are co-located with an entity. If this is the case we might propose modelling such relations by a zero vector. To recall, zero vectors are defined mathematically as having no magnitude (length, distance) and as having arbitrary direction. In the present case we may stipulate the direction as pointing from the entity to the property or possession – anchoring a property or possession deictically to its possessor and also in principle maintaining the foregrounding and backgrounding relation. Note that in the non-symmetrical case (45) above, the conceptual frame of *weigh* specifies just one direction, while *resemble* in (46a) and (46b) has alternating foregrounding (compare Langacker 1991: 311–12).

3.3.6 Ditransitive alternation

Ditransitive constructions are found across the world's languages (for an overview see Siewierska 1998, Croft 2001, Malchukov *et al.* 2010). Their common characteristic is that they enable their users to communicate about

---

[18] Foregrounding of the property in sentences like (45) and (46) is not impossible, but it has to be done by nominalisation (the weight of the package was two kilos) or by cleft constructions which maintain the property or state as grammatical direct object: *what the package weighed was two kilos, the person Bill resembled was his brother*.

3.3 Force vectors and transitivity

events that involve a relation between three participants (propositions with three 'arguments' in logic-based versions of semantics). The prototypical case is usually taken to be that of transfer verbs – the type of verb discussed above in Section 3.3.2. Certain verbs of this type, in the languages where they are found, display an alternation known as 'dative alternation' (or 'dative shift'), which in English appears as a 'double object' construction, instanced in

(47) a [= (33c)]   Jake taught the code to Bert
      b   Jake taught Bert the code.

Sentence (47a) repeats the construction illustrated in sentence (33c). Sentence (47b) illustrates dative alternation. While it is often said that sentences of this type are semantically equivalent, this is only true within a logic-based semantics. The fact that languages develop this kind of alternation at all is evidence that there is some cognitively relevant need for it.

The DST explanation is that this need is attentional focus, expressed, in English, by means of linear word order. Yet the common event schema is indeed the same: as in all the sentences (33), Jake is the agent of a force that moves (physically or metaphorically) some entity to another entity location, here Bert. Figures 3.13a and 3.13b below are an attempt, based on the natural

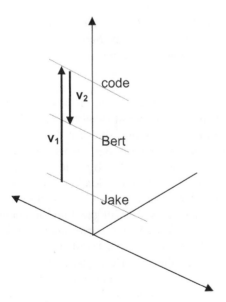

Figure 3.13a Conceptual structure of event in (47b)

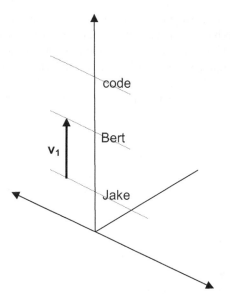

Figure 3.13b Combining the vectors in Figure 3.13a

vector principles of DST, to integrate conceptual event structure and attentional refocusing.

In Figures 3.13a and 3.13b, the entity *Jake* is attentionally 'closer' than *Bert*. In contrast with sentence (47a), the entity *Bert* is 'closer' than *code*. Figure 3.13a starts with the conceptual structure of the event modelled in terms of a force vector v_1 and a movement vector v_2. Jake is the agent of an action that causes the transfer of the code to Bert. This is merely a rearrangement of the configuration in Figure 3.10a, brought about by reordering the participant entities on the *d*-axis. If we now follow vector logic, the combination of v_1 and v_2 results in the configuration shown in Figure 3.13b. The force vector v_1 remains, but is now directed to *Bert*, as we need in order to model the dative alternation construction (47b). The sentence and the proposed DSM foreground the directed interaction (a kind of force) between the agent Jake and the interactant Bert. The entity *code* is attentionally distal relative to S – and also 'stranded', as it is in the English grammatical construction. Figures 3.13a and 3.13b make the combination of conceptual event structure and attentional structure in S's coordinate system appear sequential. It is likely that this is not the case. However, it must also be noted that vector modelling, despite its natural basis, remains a formal methodology for capturing only a part of the complexities of the cognitive processing of linguistic form.

3.3 Force vectors and transitivity

Another frequently discussed example of ditransitive alternation, locative alternation, occurs with 'spray/load' verbs (Levin 1993: 118):

(48) a The lads loaded logs onto the lorry
      b The lads loaded the lorry with logs.

Here too elementary vector combination may well apply, although the details are slightly different.

Figure 3.14a is a DSM for (48a), where the linear sequence of referent nouns in the sentence corresponds to the attentional 'distance' relative to S on the d-axis, the force vector v_1 corresponds to the verb's forceful action and the displacement vector v_2 corresponds to the directional preposition *onto*. Figure 3.14b is a DSM for (48b) resulting from refocusing the three entities and from replacing *onto* with the position preposition *with* (and position vector v_2). Now, if again we follow the logic of vector combination, we can get to Figure 3.14c.

In Figure 3.14b, the vector v_2 is redundant since it is parallel with part of v_1 (in vector terminology it is 'linearly dependent'). This means we can combine the vectors as in Figure 3.14c, resulting in foregrounding the lads' agentive interaction with the lorry and the resulting spatial relationship between lorry and logs. This is a very tentative modelling solution, using elementary vector logic that is appropriate to the conceptualisations associated with sentences (48).

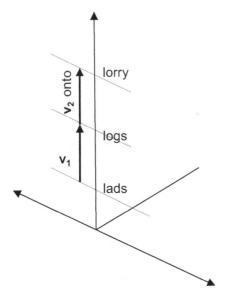

Figure 3.14a Example (48a) The lads loaded logs onto the lorry

98 Distance, direction and verbs

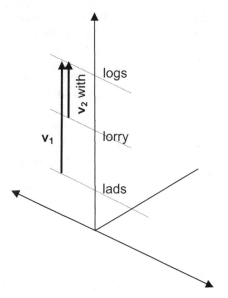

Figure 3.14b Example (48b) The lads loaded the lorry with logs

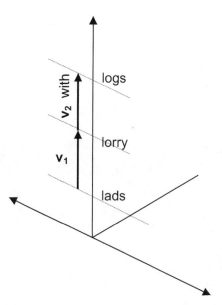

Figure 3.14c Combining vectors in Figures 3.14a and 3.14b

3.3.7 Transitivity and the passive construction

Passive and active constructions facilitate attentional foregrounding of (a) an entity that has undergone some action and (b) the state resulting from that action.[19] The resultant state can be a physical location, but states, like properties, are also treated in DST as conceptual locations, as argued above (Section 3.1.2). Given that the meanings of active and passive constructions appear to speakers to be linked, syntactic theory has come up with many accounts of the phenomenon. Generative grammar explains active and passive constructions as related by a 'transformation', with the active construction being the starting point. In that theory a transformation – or in its more general form the principle 'move alpha' – specifies a rearrangement of a sequence of symbols, where rearrangement is understood as a purely formal, i.e. meaningless, operation. Here, however, I treat rearrangement as conceptually motivated refocusing, similar in cognitive function to the spatial refocusing discussed in Sections 3.3.2 and 3.3.3 above, and applied in the analysis of ditransitive alternation in the preceding section (Section 3.3.6). Refocusing effects are apparent in the following banal sentences:

(49)   a [= (38a)]   John broke the vase
        b   the vase was broken by John
        c   the vase was broken.

In (49a) and (49b) I take the relative position of John in the linear sequence of words to be a reflex of attentional focusing. There are various reasons why a speaker would want to place this element at the start of a sentence but all these have to do with kinds of attention. The most obvious reflection of the role of attention is when the agent in a transitive event structure is not mentioned at all, as in (49c). These are not the only effects of selecting the 'passive' focalisation. The *be + past participle* element in (49b) focuses attention on a state, which may be the result of a causative action. The expression of agency by means of the preposition *by* deserves some comment. In necessarily brief terms, it is relevant to the present approach to note that *by* is a spatial preposition relating some locandum to a reference point – a relation that motivates vector representation. The semantics of *by* encompasses a variety of cognitively related meanings that have evolved and become conventionalised in English over a long period of time. The basic spatial meaning concerns the spatial proximity of an

---

[19] A revision of Chilton (2009). In that paper the passive construction was presented as resulting from a reflection in the *d*-axis. However, this does not seem to be correct, since it would imply that what S is conceptualising is an alternate *viewpoint*. This is not what passive constructions do; rather they change the focus of attention of S.

entity – static or moving – to a reference entity. Given certain pragmatic circumstances being close to an entity, or passing close by it, can be linked with instrumentality, and instrumentality can be conceptually merged with agency. Something like this happened over time with the potential meaning of *by*. When it is combined with the other components of the English passive construction, it has the conventionalised function of marking agency, of relating a resultant state to its causing agent. Other languages use spatial propositions that have equally natural though different cognitive motivations.[20]

The following figures put forward a formalisation of the refocusing, stativity and agentivity effects of the English passive. Figure 3.12a is the DSM for the default 'active' linguistic communication of force events, typified by a causative verb like *break*, which denotes the causing, by an agent, of a 'move' from a whole state to a broken state. The agent is the source of the force, hence at the tail of a vector **f**. This agent is in the foreground of attention, i.e. relatively 'closer' to S on the d-axis; indeed the force applied by the agent (the entity *John*) and the impacted entity (*vase*) as a whole can be said to be foregrounded relative to the change of state itself, i.e. the 'movement' of the entity *vase* from one state to another. It might also be noted that in physical terms such a composite event has a time course – first John's action, then the change of state – and it is of course possible in the DST format to utilise the t-axis to display the temporal duration. However, it does not seem to be the case, phenomenologically, that *John broke the vase* makes us conceptualise duration. Rather, the semantics of the verb makes us think of the event as instantaneous, and this is what Figure 3.12a represents.

Turning now to the passive construction, Figure 3.15a attempts to model the central conceptual elements of any passively constructed sentence – viz. the attentional foregrounding of the impacted entity, the foregrounding of the resulting state, and the distancing of the agent.

The fundamental difference between Figure 3.15a, modelling the passive construction, and Figure 3.12a, modelling the active construction, is the relative 'distance' of the referent entities on d. From this follows the configuration of the vectors **f** and **m** – respectively, the causal force component of the meaning of *break* and its change-of-state component. The directional relations between the entities, as shown by the vectors, remain the same as in Figure 3.12a. John, who is as it were in the background, causes the vase, in the foreground, to change its state to being broken. Once again we can take this further, and tidy up the DSM, by following, though perhaps stretching a little,

---

[20] E.g. *von* (from) or *durch* (through) in German, *ab* (or ablative case) in Latin, French *par* from Latin *per* (through).

3.3 Force vectors and transitivity

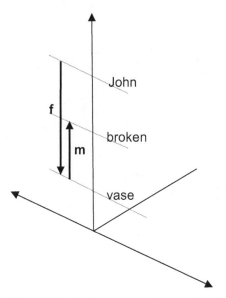

Figure 3.15a Passive construction as refocusing of entities

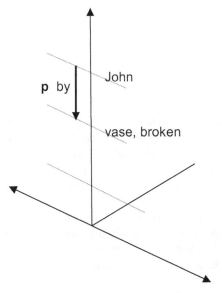

Figure 3.15b Passive construction after vector combination

102   Distance, direction and verbs

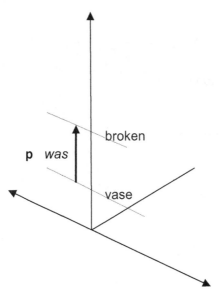

Figure 3.15c State ascription reading of (49c) The vase was broken

the elementary operations of vector combination. The result is shown in Figure 3.15b.

Figure 3.15b is derived in the following way. We allow the addition of the force vector **f** and the movement vector **m**, and also let **m** 'move' the entity *vase* to its new location (the state of being broken). The addition leaves **f** of Figure 3.15a pointing now to *vase* in its changed state. The vector **f** still represents John's application of force but this is now understood in terms of the conventional semantics, essentially spatial, of the preposition *by*. This suggests that the connecting vector is now a position vector, **p**. In fact, as noted above, the semantics of this preposition are variable and influenced by the construction in which it occurs, so what is marked as **p** should be understood, adopting a conceptualist perspective, to be both spatial and indirectly causative, modelling something like: *the vase's broken state was caused by John*. This seems to me to be what the passive communicates: it does not directly model the causal *event* of breaking; it models a resultant state, the cause of which may or may not be given by the S in the *by*-phrase. While Figures 3.15a and 3.15b both seem to be relevant it remains unclear how the relationship should be understood in cognitive or in processing terms.

Finally, consider the two meanings that can be associated with (49c): either the state of the vase is being asserted or it is being asserted that there was an event in which the vase was broken by an unmentioned agent. In line with the analysis of property ascription sentences in Section 3.1.2, the stative meaning of (49c) is modelled by means of a position vector, as shown in Figure 3.15c.

The DSM for the eventive meaning of (49c) has the same structure as Figures 3.15a and 3.15b, with the simple difference that there is no entity label for the point that is the coordinate for the tail of the vector, i.e. the source of the directed force vector **f**. This corresponds to the fact that one's mental representation of the eventive meaning of (49c) necessarily includes force from a source. It is thus misleading to call such sentences 'agentless' or describe them as having 'deleted' an agent. In the mental representation there is still agency: it is just that the agent is not lexically labelled.

3.3.8 Creative transitives: using the m-axis

To conclude this chapter, we open up another dimension in DST in order to look for a way to model verbs of creation (Levin 1993: 172–6), specifically a class of verbs known as *build*-verbs. Consider a man realising his dream:

(50)   James built a shed at the bottom of his garden.

The odd thing about *build*-verbs is that they are transitive: they have a grammatical direct object, sometimes described as an 'effected object'.[21] But conceptually, it is difficult to say what it means for an agent to exert an action upon an object that comes into existence only as a result of that same action. Stated in this way, we appear to have a paradox. The shed in (50) does not actually exist *as* an impacted target of *build* – even if we say that the shed exists at the end of the building process, it is still strange to say that it is impacted by the building at the end of the action and it certainly does not exist at the beginning, and we still want to ask how a transitive verb can *affect* a not-yet-existent entity. Except perhaps in the mind of the agent. We can in fact use this point within the DST format, for we are working within a deictic space that is, precisely, designed as the cognitive–linguistic space of a situated mind, the mind of the subject speaker S. What we have to do is propose an account of how a transitive verb can *effect* an object. In Figure 3.16 we use the full three-dimensional deictic space, including the temporal axis and, importantly, the *m*-axis which makes it possible to

---

[21] Some verbs that can be placed in the creative category display other oddities. For example the well-known case of English *paint* permits three kinds of conceptualisation: *Jane painted her bedroom wall green, Jane painted a picture, Jane painted the vase of flowers*. DST has the potential to deal with these different conceptualisations using the approach suggest here for *build* and the approach developed earlier for change-of-state conceptualisations.

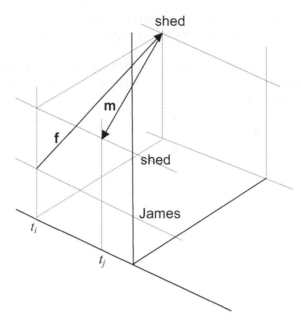

Figure 3.16 Role of the *m*-axis in modelling *build*-verbs

incorporate reference to non-existent entities – i.e. *irrealis* entities, as we seem to need for creation verbs. Moreover, because we can integrate the *t*-axis and the *m*-axis, it is possible to model the concept of emergence over a period of time.

In this DSM James is the source of some creative energy focused on an *irrealis* entity *shed* – a shed that does not exist qua complete physical object at time t_i but is perhaps imagined by James. The shed 'comes into being' (the quasi-spatial deictic verb is telling here), or 'emerges', over time, and is *realis* at time t_j and at the time of speaking.

The aim of this chapter has been primarily to explore the possibilities of using a standard part of geometry – geometric vectors – for modelling certain aspects of the meaning of verbs. In doing this we have moved from the obvious application of vector geometry to spatial prepositions into the more abstract realms of properties, changes of state, mental processes and causation. The hypothesis underlying this move is well known in cognitive linguistics – that physical embodied experience 'grounds' apparently abstract thought and language. For the cases we have examined, I have argued that vectors, in reference frames (coordinate systems), i.e. in the deictic space as defined, offer a unifying means of analysis and description.

3.3 Force vectors and transitivity

The chapter has concentrated on the particular grammatical constructions into which particular verbs enter – their grammatical frames. The association between the meaning of a verb and its constructions is not accidental. The last example that was considered – the creative verbs illustrated in Section 3.3.8 – shows how all three deictic axes may be relevant to the meaning of verbs. In particular the example shows the importance of the modal dimension of what it is that particular verbs mean. This dimension was neglected in the chapter in order to focus on the connection between types of verbs and particular grammatical constructions. Much attention in semantics has been given to the kinds of temporal meanings that particular types of verbs incorporate in denoting certain kinds of natural happenings and states of affairs. This is the dimension of verb meaning that has come to be known by the German term as *Aktionsart* (action type) or as 'lexical aspect'. This part of a verb's meaning interacts with tense forms in ways that produce complex effects. These two dimensions of verb meaning – lexical aspect and the tense forms operating on them – are the subject of the next chapter.

4 Event types and cognitive operators

Human minds get to grips with reality by way of cognitive (and affective) categories and processes; human languages schematically encode some of these categories and processes. The terms 'categories' and 'processes' here are intended to suggest the broadly static and the broadly dynamic aspects of reality that humans apparently experience and communicate about. This chapter concerns a particular phenomenon of human language-related conceptualisation that has to do not only with 'things that happen' but also 'the way things are': happenings and states, significant for humans, that are experienced, categorised and structured, not copied off a physical world 'out there'. This does not mean that happenings and states do not determine the structure of human conceptualisations of them, but such conceptualisations are not simple, complete and exhaustive copies.

Verbs denote kinds of states and changes that humans experience through time in the physical, social and mental world. Though the real-world states and changes have their own material unfolding in space-time, human languages attest to human ways of conceptualising them, choosing some features and phases while neglecting others. Human languages encode such conceptualisations primarily, though not exclusively, in the grammatical class of verbs. Verbs come with their own conceptual frames that include the temporal concept relating to the kind of real-world change or state they are associated with. Each such temporal concept has characteristics deriving from the way material states and changes, of indefinite variety, are conceptualised. In discourse, the character of the temporal concept can also be influenced by other elements represented variously in the clause by NPs, PPs and adverbials.

The many different temporal conceptualisations relating to states and changes, and expressed through verbs, are often referred to in the linguistics and philosophical literature by the German terms *Aktionsart* ('action kind') or 'lexical aspect'.[1] Competing classifications of *Aktionsarten* have been

---

[1] The term *Aktionsart* was used in roughly its modern technical sense by Karl Brugmann (1885) in his analysis of Greek verbs.

developed on the basis of an influential paper by Zeno Vendler (1957). The most commonly used and discussed *Aktionsart* typologies derive from Ryle (1949), Vendler (1957), Kenny (1963), Comrie (1976), Dowty (1977) and Bach (1986), among others. Since we are not concerned with the theory of *Aktionsart* (or lexical aspect) per se in this book, I will not go into these and subsequent discussions. There is general agreement on the relevant conceptual distinctions and broad categories that are involved, and the ones I refer to for the purpose of investigating how the geometrical approach of DST can capture *Aktionsart* conceptualisations are as follows.

The basic distinctions are between states (extended in time but with no dynamically differentiated phases), conceptually punctual or momentary events, and processes (extended in time but with internal dynamics). A further distinction is between telic (goal-oriented) and atelic (non-goal-oriented) event concepts. It is also important to note that event concepts are not completely encoded in verb semantics but are expressed by particular sentence forms in particular discourse occurrences. Some authors use the term 'eventuality' to include all types in the classification, resulting in a distinction between 'eventuality' and particular types of 'eventuality' called 'event' and 'action'. In this chapter, however, I shall avoid the term 'eventuality' and use the term 'event type' as a generic term corresponding to the notion of *Aktionksart*.

It is important to distinguish *Aktionsart* (i.e. 'event type') from 'aspect' (or more explicitly, 'grammatical aspect'), although the two terms denote conceptually related phenomena. The distinction is made clear in the pair of terms 'grammatical aspect' and 'lexical aspect', the latter being frequently used, as noted above, as equivalent to *Aktionsart*. Thus several equivalent terms (*Aktionsart*, action-type, event type, eventuality, lexical aspect) are found in use to refer to the broad notion of event- and state-concept coded in verb meanings. In distinction, the term '(grammatical) aspect' refers to operations on such verb meanings, operations that occur through a speaker's choice of verb morphology, auxiliaries and adverbials. In broad terms, aspect and tense impose a focus and a viewpoint, respectively, on an event type (equivalently, on an *Aktionsart*, action-type, eventuality type, lexical aspect type).

Accordingly, the present chapter is not only concerned with proposing conceptual schemas for event types. It is also concerned with modelling two kinds of cognitive operators that can be applied to these schemas. These operators correspond to the two main conceptual effects prompted by the English simple present tense and the English progressive. In principle, these two operators apply to any and all event-type schemas – though, as is well known, certain combinations do not appear natural, or require extra processing effort.

The overall objective is to explore the scope of DST-style geometrical modelling, to see to what extent it may unify our approach to event types and verb semantics and shed further light upon human conceptualisation, through language structure, of happenings and states of affairs as humans experience them.

Within the field of cognitive linguistics Croft (2012) offers a detailed study using diagrams with some geometric elements. The approach put forward here is formulated in terms of the DST geometry discussed in earlier chapters. Event types (*Aktionsarten*) are conceptualisations that can be modelled in terms of a t-axis and a d-axis. Some specific event types that verbs denote also require, as will be seen, the m-axis. The event-type schemas are not, however, modelled directly within the fundamental deictic space itself, because they are stable conceptual schemas that are stored in long-term memory along with phonological and morphological information. Rather, it has to be assumed that they are integrated in the fundamental deictic space during the process of communication.

4.1 Temporal aspects of happenings: event types

In this chapter the diagrams I use focus on the temporal dimension: they continue to use geometric vectors but do not use the full three-dimensional deictic space. This is because our concern here is with the stable semantic properties of verb meanings, the ones that relate to the conceptualisation of the time course of the types of processes and states that different verbs denote. These diagrams I understand as a kind of conceptual schema associated with each verb. When a verb is integrated in a sentence, and modelled in a full three-dimensional DSM, we can think of these event-type schemas as part of the labelled verb. They concern time but they are not in themselves deictically instantiated until they are tensed.

4.1.1 State schemas

Let us consider first some examples of the event-concept 'state':

(1) a Jane was in London
     b Jane knew the words
     c she was happy for three weeks.

Sentences of this kind may set temporal bounds to the state (as in 1c) or may not set bounds; however, the conceptual structure of state event types does not inherently include bounds. Many state verbs do not, at least without specific context, accept time-period adverbials. For example, *know* is a state verb (by a commonly used test, it cannot normally be in the progressive tense

4.1 Temporal aspects of happenings

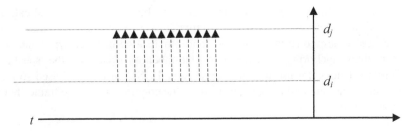

Figure 4.1 Geometric schema for states

form in English). Nor can it (without some special context) take a period of time expression, as in *?Jane was knowing the words for a week*. If a state verb does not take such an adverbial, there is no obvious inference that the state had a beginning and an end, unlike the process types discussed in Section 4.1.2 below. The most basic state-type relation is spatial location (1a) and it is relatively uncontroversial that spatial location can be given by a vector. In this simplified picture, the referents being located might be Jane and London, as in (1a). However, DST also uses the localist (and cognitive linguistics) hypothesis according to which such relations are the basis for non-spatial concepts too, among which can be counted properties.

The predicate *know* in (1b) is also analysed as a directional relation 'locating' a non-spatial referent, and properties are treated in the same way in (1c).

States, such as those instantiated in (1), can be given a geometric representation along the lines of Figure 4.1. Since we are concerned with the conceptualisation of event types and not with their anchoring in the space-time-modality space of the speaker, Figure 4.1 does not use the three-dimensional DSM of the kind used so far.

The state schema consists of an unbounded set V of vectors $\{v_1, v_2, \ldots v_n\}$ that are parallel to one another, and have the same length and direction. They have to be thought of as 'filling' the space: each vector lies at a geometric point. States, such as those instantiated in (1), can be given a geometric representation along the lines of Figure 4.1. The space they occupy is arbitrary and can be extended; the space is unbounded in this sense. Any arbitrary vector from the space is equal to any other (except for its coordinate on the t-axis): the space is internally homogeneous in this sense and corresponds to a well-known defining feature of state-type situations – namely, that any subpart is identical to any other. The stative schema is shown in Figure 4.1 with an arbitrary t-axis, and arbitrary d-coordinates; a modal dimension is not involved in the schema for states. It is important to note that Figure 4.1 is in part really only a schematic picture – it shows *sample* vectors but is to be understood as indicating a continuous unbounded space of vectors at every point on t.[2]

Since we are concerned with the conceptualisation of event types and not with their anchoring in the space-time-modality space of the speaker, Figure 4.1 does not use the three-dimensional DSM of the kind used so far. For states, no modal dimension is need for the cognitive predicate frames held in long term memory.

4.1.2   Non-state schemas

I shall refer to non-states as processes. Processes resemble states in being extended, but states are homogeneous throughout their extension. Further, the non-homogeneous processes can be subdivided into accomplishments and achievements. Accomplishments are processes that cannot be said truthfully to have happened until they are completed (cf. what has been said concerning 'creative' verbs in Chapter 3, Section 3.3.8). Examples are: *John drew a circle, Dave painted the door, John walked from Oxford to Woodstock, Tony decided.* The drawing of a circle, swimming from Dover to Calais, etc. can be contrasted with sentences like *John was drawing, Jan ran*, etc., in which no goal (in spatial and metaphorical senses) has to be reached for it to be true to assert them. The reason is that accomplishment processes consist of subparts (or phases) all of which are required for the accomplishment of the process. Achievements are also extended processes but focus on the final phase, in many cases inherently conceptualised as a punctual event. Examples of achievements are: *Mary arrived, Hillary reached the summit, Jill won the race.* With respect to these inherent semantic properties of verbs and the situations they denote, tense and aspect operations have various effects, including treating one type as if it were another, as the case when punctual events are treated in discourse as if they were extended, and extended processes as if they were punctual events.

4.1.2.1 Atelic processes

The conceptual distinction between telic and atelic (or bounded and unbounded) cross-cuts the non-state categories. Telic processes have a resultant in the type of process a verb denotes, for example a new location, state

---

[2] Though this cannot, strictly speaking, be shown graphically, it would be easy to capture in an algebraic equivalent of the graphical form of the discourse space. In formal terms, each vector in the set (which is a plane with an infinite number of points) has the same length and direction and each differs from the others only by its temporal coordinate. Each t coordinate can be defined by a factor in terms of any of the others. The set is a vector space consisting of linearly dependent vectors. This formulation is equivalent to the classic truth-conditional approach to lexical aspect: if a sentence S is true for some interval time I, then it is true for all subintervals of I (cf. Dowty 1979, 1986).

or product. An atelic process has no natural completion. The inherent semantics of particular verbs may denote inherently telic processes but in practice, because verbs appear in particular constructions (particular contexts too), whether the hearer has a telic or an atelic conceptualisation depends on the construction, especially morphological tense, and the other lexical items in it. For example, *Fred built a shed in a fortnight* is telic but *Fred was building a shed last week* is atelic, as is *Fred built sheds for a whole month*.

Atelic processes: semelfactives

Semelfactive event types include those such as *cough* and *knock* that can occur once (hence their name), have no *conceptually significant* duration (of course, in reality all events have *some* duration) and are thus temporally bounded. Being bounded should imply that semelfactives are telic but the usual *in*-test for telic meanings does not work: **John knocked in a second*. However, the atelic *for*-test does work, provided the sentence is understood to denote a series of knocks extended over time, as can be seen in (2b) and (2d) below. In general, semelfactive events can be single or consist of a series of the same event indefinitely extended in time. Many sentences containing semelfactive verbs will be ambiguous between the singular and the serial meanings, as in (2a) and (2c):

(2) a the judge coughed
     b the judge coughed for five minutes
     c John knocked
     d John knocked on the door for five minutes.

Certain event types involving actions that last some critically short period of time seem to be inherently conceptualised as points. Langacker calls them 'punctual verbs' (Langacker 1991: 209). In geometric vector terms it is possible to model the non-serial meaning illustrated in (2) as a vector in the plane (d, t) that has no temporal duration, i.e. occupy a geometric point on the t-axis, as shown in Figure 4.2. To do this, we have to allow that a unit geometric vector can stand for all kinds of event types, whether or not directionality is salient and whether or not there is a conceptualised entity that is an undergoer of some sort.

The example (2d) includes an expression (*the door*) denoting the target of the action of knocking. Significantly this is not expressed as a grammatical direct object but a prepositional phrase, suggesting that the target is *conceptualised* as a location of the knocking, rather than as some thing that is impacted (though in reality it is, of course). The verb *knock* is similar to semelfactives like *cough* in not having an intrinsic entity that is a direct undergoer of the action.

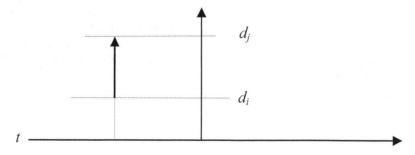

Figure 4.2 Semelfactive event

If we were to try to draw a diagram for the serial meaning of the sentences in (2) we would have a diagram looking misleadingly like that for statives (Figure 4.2): an unbounded set of equivalent vectors. The important difference is that the diagram for serial semelfactives includes intervals between the time points of each vector, whereas for states there are infinitely many vectors filling the plane (something that we cannot draw, as noted above). Still, that this geometric approach suggests some similarity between states and serial semelfactives is interesting since it corresponds to the intuition that an unbounded set of repeated semelfactives is indeed rather like a state, despite their dynamic quality.

Atelic processes: activities
In the following,

(3) a the dragon is sleeping
     b Lance is pedalling,

sleep in (3a) has a natural onset and termination but is not conceived as having phases (except by experts), and in (3b) *pedal* can be extended in time but is conceived as internally cyclical (like *run, walk*).[3] Intransitive verbs such as *sleep, ponder, sing* (but not *sing a song*) are unbounded events in which there is continuity but no change in the participant: a labelled vector (representing the semantic properties of the verb, *sleep, ponder*, etc.) maps a discourse referent into itself (Figure 4.3). Transitive

---

[3] This example may be felt to be the same as semelfactives. However, *pedalling* is a continuous activity that is cyclic in the strict sense, not one in a series of actions separated by time intervals.

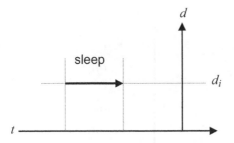

Figure 4.3 Homogeneous activity

verbs that denote homogeneous durative activities (e.g. *wear a hat*), which similarly require the present progressive, can be modelled as a continuous set of vectors in a continuous vector set like the one in Figure 4.2, with the difference that the set is bounded (the wearing of hats is limited to bounded time periods, like sleeping). All homogeneous activities are thus state-like to the extent that they involve continuity. In fact, there is uncertainty in the *Aktionsart* literature as to whether *sleep* and similar verbs (e.g. *stand, lie, hang*) are activities or states (compare Dowty 1979, Carlson 1981, Bach 1986, Michaelis 1998, 2004, Croft 2012). They are unlike states in requiring in English the progressive tense form, and therefore should seem like processes conceptually. We can also ask *What is the dragon doing?* And grammatically answer *He is sleeping*. However, we do not seem to think of (3a), for example, as denoting a dynamic process. The contradiction between categories need not perhaps be resolved, nor a new category invented, since the uncertainty itself may point to actual conceptual uncertainty.

Sentence (3b) (cf. also *walk, skip*, etc.), on the other hand, involves more than one *discrete* subaction (specific cyclical leg movements). Each of these is conceptualised as punctual.

Telic processes: accomplishments

The two subtypes of telic process (accomplishments and achievements) can also be represented geometrically by exploiting simple properties of vectors in a spatio-temporal coordinate system. Figure 4.4 shows the general case for accomplishment types.

Unlike states, the process vector is a set **V** of linearly dependent vectors $\{v_1, v_2 \ldots v_n\}$ lying in a straight line with end points that may or may not become specified in the clause or discourse. Such a set can be understood as a path consisting of an indefinite number of smaller paths (displacement

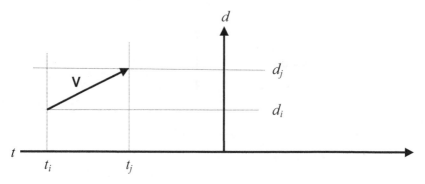

Figure 4.4 Process type: accomplishment

vectors) with the same direction. The participants specified at (d_i) and (d_j) can be conceptualised as 'theme' and 'goal' with respect to this type of vector. Unlike states, extended processes have boundary points in their basic semantics, even though they can be 'viewed' as uncompleted when given 'aspect' by linguistic forms such as the English progressive. Similarly, each vector in the set, though linearly dependent on the others, has a different position. This corresponds to the well-known heterogeneity criterion of process event types, viz. that no part of the process at any proper subinterval is equivalent to any other (it occupies a different position in the overall process).

(4)   a Jane walked from London to Manchester
        b Jane sang a song.

Processes not involving physical displacement of a discourse referent (e.g. 4b) can be understood on localist assumptions. The sentence *Jane sang* would be a process without a goal, akin to states, but sentence (4b) denotes an accomplishment, of which the 'goal' is the song; the process of singing a song is accomplished only at the point (d_j, t_j).

At any discursively instantiated point t_k, $t_i > t_k < t_j$, the process is not complete. There is also a significant implication in the vector formalism. The orthogonal projection of the entire vector onto the d-axis (or any other specified vertical line intersecting with the t-axis) is a vector with no temporal extension: that is, it has the configuration described above in Figure 4.2. This is an important point, because one of the effects of tense–aspect operators in discourse is to transform extended processes (and sometimes states) into punctual events (application of simple past), and conversely (application of past progressive or imperfect). In general, the whole of an accomplishment process (unlike, as we shall see, achievement processes) can be projected as an event.

Telic processes: achievements

Achievement type processes, such as *deciding, arriving, remembering, dying, reaching the summit* etc., all have in common that they denote events that 'come into a state'. Many analysts treat them as inherently punctual (e.g. Lyons 1977: 712, Rothstein 2004). However, there are many examples that, on intuition, presuppose a process prior to a point. Connected with this observation, some of them are happy when modified by the adverb 'gradually' (cf. Taniwaki 2005). Such verbs pose a problem for formalisation in the conventional formats because they involve conceptual 'focus' on the end phase, which may or may not be a point, in the denoted process. In Langacker's terminology, the end point is 'profiled' and the profiling can be shown iconically in his pictorial system. One way to model this situation in DST format is as follows, and it is novel compared with classic formal–semantic treatments.

Achievements are in effect treated in the present proposal as a special case of accomplishments, but unlike accomplishments they consist of just two linearly dependent vectors, one of which is relatively shorter than the first or may be a zero vector. The reason for saying 'one of which is relatively shorter than the first or may be a zero vector' is that, as claimed by Taniwaki (2005), achievement verbs probably come in several subtypes. Some are happy with both the adverb *gradually* and the adverb *immediately*; some prefer one or the other only. One can say *Jane gradually/immediately recognised John's voice* but one cannot say *Jane gradually found the key she'd lost*, though *the research team gradually discovered a cure for the disease* is acceptable. It seems to me, though individual intuitions may vary, that *he gradually reached the summit* is possible also.[4] Hence, the present account leaves open the possibility that the end phase is either (i) a short vector relative to the first phase vector (this phase remains) or (ii) a zero vector. I think it should be hypothesised that the presupposed first phase remains and that it is the end phase that can be either a short vector or a zero vector. Since zero vectors have zero length and no direction, they are exactly appropriate in the model for representing punctuality.

Consider now the example

(5)   Hillary reached the summit.

The model I propose has a vector (whose starting point may or may not be indicated in discourse) composed of two component vectors, corresponding to a presupposed period of movement followed by an end phase, as outlined in Figure 4.5.

---

[4] Note that 'he immediately/suddenly reached the summit' is also odd, though evidently punctual, perhaps because it is redundant, if *reach* is inherently punctual.

116   Event types and cognitive operators

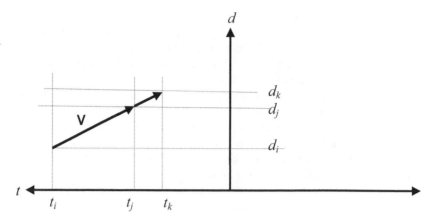

Figure 4.5 Process type: achievement, sentence (5) Hillary reached the summit

Non-physical achievement concepts denoted by verbs like *decide* are analogous to physical achievement meanings: there is a period of cogitation 'on the path' to the 'goal' denoted by *decide*. After all, one 'arrives at' a decision.

We are treating achievements, like the other event types, as independent of the kinds of temporal perspective that are introduced by tense and aspect. However, tense and aspect markers, often in combination with particular lexical content and with context can certainly make a difference to the precise conceptualisation involved. Thus, although achievement verbs all focus on an end phase of a process, the end phase is not in all actual cases always conceptualised as instantaneous. True, it seems odd for some achievement situations (e.g. *Mary arrived*) to ask 'How long did it take?' and also odd for many achievement verbs to take progressive aspect. But many achievement verbs are readily interpretable on both counts in certain cases. The second ('terminal') vector may be the location for the progressive operation, under certain discursive conditions, making it possible to conceptualise a meaning for sentences like *the train is arriving at platform four*.

4.2   Tense forms as cognitive operators: instancing and presencing

Event schemas are conceptual constructs arising in human minds from the point of view of human experience of the physical and human world. Languages select and stabilise conventional conceptualisations in conceptual frames associated with particular lexical predicates. In addition, by means of a range of morphological realisations of tense and aspect, languages express another kind of conceptualising activity of the mind related to time. I propose here that tense–aspect verb forms in English constitute cognitive operators

4.2 Tense forms as cognitive operators

applying to event-type schemas of the kind described geometrically in the preceding section. The English *simple present* and the English *progressive* forms, though generally called tenses and thus generally linked with different perspectives on – indeed different experiences of – time, are best understood as two kinds of cognitive operators that combine with event-type schemas in various ways and in turn combine with time-line concepts. An important element in the theoretical framework sketched in this chapter is the relativity of reference frames.

We have a set of lexicalised conceptualisations of event types on the one hand, and on the other hand two key cognitive operations that can be applied to them. To some of these event types the conceptualisation associated with the present tense applies more readily than that associated with the progressive. This is what gives rise to the categorisation by analysts into stative (cf. Langacker's imperfective) and non-stative (or dynamic or Langacker's perfective) categories. But as is well known, there are exceptional uses: e.g. *Mary believes in ghosts*, *Mary is believing in ghosts at the moment*. It is therefore preferable to speak of a particular verb as being *prototypically* stative or non-stative (and subtypes) and to allow for either simple present or progressive forms to potentially operate on all event types.

Many attempts have been made to characterise the conceptualisations associated by linguistic convention with simple present and progressive forms in English (and other languages, although there are important differences cross-linguistically). Such characterisations have been carried out in cognitive grammar in much descriptive detail, partly by means of pictorial diagrams and partly through metadescription in ordinary English (Langacker 1991, 2001, 2011, Michaelis 1998, 2006, Brisard 2002, are important examples). The approach I am putting forward is tentative and not intended to be complete. The DST model is stated in terms of conventional elementary geometry. I want to see whether and how much this geometric modelling can tell us, and make more precise, concerning the semantics of the English simple present and the English progressive forms.

In this section I outline two geometric transformations of the base reference frame R which, I am suggesting, correspond to the conceptualisations prototypically brought about by the simple present and the progressive verb forms, at least as found in English and as hitherto described in many places in the linguistic, particularly the cognitive–linguistic, literature.

4.2.1 Instancing operator (simple present tense)

The core of this operation is the selecting of an instant coinciding with the instant *now*. Now is the subjective awareness of time, emerging from short-term memory and short-term forward planning: it is what I have termed 'peripersonal time', within which there is subject awareness of an 'instant' of consciousness accompanied by deictic awareness of self in relation to an environment in the recent past and the same self as it is anticipated in the near proximate future. There may be here a sense of moving with a flow of time, the deictic past and future moving with the self. There is, however, another way of conceptualising what is *now*. Abstracting from the moving instant is the concept of 'timelessness' – what is now is always. Abstraction is also generalisation, categories and truth. The English simple present tense form is usable in both these cognitively linked conceptualisations, depending on the associated event-type schema and combined with contextual and cultural input. The term 'instancing' is chosen with this double meaning in mind: it combines the temporal instant with the abstract and generic that an instance belongs to.[5] There is a seemingly paradoxical connection between the subjective temporal instant *now* and the objectivising, generalising and categorising 'instance'.

What then is *now*? A conceptualisation of *now* emerges from the three-dimensional abstract deictic space, which I have described. Geometrically, the point (0,0,0), the origin, defines *now* for S. What does this mean in conceptual terms? For the moment I will not consider the notion of speech time, which is important for cognitive grammar accounts of the present, and focus on conceptualisation as distinct from the physical act of speaking. Thus (0,0,0) is a point in consciousness, but not a simply temporal one. It integrates consciousness of the present, maximal epistemic certainty and maximal attention. This is perhaps equivalent to what is called 'immediacy' by cognitive grammarians. It is not purely 'modal', since it involves consciousness, attention and the temporal.

It is important to note, however, that (0,0,0) is a reference point, the deictic origo S. It is logically possible, and in fact the usual situation, to have (d, t, m) specified for relatively distal points on the d-axis and m-axis, while t remains 0, as for example: *it is possible that carbon emissions cause climate change*. The later expression would have coordinates on the d-axis for *carbon emissions* and *climate change*, with a vector relating them, all located on a plane positioned at the midpoint on the m-axis, reflecting the modalisation effect of *it is possible that*.[6] The vector for *cause* is at $t = 0$. We shall consider such uses of the present tense form below. The crucial point is that in the DST model, this type of expression has *no dimension* on the t-axis, i.e. no temporal

---

[5] In French *un instantané* is a 'snapshot'.
[6] The labels for the noun phrases would be on the (d, m) plane at $m = 0$, assuming that S believed carbon emissions and climate change exist. Their coordinates run through the deictic space to the plane at the midpoint on m. The vector relating the two noun phrases is what is modally uncertain.

duration. This is a useful consequence of the geometry and corresponds to what Dummett, following Frege, calls the 'tense of timelessness' (Dummett 2006: 108).

The zero-point concept of time does not, however, correspond to all the conceptualisations that are set off in the mind when the English present tenses (whether simple present or progressive) are used. As has been seen, the DST space includes a temporal peripersonal region, which 'surrounds' the point (0,0,0). Like peripersonal space, I take peripersonal time to have a basis in neurological and physical processes, and to have a basic minimal span. However, both spatial and temporal spheres around the self can be extended depending on context and activity. The peripersonal time span consists of points, of which one is the origin, the zero point; the peripersonal time-span contains the *now* instant located at the origin. These two notions of *now* are thus not separate and it is not surprising that both are needed to give an account of the conceptualisations prompted by the present tense forms.

4.2.1.1 The instancing operator, operating on a state schema

Figure 4.6 lets us visualise the conceptual operation associated with the simple present. By way of example, the vectors stand for the predicate *know*, and the d_i and d_j coordinates for *Mary* and *answer*, respectively, in a stative sentence such as (6):

(6)   Mary knows the answer.

The hypothesis is that the cognitive schema for an event type is retrieved from long-term memory and 'inserted' in the three-dimensional deictic space – that is, it is used in an utterance. We need to define the simple present (instancing operator) as selecting the vector in the state schema set that coincides with $t = 0$.

This is also $m = 0$, the point of maximum modal 'closeness', i.e. epistemic certainty. But the single vector or the subset of vectors is identical to (has the same length and direction as) every other vector in the unbounded set of the state schema, except that each vector has a different coordinate on the t-axis. It can be seen that the instancing operator, as described here and in Figure 4.6, corresponds with the important point, made many times in the literature on tense (e.g. Michaelis 1998, 2006), that the simple present selects an 'instance' of a state predicate that is equivalent to all other instances over time. We also need to understand the result of instancing operator as neglecting all vectors except the one at $t = 0$. This result can be thought of as a 'compression' (cf. Langacker's notion of 'contractibility') and it is this outcome of the operation that is important. We need to say more about this compression operation when we consider how the instancing operator applies to *non*-stative schemas in the next section.

120   Event types and cognitive operators

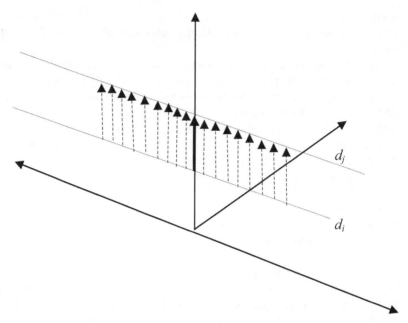

Figure 4.6 Instancing (simple present operator) operating on a state schema

The instancing operator is activated by the English 'simple present tense', which is not just a tense in the sense that it relates to time points relative to the speaker (or other reference point). Its function is primarily to produce instancing within a time-frame. As is well known, the simple present tends to correlate with semantically stative verbs, though non-prototypical uses have to be accounted for. This does not mean that the instancing operator arises from stative verbs; here I am exploring it as an independent cognitive 'perspectivising' phenomenon. So what happens if the instancing operator is applied to non-stative event schemas?

Let us assume that, in some stretch of discourse, a tense operation locates an event-type structure (called up from long-term memory) of some verb at $t = 0$. For stative verbs the simple present suffices (*Jane knows the words* vs. **Jane is knowing the words*). The progressive form is not an option except perhaps under very special discourse conditions. In DST terms, the coordinate at $t = 0$ coincides with one vector in the unbounded continuum of vectors representing the untensed string *Jane + know + the words*. Since this vector is equivalent to any other in the state schema set and also there is in fact no progression, the progressive aspectual operator does not 'make sense'. We can see why if we assume that what the progressive does is pick out a subphase of a process that involves progressive change over time. But there is

4.2 Tense forms as cognitive operators   121

no change in stative event types, so the progressive will only pick out a subset of similar states, and is thus conceptually vacuous.

4.2.1.2 The instancing operator, operating on a process schema

We can now consider the potential effect of instancing, so to speak, the process type schematised in Figure 4.6 above. The simple present cannot naturally apply to a sentence containing a process-type verb, as (7) shows.

(7) a What is the decorator doing in there at the moment? *He paints the door.
     b What is the decorator doing in there at the moment? He is painting the door.

A geometrical account is consistent with this intuition. The instancing operation consists in inserting the event schema in the deictic space at $t = 0$, the origin, equivalent to S's *now*, in such a way that this point falls within the span defined by the projection of the end points of the process vector onto t. The consequence is geometrically quite different from the case of instancing a state, both geometrically and conceptually, as is obvious in Figure 4.7(i) and (ii).

(i) Stative schema
(ii) process schema
(iii) result of instancing operation

In (i) we have, by application of the instancing operator (simple present), the selection of one instance of, or the peripersonal subset of, the indefinite vector set of the stative schema, at $t = 0$, as already discussed. In (ii) we also apply the simple present instancing operator, this time to a process schema (cf. Figure 4.4 or 4.5). This is unsatisfactory because the vector at $t = 0$ merely cuts the transverse process vector and defines a single point. Since points have no dimensions, this point has no time dimension. But the event schema for process has a vector defined over time. The instancing operator produces a point that has no extension on the t-axis and also no extension on the d-axis. The application of the instancing operator (simple present tense) produces nothing that corresponds to linguistic intuitions. The geometry is therefore consistent with linguistic intuitions about (7). But things are not quite so simple.

The applicability of the instancing operator is dependent on extra lexical material in the sentence and/or the context of utterance. If S is reporting a process at *now*, the verb form is present progressive, say, *Mary is writing the report at this very moment*. But as is well known, simple present forms, such as *Mary writes the report*, can readily be given a meaning, one indeed that may have nothing to do directly with time *now* but can also refer to time future or time past or to a temporally periodic (i.e. habitual) event. Thus we need a transformation that will turn the process schema into a configuration that will correspond to, for example, *Mary writes the report next week* or

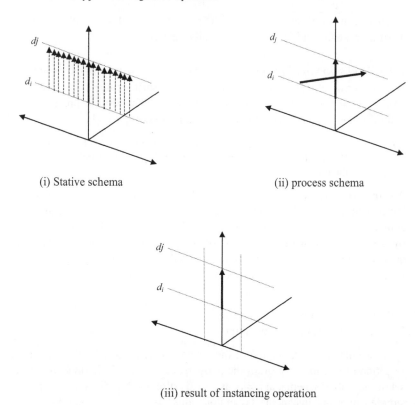

(i) Stative schema   (ii) process schema

(iii) result of instancing operation

Figure 4.7 Instancing (simple present operator) operating on (i) a stative schema and (ii) a process (accomplishment) schema, resulting in (iii) an 'instance' or 'instant' corresponding to the simple present tense form

Mary writes the report every Thursday. That is to say, we need a transformation that turns a process vector into a vector without time duration so that it looks like a stative verb vector (see Figure 4.6). In fact, all we need do, to turn a process vector like that depicted in Figure 4.7 into one like the vector that is selected (i.e. instanced), in Figure 4.6, is to specify that its end points both have time coordinates $t = 0$. This does in effect mean setting up a separate kind of instancing operator for simple present, one that applies in certain cases to the event schema of process-type events. This is perhaps not so ad hoc as it might seem, since the instancing operator as initially defined (for state schemas) also specifies that end points are at $t = 0$. The difference is that the application to the stative schema *selects* one vector from an unbounded set, whereas application to a process schema *transforms* the process vector. It does this by a sort of contraction transformation – imagine the ends of the *t*-coordinates of the process vector coming closer together, until this 'diagonal' vector is pointing straight up, like a state vector.

There is a further consideration: how to interpret the line vector thus produced. So far we have suggested that because it is geometrically timeless we have a useful account of the 'timeless present' that is evident in general validity statements, such as *light travels at 299792.458 km/s*. Here the verb *travel* cannot be instanced in an utterance that refers to an event at S's *now*. The simple present is used to conceptualise timeless properties, that is, properties that are conceptualised as timeless (in objective reality they may not be timeless). As we have seen the instancing operator acts on temporally unbounded sets of vectors that ascribe a property or condition to an entity and abstracts from that set one instance. It is the underlying conceptual abstraction that makes the simple present 'eliminate any temporal restriction where timelessness or eternity is part of the thought'.

It is, however, additionally useful to let the point zero where linear vectors occur through instancing be understood as a temporal 'instant'. Instants are the smallest cognitively (ultimately neurologically) possible duration of the epistemic span, corresponding to *now*. Utterances referring to 'instants' occur in on-the-spot narration when the speaker refers to an event of minimal duration, e.g. *Jones passes the ball to Smith*. It seems that we need both the timeless zero point understanding (for general validity claims) and also the minimal moment, or instant, understanding for other kinds of meaning that are cued by the simple present.

Homogeneous atelic activities like *sleep* (e.g. (3a) *the dragon is sleeping*) cannot report a present situation with the simple present tense: **the dragon sleeps*. This is because of the inherent boundaries in the event type structure: a sleep period cannot be contained in a point. Present progressive and past progressive are possible. Again, this is because a function of the progressive is in some way to invoke a whole process-type frame and simultaneously extract or 'focus on' some subinterval of that process.

For heterogeneous activities the simple present is also not possible (*Lance pedals, *Jim walks, etc.), but for slightly different reasons. Since the simple present denotes a point, it can at most coincide with only one of separate sequential leg-movement cycles that constitute *pedal*, *walk*, etc., but *pedal* and *walk* involve a sequence of several cycles. Simple present is not compatible with these event types, presumably because it would pick out only one element of the activity, one full cycle only, which does not constitute what we understand by *walk* or *pedal*.[7] The present and past progressive are compatible with this kind

---

[7] Michaelis (1998: 47), following Dowty (1977), argues that the simple present picks out a point within the cycle, whereas here I am treating a whole cycle as (conceptually) a punctual event.

of action type: again this implies that the progressive calls up at least a sub-interval of a process in which a number of cycles occurred. In the case of heterogeneous activities, this means, in DST terms, a subset of discrete vectors.

Semelfactives (or 'punctual verbs') also cannot occur in the simple present: *Mary coughs. If the simple present's job is to pick out a punctual vector from a *continuous* (i.e. homogeneous) set, then this is not possible with semelfactives: the input is not appropriate since coughs, blinks, shrugs and the like are conceptualised as isolated events at a single time point.[8] Unlike *walk*, etc. they are not part of a continuous physical movement with cyclic subphases, but are a discrete movements, with intervals of time between each movement if they occur in a series. The present and past progressive cannot be used to report the ongoing event of a single semelfactive event. Rather, application of the progressive operator prompts a conceptualisation in which discrete spasmodic events succeed one another in a series. Thus: *Mary is coughing* cannot without effort be understood as Mary being in the middle of one tracheal spasm but as in the process of a series of separate spasms. Again, this must be because the progressive opens, so to speak, a temporal window on a subinterval of some process.

Nor can accomplishments and achievements be combined with the English simple present tense form:*John draws a circle*, **the train arrives*. Since the simple present is a point on the *t*-axis it is not compatible with processes, which are inherently time-structured. Accomplishments are compatible with the present progressive, but it is often said that achievement verbs are incompatible with the progressive. However, this is not entirely the case since discourse conditions can be imagined for *the train is arriving*, for example. This follows from the way we have defined achievement event type above, with the distinguishing condition that progressivisation only operates on the second vector.[9] Simple past collapses just the second vector into a point, which represents completion of the whole process: we consider how this works in the next section.

Let us now turn to the operation of the cognitive operation which is associated with the English progressive verb suffix *ing*.

4.2.2   The presencing operator (progressive tense forms)

There have been many descriptions of the progressive tense form (*be* V+*ing*) in English, in effect of what I am calling specific cognitive operation conventionally associated with the progressive tense morphology of English and their analogues in other languages. Several key characteristics receive regular attention.

---

[8] 'Mary coughs' can of course mean that she habitually coughs once or repeatedly (e.g. on getting up in the morning), but this is not the sense under consideration here.
[9] The second vector cannot in such instances be zero.

4.2 Tense forms as cognitive operators

(i) The progressive tense form is often said to place the speaker located 'close to' a process, and is often described in spatio-visual metaphors as 'zooming in', imposing a 'viewing frame' or taking a 'snapshot', in all cases losing sight of the end points of the process (see, for example, Lakoff 1987, Langacker 1991, 1999, Michaelis 1998, Brisard 2002).

(ii) Another, overlapping, account says the progressive tense form is an operation that locates the speaker 'inside' a process, presenting it 'from the inside', temporally 'elongating' it 'from an interior viewpoint', or taking an 'internal perspective' (Langacker 1991, Frawley 1992, Kamp *et al.* 2011).

(iii) In addition, linguists sometimes claim that the progressive tense form induces a stativisation (homogenisation) effect, or serves to 'arrest' the progression of a process towards its completion (Michaelis 1998, Langacker 1999, Kamp *et al.* 2011). This idea raises some questions of consistency and is somewhat different from points (i) and (ii), as pointed out in Chilton (2005). It possibly arises from the *be* element of the present progressive. However, I will not explore this further here but focus on points (i) and (ii), which involve spatial descriptions that can easily be modelled in the geometrical format of DST.

The proposal is that the progressive can be modelled as a translation on d of a copy R' of the base axis system R, as shown diagrammatically in Figure 4.8. The system R' can be thought of as 'virtual' reality conjured up by the

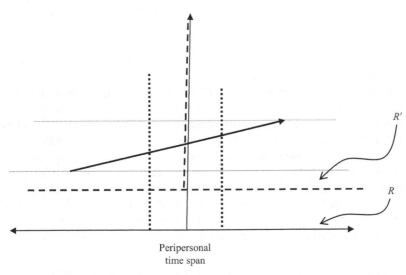

Peripersonal
time span

Figure 4.8 Presencing in the (d, t) plane, applied to a process event

semantic character of progressivisation, a notion that plays an important role in DST's modelling of many linguistic phenomena (see also the idea of 'virtual' representations in Langacker 1995, 1999, 2001). To capture the intuitions reported in (i) and (ii), R' is positioned at some point t and at some point on the d-axis within the interval defined by the vector(s) of the event schema (here state or process) and R' is located on the (d, t) plane at $m = 0$. This captures, geometrically, the core notion that the presencing (progressivisation) operator 'zooms in' and puts you 'inside' and 'close to' the event. In fact, the visual analogy of a close-up photo shot holds well, since close-ups leave the peripheral regions out of focus. It should be noted the use of a transformation introducing a reference frame R' carries the implication that conceptually the progressive form has an effect rather different than the instancing operator, suggesting that progressivisation (presencing) introduces a new reference frame, even a separate 'world'. I shall not try to justify this further here, except to note that it is not inconsistent with points (i) to (iii) above.

How do we capture the notion that earlier and later phases or endpoints, are not 'in view', i.e. that the *ing* operator focuses in on a segment of the event? Recall that the base coordinates space in DST includes not only an origin point, equated with S, but also a peripersonal time-span. Whereas the instancing operator uses the d-axis at time $t = 0$ and results in a single vector (an instant or instance) located at the origin of the deictic space, the presencing operator imposes a peripersonal time span on the inserted event-type schema: a kind of 'windowing' operation.[10] Figure 4.8 shows how a peripersonal span applies to an inserted event schema. The diagram is misleading to the extent that the start and end point of the event schema need only be specified as earlier and later than the peripersonal boundaries. This accounts for the common description of the progressive tense form as bringing about the effect of obscuring end points or being 'internal' to an event. The peripersonal span itself is flexible. Figure 4.8 in fact shows two effects of the presencing operation. The peripersonal windowing on the event schema is primary. But what of the 'zoom' effect? I suggest that this is a secondary effect, due to the loss of focus on peripheral end points. Geometrically, this effect can be understood as a translation of a copy of the base axis system, as shown also in Figure 4.8, where the translation is shown as a shift 'close to' whatever the agent label may be on the d-axis.

Rather than label the conceptual effect I am trying to describe here 'progressive', the traditional term for the English present tense form $be + ing$, I have introduced a new term, as I did also for the 'simple present' tense form.

---

[10] This proposal differs from Chilton (2007), in which the segmenting of the process vector was achieved by way of a 'viewing angle'.

4.2 Tense forms as cognitive operators

The term 'instancing' was introduced in an attempt to capture the ambiguous nature of the English simple present, deliberately playing on the meaning 'instant' and 'instance'. The term 'progressive' does not seem wholly helpful in capturing the conceptual particularity linked with the English 'progressive' tense form, and speculatively certain analogous expressions in certain other languages. In other writings (e.g. Chilton 2013) I have used the term 'windowing' because this seems appropriate to the 'viewing arrangement', which occludes parts of the process viewed outside the 'window'. But views through windows can be distant, whereas the *ing* form seems associated with 'close up' conceptualisation of an event. I have, therefore, preferred the term 'presencing', which suggests presence in the proximity of the event in one's personal 'space', as the geometric model also suggests, and such closeness in turn implies that earlier and later phases are not in view.[11]

4.2.2.1 The ing *presencing operator, operating on a state schema*
Applying the presencing operator (by means of the progressivising *ing* morpheme) to a state event schema (activated by a stative verb) will give us sentences such as (6):

(6)   *Mary is knowing her lines, *Mary is understanding the problem, etc.

Out of context such progressivised sentences appear unacceptable. For some reason the conceptualisation attached to the progressive tense form (presencing operator) is incompatible with the conceptual event-type schema attached to the predicates *know* and *understand*. Figure 4.9 gives the result of applying the presencing operator (Figure 4.8) to a state event schema (Figure 4.1).

First, we can make geometrical argument that explains why applying presencing (progressivisation) to a state schema does not work well. As we have defined the progressivising (presencing) operation, the peripersonal space picks out a bounded set of vectors; but just as for the entire *un*bounded set that corresponds to the state event schema, each vector is equivalent to the other, so collapses into one vector (as above, Section 4.2.1.2, Figure 4.6). This makes the peripersonal width imposed by the presencing operator redundant and irrelevant; the result is the same as the instancing operator – it is as

---

[11] There is another motivation for this choice of term, viz. an allusion to Heidegger's concept *Anwesen*, generally rendered 'presencing' or 'presence' in English translations. The basic sense of German *anwesen* is something like 'be at', comparable to Latin *prae-esse*, 'be in front of', the source of English 'presence'. It is relevant that McGilchrist (2010) thinks 'presencing' experiences and concepts are affairs of the right cerebral hemisphere. Whether what I've called the instancing operator is a left-hemisphere activity is another question, but it is worth noting that the instancing operations as I have described them include conceptualisation of abstract generic compressions, a typically left-hemisphere operation..

128   Event types and cognitive operators

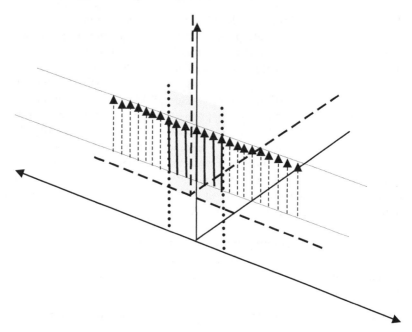

Figure 4.9 Effect of *ing* presencing operator on a state event schema

if the instancing operator had applied. The other component effect of the presencing operator is the zooming in effect (translation of R'). What can be said about this is that it is also conceptually irrelevant or inconsistent with the concept of state: one can be 'closer' to an action but it is unclear what being closer to a state – especially a state of mind of the kind denoted by the verbs *know*, *believe* and *understand* – would be.

Second, we can probe what the application of presencing to states in conceptual, phenomenological, terms. It seems that 'zooming in on', 'taking a closer look at' a particular subpart of a continuous state tells you no more than zooming in on any other subpart. Having a sense of being 'internal to' a continuous state does not have any obvious interpretation when stated in the abstract in this way, but may well have an conceptual interpretation in some special context. For instance, the default for the verb *believe* is that it requires the simple present tense form and is problematic with the progressive:

(7)   a Mary believes it
       b *Mary is believing it
       c Mary believes it at the moment.
       d Mary is believing it at the moment.

4.2 Tense forms as cognitive operators

Without going into all possible conceptual response to these sentences (7a) and (7b) contrast starkly when presented outside of a context. Progressivisation of a state seems to be fine when temporally relativised, as in (7c), so long as *at the moment* in that sentence is understood as some pragmatically extended peripersonal time-span; it is odd if you try to imagine *at the moment* as an instantaneous unextended point: *believe* cannot be punctual or timeless. It is important to note that even (7b) is an example of a sentence that is imaginable, given extra lexical and situational context, as in (7d). Now it is not easy to produce evidence as to what the conceptual distinction actually is – what the experience of reading one version of that English sentence rather than the other – does in one's consciousness. My own sense is that the progressivisation (presencing effect) of (7d) in some way involves a sense of being closer to Mary's personal experience, as well as isolating a subperiod of a state. In addition, because progressivisation is associated with dynamic events (7d) seems to implicate intentional action on Mary's part.[12]

Similar but not identical arguments can be used to make sense of the much-discussed alternation between simple present (instancing) and progressive (presencing) that is found with the verbs *stand* and *live*:

(8) a the statue stands in the hall; Bill lives in Chicago
     b the statue is standing in the hall; Bill is living in Chicago.

In such cases, the simple present applies naturally because instancing examples like (8a) ascribe properties and properties are conceptualised as permanent, i.e. timeless. In certain contexts, in particular those concerning animate entities, this kind of conceptualisation may be better termed 'habitual'. The application of progressivisation (presencing) in the (8b) sentences involves the imposition of the bounds of peripersonal time on a continuous unbounded state schema. The implicature is that the state is not unbounded or permanent: the timeless becomes the temporary.

4.2.2.2 The presencing operator operating on a process schema

The default use of the presencing operator produces examples like (7b) above. In Figure 4.10 the schema for a process (cf. Figure 4.4 or 4.5) is inserted in the deictic space, at $t = 0$, for illustration, though it could go anywhere along the t-axis. Imposed on the process schema is the effect of applying the presencing operator, analogous to Figure 4.9. We have a second axis system R' translated to a position on the d-axis that is 'close' to the event schema, satisfying the intuitions that one of the conceptual effects of the progressive may be that S is sensed to be 'closer' to the action.

---

[12] A pragmatic analysis involving implicature is clearly appropriate here.

130   Event types and cognitive operators

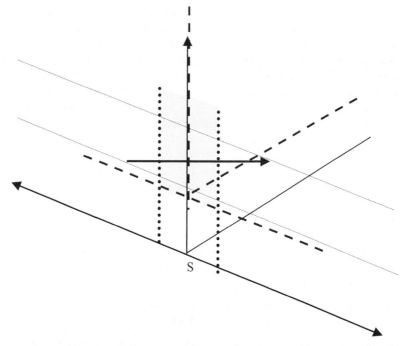

Figure 4.10 Insertion of process schema and application of presencing operator

The more salient effect of the presencing operator here, under the definition of it that we have given, is the imposition of pragmatically determined peripersonal time boundaries. Figure 4.10 shows how this operation 'zooms in' and cuts out a piece of the whole process schema, an event still in progress for S.

The overall effect of presencing on a process schema – namely, translating of a copy R' of S's base coordinate system R – is to produce a virtual viewpoint that gives rise to temporal bounding within the process schema. This effect also seems to imply that the full process schema does not disappear. It seems to remain in the mind but cognitively backgrounded.

A related implication of the progressive tense form (presencing operator) should be noted. The end points are 'out of view' from S's virtual viewpoint. But there is an asymmetry. In sentences where presencing is operated on a process schema nothing is asserted as to the *completion* of the process, though it is presupposed that the process schema itself has a *beginning*. Strictly speaking therefore the diagram should be revised to take account of this cognitive limitation and asymmetry. But it should also be noted that this asymmetry is a function of the mind–brain's relationship with the future (subjective or objective): it is modal, it is unknown for certain (although it can be conceptualised through language as certain).

4.3 Instancing and presencing in the past

To conclude this chapter, let us return to the reference frame properties of the presencing operator (progressive tense forms), as I have defined it. Let us consider the short discourse in the following:

(9) a Dave was painting the door. Mary arrived
 b Mary arrived. Dave was painting the door
 c Dave was painting the door when Mary arrived
 d When Mary arrived Dave was painting the door.

The first point to note is that the auxiliary *was* locates a reference time at some point (pragmatically determined), call it $-t_j$, relative to S, and that this is anaphorically related to the time denoted by the simple past form of *arrived*. To be more specific, the relation in such cases is such that the time point denoted by the *ed* form is included in the interval that is denoted by the *ing* form. The question is how the DST formalism can show this, together with the other effects mentioned above.

Let us assume that a discourse processor calls up an entire event schema and then applies the progressive (presencing) operator. It may be the case that the processor first inserts the schema in the ongoing discourse representation (as suggested in Figure 4.11, which is the DSM for the clause *Dave was painting the door*), operates the progressive (presencing operator), and then allows the full schema to 'fade', or to be 'backgrounded'. In the present state of knowledge it is not clear how to characterise this precisely, but foregrounding/backgrounding effects are in any case required in numerous aspects of discourse processing. Provisionally we assume that the event schema for *paint* in example (11) is inserted in the deictic space with coordinates on the *d*-axis corresponding to *Dave* and *door*. The modal coordinate is 0. How is the semantic schematic for *paint* positioned on $-t$?

We know that $-t_j$ must occupy some point within the interval set up by the progressivising operation and that the latter must be a proper subinterval of the total time course of the *paint* process. Inception of the painting process may or may not be given pragmatically or by the discourse and at this point in the discourse processing the processor cannot know if Dave goes on to finish ('accomplish') the painting, though subsequent discourse details may decide the shape of subsequent DSMs (e.g. *Dave put down his brush; he finished the job before making a cup of tea*, etc.). Thus in Figure 4.11 (and also Figure 4.5) the *t* coordinates of the end points for the process are arbitrary. The relative position of $-t_j$ within the *paint* schema assumes that the default is that $-t_j$ is

132   Event types and cognitive operators

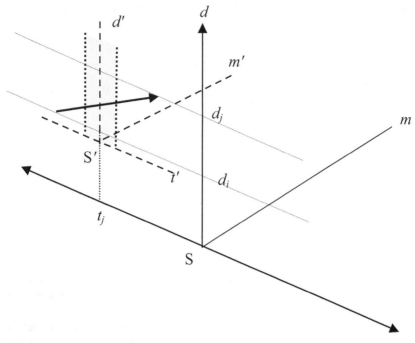

Figure 4.11 Progressive (presencing) in the past relative to S

in the centre of the frame, unless the discourse indicates proximity to either inception or conclusion. Some justification for this comes from the English spatial–metaphorical expression *Dave was in the middle of painting the door*.

The proposal is that the past progressive can be modelled as a geometric transformation of axes – specifically a translation of axes as shown diagrammatically by the dashed axes labelled d', t' and m' in Figure 4.11. The d'-axis of the shifted coordinate system is aligned with $-t_j$; the origin – i.e. the deictic centre for the speaker S – is shifted to a point on the d-axis, remaining at 0 on the m-axis. The translated origin is labelled S'.

In this chapter we have concentrated on the interaction between the event-type schemas (associated with the meaning of each verb) and two grammatical tense forms. We have anticipated a more detailed exploration of the meanings of tense forms and the role that the shifting of the base reference frame in the deictic space plays in the meaning of tenses. The next chapter takes up these matters.

5 Times, tenses and reference frames

> ... it is not strictly correct to say that there are three times, past, present and future. It might be correct to say that there are three times, a present of past things, a present of present things, and a present of future things. Some such different times do exist in the mind, but nowhere else that I can see.
>
> St Augustine of Hippo, *Confessions*, xi, 20

The relationship between tense forms and the awareness of time is not straightforward. Present-tense forms are used in ways that do not always refer to the speaker's present, and often refer to the speaker's future; both past and future tenses may be used not only to refer to times relative to past times and future times relative to the speaker's present but also to express epistemically modal meanings ranging from absolute certainty to mere possibility to counterfactuality. This chapter, like the last, considers only the two English tense forms known as simple present and present progressive. I shall not be concerned specifically on the tense forms known as simple past tense, or with the 'perfect' tense forms using the auxiliary *have*, or with the complexities of future reference of the tense form *will* + *verb*. These are all complex and interesting areas but this is not a book about verbs or tenses; rather, its concern is to explore the possibilities of the geometrical approach developed so far. The central puzzle to be addressed is how we can model the use of present-tense forms to refer to past and future. To glimpse something of the overall complexity of the associations between tense morphology and temporal conceptualisation it is useful to take an overview.

Table 5.1 summarises the key temporal–deictic uses that appear to combine with tense form (morphology and constructions). The columns headings *now*, *past* and *future* are intended to refer to the zones of S's experiential time built into the bidirectional deictic *t*-axis of the deictic space. The left-hand column lists English tense forms. The cells contain illustrative expression of the relation between these tense forms and experiential time zones. It is clear that the tense forms associated conventionally with deictic present, past and future, are not in one-to-one correspondence with these three temporal concepts. The table leaves out some important uses of tense forms – in particular uses of simple past tense and past perfect tense in conditional sentences

Table 5.1 *Correspondences between present-tense forms and deictic time reference*

Tense form/ operator	Now and generic	Past	Future
Simple present *(instancing operator)*	Earth revolves around the sun Henry knows the answer (statives) Henry walks to work The statue stands in the hall Rooney heads the ball	1520: Henry sails to Calais The king waves from the balcony in 1945	Henry visits Calais next week (non-statives)
be to		It is April 1520 and Henry is to visit Calais	Henry is to visit Calais next week
Progressive *(presencing operator)*	Henry is visiting Calais at this moment (non-statives) The statue is standing in the hall	Henry is sailing to Calais. Suddenly a storm blows up.	Henry is visiting Calais next week (non-statives)
be going to		It is 1520. Henry is going to visit Calais but is prevented.	Henry is going to visit Calais
Modals			
will	Henry will be in Calais now	Henry will have been in Calais last week	Henry will visit Calais next week
may, might, must			Henry may/might/ must visit Calais
present perfect		Henry has visited Calais (?last week)	
past perfect		Henry had visited Calais earlier	
simple past		Henry visited Calais last week	

(we return to these in Chapter 6) – but the uses of *will* that are included in the table show the close interrelationship between time and epistemic modality. The aim of this chapter is to show how DST can offer models for some of the complexities involved.

What is called 'tense', expressed by way of verb morphology and by means of auxiliary verbs, is used to locate events in time relative to the speaker; but it is not the only means of doing so, since prepositional phrases and adverbs can also pick out times deictically relative to S, and in such cases it is not even necessary to have the so-called present, past or future tenses line up one-to-one with the relative time that S intends to refer to. For example, S can use some form of the present tense when referring to an event in S's past or future. Furthermore, there are alternate perspectives on events – the conceptual operators of instancing and presencing discussed in the last chapter, which are also expressed, in English, by means of verb morphology and auxiliaries. In this chapter we are concerned not primarily with the interaction of tense forms with event types (*Aktionsarten*), as we were in the last chapter, but with the interaction between the linguistic tense forms and the conceptual temporal space, represented in DST by the bidirectional deictic *t*-axis.

The analysis attempted in this chapter, like those of a number of other scholars, builds on Reichenbach's (1947) framework and uses insights from Langacker (1991, 1995). Description of English tenses requires time points: the point of speech, event point and reference point (Reichenbach 1947: 287–8), with reference point often being defined in the discourse or pragmatically rather than explicitly in the sentence. In broad terms the DST account I develop follows Reichenbach in treating the three points in terms of relationships and ordering on the time-line. However, because the vector space includes direction as well as distance, DST says more about the nature of these relationships, and in particular it is able to incorporate 'viewpoint' (Langacker 1995, Michaelis 1998, amongst others). Further, DST carries with it the geometric idea of transformation of reference frames: this will be an important element in addressing the puzzle of the non-correspondence between times and tenses.

5.1 A present of present things

Now is a subjective experience or an intersubjectively and culturally agreed-upon conceptual category – 'the present'. The past and the future are subjectively understood as relative to *now*. As already discussed in Chapter 4, in the DST framework, S experiences 'the present' either as a point or as a peripersonal extension and normally experiences it as 'real', more real than the past and the future, which may be felt as less real than the past. A more abstract way of putting this is to say that *now* is experienced as maximally certain on a scale of epistemic modality. The relativity of the conceptual constructs of 'the past' and 'the future' to the *now*, where S is situated, is what makes reference to these 'times' deictic. This section is concerned with the effects produced by the use of the simple present tense and present progressive (*ing* form) in combination with various meanings expressed by particular verbs. As Chapter 4 has already explored two major types of present-time conceptualisation (instancing and presencing), this section is confined to summarising some extensions of these two types found in certain particular uses of the simple present and the progressive *ing* forms.

The English simple present does not always refer to the present time of the speaker, the speaker's *now*, though it generally includes it. In fact using the simple present to refer to a present instant is a rather specialised usage. It is generally recognised as conceptually connected with the following: states that include but are not restricted to *now*, generic concepts, habituals and narrative instancing (e.g. the 'instant' narration of audio commentators).

Frege thought of the simple present tense in terms of timelessness and time-setting:

> The present tense is used in two ways: first, in order to give a date, second, in order to eliminate any temporal restriction where timelessness or eternity is part of the thought. (Frege 1956 [1915]: 296)

By 'giving a date' Frege can be understood as meaning that an utterance of a sentence implicitly locates the time at which the uttered content of the sentence is asserted to be true. There are other, more language-oriented ways we can understand this in the DST framework. The present tense can be used to refer to a past event indicated in the discourse, perhaps with a date expressed in an adverbial adjunct and we below shall look at such cases in terms of the transformation of the base reference frame R.

Frege's idea of timelessness is important here because it characterises the abstract generalising use of the simple present, and is related to the logical notion of universally quantified propositions, which are timelessly true (or false). The important point is that treating the present tense as expressing temporal *presentness* (as opposed to temporal distance) is integral with epistemic certainty. And epistemic certainty inheres *both* in perceptual (therefore temporal) immediacy *and* in the conceptual transcendence of temporal boundedness. This characterisation emerges naturally from DST's geometric approach; it also seems phenomenologically natural. This way of understanding the meaning of what it is that the simple present tense corresponds to is similar to revised ideas about the English present tense found in cognitive grammar (Langacker 1991, 1995, 2001, 2011, Brisard 2002: 262ff).

We have already seen in Chapter 4 that although simple present (instancing operation) is the default for verbs denoting states and progressive (presencing operation) is the default for verbs denoting dynamic processes when S refers to them in S's present *now*, these default operations can be reversed. Applying simple present to dynamic process verb meanings produces the conceptual effect of atemporal generality, which can be understood in the DST geometric framework as a point on the t-axis.[1]

---

[1] I do not consider progressive present in this section, since events in the present are typically processes that are by natural default referred to in the progressive. They have been covered in Chapter 4 and it is the apparently discrepant use of the simple present that bears further discussion.

5.1 A present of present things

The following sentences all use the simple present. Only (1b) involves a stative verb that, for the reasons described in Chapter 4, occurs by default with the simple present. In example (1d) the verb *stand* is semantically open to alternate construals by either the instancing or the presencing (*is standing*) operator. The point of grouping these examples is to consider further the conceptual similarities emerging from the application of instancing in all cases.

(1)   a   the Earth revolves around the Sun
        b   Henry knows the answer
        c   Henry walks to work
        d   the statue stands in the hall.

The simple present in example (1a) is not understood as a temporal instant or a temporal zone around the self. It seems rather to be understood as Frege's timelessness present. In DST we have proposed capturing this geometrically by 'compressing' the vector for *revolves* to $t = 0$ and simultaneously locating it at $m = 0$ – thus defining a point without duration that is also maximally certain for S. This is an abstract certainty that is different from the experiential certainty of being 'close' to some represented dynamic process.

This kind of characterisation of the conceptual implications of the simple present might seem inappropriate to (1b). The dissimilarities disappear if one thinks of both sentences as predicating permanent properties. In understanding (1b) it seems likely that there is a tendency to assume that Henry does not stop knowing the answer, and the time at which he started knowing it is conceptually not represented at all. One of Henry's properties is that he knows the answer. The DSM for (1a) and (1b) are the same on this account (see Figure 5.1).

This kind of timeless epistemic certainty concerning essential properties is not limited to culturally accepted scientific certainties but is extended to personal experiences, opinions, philosophical claims, etc.: *broccoli is disgusting*, *Syd is a liar*, *politics is the pursuit of war by other means*, and the like. The DST format relativises all such cases to S's cognitive–deictic space; they are not modelled as objective (for the modeller) truths. The key conceptual feature is timelessness, which distinguishes this kind of temporal–epistemic certainty from that expressed in e.g. *the decorator is painting the door*, *Henry is sailing to Calais*, where the progressive tense form expresses epistemic certainty that is time-bound and based on perception or perceptual evidence. Such statements as those in (1) are *timeless*, that is, temporally universal facts that hold true at the present moment, even if S does not perceive them. Only the simple present expresses this: *Syd was a liar* is not taken as a fact true as of speaking, and is thus not temporally universal.

As proposed in Chapter 4, DST models the timeless validity meaning of the present tense as in Figure 5.1.

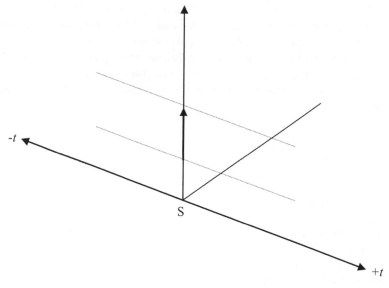

Figure 5.1 Timeless simple present for (1a) and (1b)

In sentences (1b) and (1c) the simple present tense form produces different conceptualisations. In the case of (1b) the schema for the stative *Aktionsart* combines with the instancing operator (expressed by the simple present form) to produce a timeless vector relating Henry and the answer. Sentence (1c) combines the instancing operator (again via the simple present tense form) with a process schema, producing, again, a timeless vector. Since processes cannot be conceptualised as timeless or as instants, they are conceptually processed as timeless *properties*.

The concept 'habitual', which comes to mind in understanding (1c), can be seen to be closely related to the concept 'timelessly valid' which comes to mind in processing examples like (1a). Both are so to speak removed from particular time reference. Even in (1c), Henry's walking is conceptually rendered timeless by the instancing operator (simple present). The difference between a human individual's walking and the Earth's orbiting is settled by way of additional conceptual processing drawing on the background knowledge triggered by the semantic frame of the two respective verbs and of the nouns. Now this explanation requires further detail that cannot be undertaken here. But it should be noted that not only do the *Aktionsart* schemas and semantic frames for *revolve*, *walk*, *Earth*, and *Sun* come into play but also, in the case of (1c), frames concerning social knowledge and linked to expressions such as *walk to work*. This particular frame will contain

5.1 A present of present things

knowledge about prototypical work routines in a particular human society and it is some such frame that makes (1c) 'habitual' in the sense that it evokes a series of repeated actions based on a daily cycle. It is possible that this is the case for all English expressions using the simple present that are understood as 'habituals'. This is perhaps the reason why we do not think of (1d) as being a habitual, although there are plainly some similarities with the (1c), as well as more abstractly with (1a.): statues are not framed as having habits. Such additional semantic material is not part of DST, which is set up to model only the abstract deictic framework (here especially the temporal and epistemic dimensions of it) that is common to the meanings of the sentences in (1).

Both sentences (1a) and (1c) in this sense express timeless properties. Habituals therefore can be regarded not as a separate class of event types, but as a result of the cognitive instancing operator that 'squeezes' temporally extended process schemas into state-like, temporally unextended vectors at $t = 0$. The position at $t = 0$ is a truth condition of the sentence for S at $t = 0$ in a particular reference frame modelled by a particular DSM – the zero t in such diagrams is S's current reference frame relative to the universal space-time reference frame.

The most general way of stating the proposal here is that a proposition predicating a habitual action or state of some argument represents a conceptualisation according to which that predication is a temporally unbounded property of that argument.

In order to account for the use of the simple present to denote the habitual meaning and the general validity meaning, Langacker (1991, 2001) and Brisard (2002) are obliged to postulate an extra mechanism. Langacker (1991: 264–5) adopts a distinction between 'structural' and 'phenomenal' knowledge taken from Goldsmith and Woisetschlaeger (1982) and uses the idea of 'structural' world knowledge to account for both habitual and generic sentences. He then proposes a schema to describe an ad hoc 'rule' (really an iconic diagram) by which the simple present turns a perfective (processual) verb into a 'derived' configuration in which what is expressed by the particular processual verb is now incorporated in 'structural' world knowledge, in effect rendering it imperfective. I want to suggest that, while compatible with Langacker's insights, the DST approach follows automatically and more simply from its geometrical principles. It does imply, however, that the habitual meaning emerges from pragmatic processing: (1b) and (1c) have the same schematic conceptual form, but (1c) involves a conceptual inconsistency between the process *Aktionsart* schema and the instancing operator (simple present tense) that is resolved by introducing the concept of habit.

The cases of simple present I have considered so far are somewhat different from cases where an event that is to some sufficient degree temporally concurrent with the speaker's speech time – as in sports commentary:

(2)   Mary bowls. Alice races to mid-wicket.

Such uses are not restricted to this particular genre but would also be natural in the following:

(3)   The surgeon picks up a scalpel. She makes an incision.

Cases (2) and (3) are not essentially different from conversational narrative uses of the simple present:

(4)   He was just walking down the street. This cop goes up to him and asks for his ID.

The example in (4) is, however, relativised to S's past by the tense of the first sentence. We shall look at a way of formalising this phenomenon in the next section.

Langacker's (1991, 2001) account of such uses of the simple present is appealingly straightforward in proposing that the use of the simple present in these situations is motivated by the approximate simultaneity of speaking time with the duration of the event being observed by the speaker. The DST account is the following. In an example like (2) we have process verbs that are not expressed in the progressive tense form (presencing operator), that is, they are not presented as portions of actions observed, as it were, close up. Given their event-type schema the expected default would be the progressive (*Mary is bowling*, *Alice is racing*). The use of the instancing operator (simple present) compresses the durational quality into a vector at unextended $t = 0$, as described in Chapter 4. This may be because bowling, racing to a position in cricket, making an incision, etc., though durational, can possibly be conceived as point-like relative to other types of actions. Be that as it may, application of the instancing operator (simple present) has to be understood here not in the sense of producing a single timeless vector that is then understood conceptually as an *instance* of a temporally unbounded set (as for examples (1a) to (1d)), but rather in the sense of producing an experiential *instant*. Such a temporal instant is still positioned on the t-axis at 0, but understood as the limit point of experienced peripersonal time.

In this way it is possible to construct a cognitively plausible account of uses of the simple present with process type verb schemas, one that is to some degree at least consistent with the geometrical account. But the main question is how to account for uses of the simple present that do not concern the present, that is, S's present, but to past or future times relative to S in S's reference frame R.

5.2   A present of past things

What is often called the historical present occurs widely in conversational narrative and literary narrative. In Langacker's formulation the historical present is a 'radical mental transfer pertaining to the deictic centre' in which 'the speaker

5.2 A present of past things

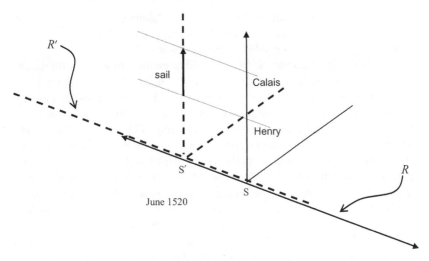

Figure 5.2 Transforming of reference frame for historical present: (5b) In June 1520 Henry sails to Calais

decouples the deictic centre from the here-and-now of the actual speech event and shifts it to another location' (Langacker 1991: 267). We can also use such a shift to account for certain uses of the present tense to refer to future times relative to the speaker (*the train arrives in five minutes*). In writing of 'displacement' or 'transposition' of the deictic centre, Bühler (1990[1934]) points us toward a geometric approach to describing uses of the present tense to refer to past events. In the DST framework, translation of reference frames on the t-axis seems like a natural solution and will be explored in this section. In both these cases, the relevant transformation is a translation of axes not of course a reflection. A second set of axes is copied which places a new $0'$ (the deictic centre) at some point $t_i > t_0$. The way this works is outlined in Figure 5.2 below.

Sentence (5a) below is feasible when there is a conversational context located in a past jointly attended to by S and an interlocutor. Sentence (5b) shows that the date-setting can be fairly precise and (5c) is linked to a photograph as well as time located relative to a reader (who may be reading it at a pragmatically appropriate (not too soon after) the date mentioned in the adjunct.

(5) a This cop goes up to him
    b In June 1520 Henry sails to Calais
    c Lennon waves from his hotel window in 1962.

Notice that such sentences can be temporally ambiguous. For example if (5b) is uttered by a twenty-first-century historian, it is present tense referring to the past relative to the historian and hearers. Equally, however, it could be uttered by, say, a privy councillor in the year 1518 planning a meeting with the French king to take place in 1520. The different possibilities are simply modelled by means of embedded reference frames.

Let us consider the case of the twenty-first-century historian's utterance. The speaker induces us to set up a second set of coordinates R', a temporal reality space in which we take the position of S'. Figure 5.2 translates a copy of R to a point on t representing the publicly agreed-upon date *June 1520*. In this reference frame the event of Henry sailing is expressed in the simple present tense form, following an application of the instancing operator to the process event type schema, as outlined in Chapter 4. It is positioned in the same way and with the same conceptual effects, at $t = 0$ in the deictic space, i.e. the reference frame with the usual DST axes, as described in the preceding Section 5.2.1. The 'instancised' process verb *sail* is at $t = 0$ in R', which is a copy of the base frame R, so carries the subjective deictic origo, now marked S'. In other words, S is 'seeing' an event taking place concurrently with his or her *now*, but this *now* is displaced into the past relative to the base frame and to S.

The discourse entities are labelled on the d-axis in the base space R: their coordinates run through into the virtual space R', so we have trans-frame identity. The fact that the 'ends' of the time axes t and t' do not coincide is of no consequence since the t-axes are potentially infinite. The essential point is that the vector for *sail* is simple present (the process *sail* has had the instancing operator applied to it) with respect to R' and simultaneously past relative to S in R. This is exactly what is required in order to formally model a present tense form that refers to a past event.

In this example the instancing operator has applied to the process verb *sail* (as in examples (1) above), resulting in a particular construal of the event. We can think of the configuration similar to that of the action-by-action commentator or observers (cf. (2), (3) and (4)), with the process event compressed to a conceptual instant of time in which the event is completed. A connected motivation may be that process events in the past are viewed as complete when presented in the usual way by means of the simple past: *Henry sailed*. In fact, the temporal translation of R is only part of the construal potential. Once R has been shifted the event of sailing can be represented either by the using the cognitive operator provided by simple present or by using the cognitive operator provided by the present progressive as in (6):

(6)   In 1520 Henry is sailing to Calais. Suddenly a storm blows up.

In such a representation R' is positioned on the t-axis as in Figure 5.2, but is also translated on the d-axis, so that it is closer to the coordinate for the *sail* vector – in accordance with our characterisation of the presencing

5.3 A present of future things

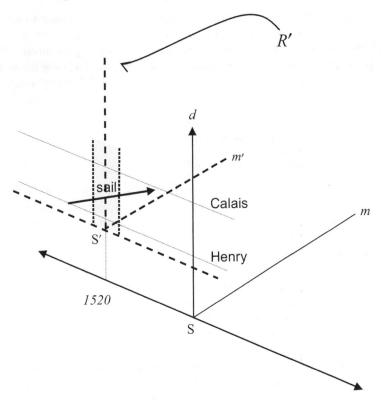

Figure 5.3 Present progressive in the past: (6) In 1520 Henry is sailing to Calais

operator in Chapter 4. The resulting DSM would in fact look like Figure 4.11 in Chapter 4, repeated above as Figure 5.3, with labelling appropriate to (6).

5.3 A present of future things

The relationship between the English present-tense forms and reference to future time is particularly complex. It is often pointed out that reference to the future has a strong modal element and this is assumed to be connected to the fact that the future is inherently unknown, though it is mentally representable because of the cognitive phenomena of intending, planning and expecting.[2]

---

[2] This section is a version of part of Chilton (2013).

Jaszczolt (2009: 38–45, 50–5, 140–1) argues strongly for a modal view of the expression of deictic future times and proposes a scale of future time markers corresponding to a scale expressing 'a certain of detachment from the certainty of *now*'. Following Jaszczolt's scheme, but excluding *be to*, *be about to* and *be on the point of*, we might arrange the following sentences in the following order:

> Henry visits Calais this Thursday
> Henry is visiting Calais this Thursday
> Henry is going to visit Calais this Thursday
> Henry will visit Calais this Thursday.

According to Jaszczolt these are graded from highest certainty (least detachment) to least certainty (highest detachment). While this scale seems intuitively satisfactory, it is not completely clear how these particular tense forms encode graded modal meanings. Jaszczolt outlines in some detail how Default Semantics, including its revised version, explains meaning as an emergent phenomenon arising from societal, cultural and other pragmatic information (Jaszczolt 2009: 132). But the tense form must also have core meaning and what is not clear is whether and how these core meanings per se encode modal gradations. I shall leave aside the question of whether the semantic structure of the different tense forms have inherent modal meaning that enables us to place them on a modal scale, and instead focus on the temporal axis as set up in DST. The question we are concerned here is how DST can model the use of tense forms that appear to refer to the speaker's present in order to refer to the speaker's future. We shall consider the following examples:

(7)   a Henry visits Calais this Thursday
        b *Henry visits Calais [future time reference, no pragmatic or lexical indicator]

(8)   a Henry is visiting Calais this Thursday.
        b *Henry is visiting Calais [future time reference, no pragmatic or lexical indicator].

These examples suggest that present tense forms cannot refer to a subjectively future time without further cognitive input. This means some explicit lexical indicator such as *this Thursday*, *next year*, *in a few days' time* is required. Note that a vague indication such as *in the future* is not enough:

(9)   a ??Henry visits Calais in the future
        b ?Henry is visiting Calais in the future.

It seems that some more precise time-indicating expression is required. This is not necessarily a lexical time expression in the clause, however. It may be a

5.3 A present of future things

mutually shared piece of knowledge in the conversational context.[3] These considerations will lead us to a detailed model of present-for-future expressions that is somewhat more detailed than the conventional accounts and which is primarily temporal rather than modal. I turn now to simple present and progressive present forms used to conceptualise times that are in the future relative to the speaker and hearer.

5.3.1 Simple present and the future

In order to account for the 'scheduling' use of the simple present in (1), Langacker (2001) makes use of a notion of 'virtual entities', understood by him in terms of a 'non-canonical viewing arrangement'. Brisard deals with the problem primarily by emphasising a modal meaning of the simple present that he calls 'immediate givenness' (Brisard 2002: 263–8). Brisard retains Langacker's idea that the simple present has a double meaning (immediate phenomenal experience and general structural knowledge of the world), but offers a different description, in which the processual verb is 'projected' (by use of the simple present) onto a 'virtual plane' equated with representation of 'structural aspects of the world' (Brisard 2002: 274). The DST approach, however, provides a way of dealing with such a matter within the overall theory of frames of reference. Lonergan proposes three kinds of reference frames: the personal, the public and the special. The latter are mathematical and physical and need not concern us here. The first corresponds with personal cognition of three-dimensional physical space:

> ...everyone has his personal reference frame. It moves when he moves, turns when he turns, and keeps its 'now' synchronized with his psychological present... (Lonergan 1957: 144)

It is the case that DST does not work with the physical space but with a personal conceptual space of a kind that specifically underlies language. To that extent DST seeks to advance beyond Lonergan's idea of reference frames. Nonetheless, Lonergan's general point and the link he makes with temporal as well as spatial deictic expressions is consistent with DST's deictic space. What is of special concern in the present context is his second class of reference frame, 'public reference frames'. These are both spatial and temporal. In the following Lonergan is primarily describing temporal reference frames:

---

[3] Jaszczolt's Default Semantics also takes account of such information in explaining semantic representations (merger representations) of utterances.

[people] are familiar with alternations of night and day, with the succession of weeks and months, with the uses of clocks and calendars. Now such relational schemes knit together extensions and durations. But they are not personal reference frames that shift about with an individual's movements. On the contrary, they are public, common to many individuals, and employed to represent the *here* and *now* of the personal reference frame into generally intelligible locations and dates... (Lonergan 1957: 144)

Now cognitive linguistics, coming from a different angle, has a complementary point to make concerning the nature of this publicly shared knowledge of times (Lakoff 1987: 68–9), drawing on Fillmore's work on cognitive frames (e.g. Fillmore 1985). Lakoff describes what English calls a 'week' as an 'idealized cognitive model' – we could equally call it an idealised public reference frame – that contrasts with for example Geertz's description of the Balinese calendar (Geertz 1973: 392–3 cited by Lakoff 1987: 68). Names of days, such as 'Tuesday', can only be defined relative to a reference frame, and this is clearly a conceptual construct:

Tuesday can be defined only relative to an idealized model that includes the natural cycle defined by the movement of the sun, the standard means of characterizing the end of one day and the beginning of the next, and a larger seven-day calendric cycle – the week. In the idealized model, the week is a whole with seven parts organized in a linear sequence ... (Lakoff 1987: 67)

Such frames (or 'models') do not exist in nature but are conceptualisations that are agreed upon by long processes of cultural coordination.

Some formal semantic accounts of the scheduling present take it that such cases arise because they have the 'connotation' that they are *predetermined* with respect to the present (Kaufmann *et al.* 2005: 90). I suggest that this 'connotation' arises because of the insertion of a cognitive frame such as the *week* frame, where the days of the week and the cyclic recurrence of weeks are in a sense 'predetermined'.

I am proposing that the notion of idealised frames can be usefully thought of in terms of reference frames: Fillmore's frames and Lakoff's models, at least those that concern time, can usefully be understood as *reference frames* that enter into relations with personal reference frames of the kind that DST models in its particular geometry. The crucial point is that *now* is a personal cognitive experience not a socially agreed-upon time in a public frame of reference: the personal and the public frames of reference have to be 'coordinated' – their separate coordinate systems have to be brought into some sort of alignment, the one relatively to the other. We shall make use of this idea in looking at the meanings of English present-tense morphology.

In DST we already have the apparatus to model time-related frames. The first step in considering how to use them to model the 'scheduling' use of the simple present tense form is to note that its use suggests a secondary set

5.3 A present of future things

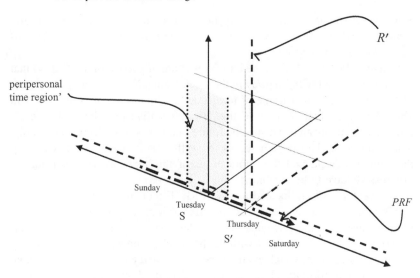

Figure 5.4 Example (7a) Henry visits Calais this Thursday

of axes R' whose origin is located at some time t in the future relative to S. Further, we may say that the origin of the scheduling axes R' is not at S', but some other referent such as a timetable or some pragmatically given 'public reference frame' (PRF), which might include a shared mental representation of a written or verbal agreement, or shared cognitive frame representing a time cycle such as the calendar or the seasons, in the encyclopaedic memory of the interlocutors.

In using the term 'schedule' here, one should not be misled into importing contemporary associations, since modern scheduling is merely a kind of cooperative planning for the future that is characteristic of all human societies from early beginnings that may very well be closely tied to the emergence of human language. More needs to be said about how DST can geometrically model PRFs but this cannot be done here. We focus on the relations between frames of reference, depicted in Figure 5.4. In this and subsequent figures, the peripersonal temporal region is shown as a shaded plane.

PRF is a public frame of reference. Because it is public there is no S defining a *now*, so there is also no m-axis representing the modal component of S's conceptualisation, and no d-axis on which S 'locates' discourse referents at subjectively relative 'distances'. It is simply an ordered set of seven arbitrarily named time periods, each corresponding to the diurnal cycle. These intervals lie on a finite directed time-line that defines a one-dimensional plane. Though inherently it is not deictically centred, it is directed in the direction of ontological time. In common with other individuals in the culture,

S holds PRF in long-term memory but also aligns it, in virtue of socially shared knowledge about 'which day it is', with his subjective conceptualisation of peripersonal *now*.

Consider Figure 5.4. First the base axes *R* are aligned with the PRF so that S's *now* is aligned with *Tuesday* in the week frame. Next we set up R', a copy of S's base *R*, a virtual frame of reference in which S's *now* is aligned with *Thursday* in the PRF. It is in R' that Henry's sailing to Calais is located, a reference frame located at a specific future that has both subjective and public reality for S, as the diagram shows. Now this future event is expressed by the simple present, and the implications of this tense as distinct from the progressive have to be explained.

The vector *visits* is located at 0 in PRF. Either this means the event of leaving on Thursday is a timeless property, or it means that it is viewed as a timeless instant – the relevance here is that the event is viewed instantaneously as a whole, in contrast with the presencing conceptualisation cued by the progressive form, which excludes beginning and end. This is an aspectual distinction but formulated in the terms just used we can see why it is appropriate to express the type of meaning under consideration.

5.3.2   Present progressive and the future

Neither simple present nor present progressive can refer to the future without extra lexical or pragmatic specification and presumably therefore additional cognitive structure. Compare the following:

(10)   Henry is visiting Calais (now)

(8a)   Henry is visiting Calais this Thursday.

In (10) the default understanding is that Henry is visiting Calais *now*, at the time of speaking. In (8a) the speaker sets up a mental space R'. But this future is located in a PRF, as shown in Figure 5.5. The deictic space model for the conceptualisation induced by this use of the progressive combined with lexical (or pragmatic) time indicator proceeds as for the case of the simple present (Figure 5.4 above).

In this new space, represented by the bold dashed axes in Figure 5.5, S (and the hearer) have a present *now* located at a deictically future time, indicated by a temporal deictic *this Thursday*, defined relative both to the PRF and to base *R*. Future *now* and present *this Thursday* are conceptually co-located. Within the embedded frame, i.e. relative to it, we have the presencing operation due to the progressive form *is visiting*. The 'up close right now' element is represented by the conceptual presencing operation within that frame. Sentence (8a) seems to bring us 'close' to the action, 'puts us in the picture'. It does not seem necessary to consider this effect 'modal'. Moreover,

5.3 A present of future things

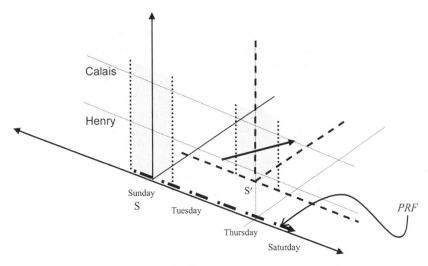

Figure 5.5 Frame shift for (8a) Henry is visiting Calais this Thursday

it is often noted that the intuited sense of such examples is that the future event is somehow viewed as rooted in the speaker's present. According to the present model, this is accounted for within the shifted reality frame R'.

How should we characterise the difference between the simple present and the progressive present referring to a future time? The elements of a description are made explicit in the notions of coordinate geometry, vectors and shifts of reference frame. The temporal component is the same for both tense forms (see Figures 5.4 and 5.5). The difference is the structure of the instancing operator (simple present) and the presencing operator (progressive present). In the former the whole event is viewed as a completed instant, but since the simple present also serves to conceptualise timeless generic properties, it carries a potential implication of high subjective certitude. This is not the case for the progressive form (presencing operator), which has the slightly contradictory effect of close-up viewing and incompleteness (as the 'ends' of the action are not 'in view'). This effect may or may not, it seems to me, result in a sense of greater or lesser certitude with respect to the simple present. It is thus not clear to me that these forms should be viewed as primordially 'modal', though this may arise by implication under the influence of various contextual factors.

5.3.3 Going to the future

We are concerned here with the 'periphrastic future' tense forms *be going to/gonna*, as in (5)

(11)   Henry is going to/gonna visit Calais this Thursday.

Such forms are discussed in terms of grammaticalisation by Hopper and Traugott (2003). In the case of *gonna*, which is only possible with a following verb (not a NP), the fact that *going* assimilates *to* indicates that we are dealing with a coalesced concept of directed motion, i.e. we are not dealing with an infinitive form *to* V, e.g. *to visit*. In terms of cognitive metaphor theory, the underlying metaphor as is well known is PURPOSE IS DIRECTED MOVEMENT (cf. Lakoff and Johnson 1980). Hopper and Traugott's (2003) explanation of the grammaticalisation of *going to* as an auxiliary expressing future time rests on a pragmatic inference of futurity from purposive action as expressed in the lexical verb *go*. However, this fails to generalise over the symmetrical pair found in a language like French: *venir de V* (literally 'come from V', i.e. to have just done V) and *aller V* ('go V'). Purpose cannot be inferred from 'come from'. While Hopper and Traugott's account is not necessarily wrong, the French example suggests that this kind of temporal auxiliary results from a direct conceptual transfer from space to time, preserving deictic relations. The DST account makes this assumption and the coordinate and vector components of that theory are well placed to model what is at issue. Spatial translation to/from a location is directly mapped onto temporal translation to/from a time point.

As suggested by the grammatical construction in question, the model in Figure 5.6 shows two vectors, a component for 'be going' and one for 'visit', following the modelling already discussed for progressive forms (presencing operations). This corresponds to the reported sense that *be going to* futures are closely connected to the present, to some process already in train at S's *now*. The main verb (here the *visit* component vector) appears as a timeless event vector at some point $t_i > t_0$ that might be specified pragmatically or, lexically, by a PP (here 'this Thursday').

Presencing operations usually leave the culmination of activities 'out of view' but in the case of the periphrastic future construction that uses them the preposition *to* specifies the end point, namely the event expressed in the main verb (here *visit*). Nonetheless, the model also captures the fact that from S's present viewpoint, the culmination is 'outside' her peripersonal space. And as for all progressive forms, the presencing operator also leaves 'out of view' the beginning point of the activity, which equally seems to be appropriate for the conceptual effects of the periphrastic future that are often described.

These features seem to correspond well with the sense that periphrastic futures refer to future events that are expected with a high degree of certainty because the speaker has experiential evidence in the present that they will take place in the subjectively 'near' future. Another way of putting this (Jaszczolt 2009: 65, reporting Eckardt 2006), is to say, in Reichenbachian terms, that E (event time) is in the future and R (reference time) is in the present and that *be going to* sentences are about 'the present time and what is imminent in it', or perhaps, one might say, 'immanent in it'.

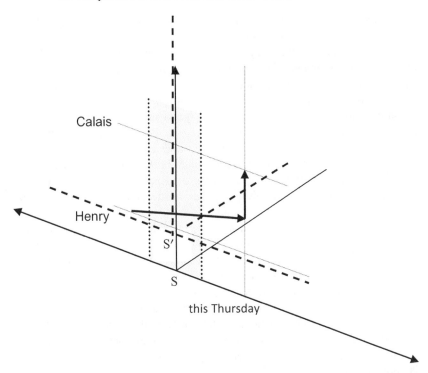

Figure 5.6 Example (11) Henry is going to/gonna visit Calais this Thursday

Note that there can be a subtle conceptual difference in *Henry is going to visit Calais next week*. The speaker S may be imagining Henry's intention, or S may be communicating her/his own prediction. DSM modelling handles this. In Figure 5.6 we have the case where S projects her own world R' in which Henry visits Calais at a future time. To model the meaning in which S is communicating Henry's intention, the origin of R' has Henry's coordinate on the d-axis.[4]

5.4 The putative future: a reference frame solution

Across languages it is not uncommon to find that a future-tense form occurs in contexts that are interpretable and that yield a cognitive effect generally described as modal, or epistemic, or as we shall prefer here 'putative'. I am

---

[4] A DSM for *venir de* constructions can be built along similar spatio-temporal lines, allowing for the fact that this French construction does not involve the progressive.

not concerned with the precise pragmatic factors that allow such understandings, merely with their conceptual structure when they do occur.

As Jaszczolt argues, the fact that the conventional future form *will* + V can express present probability may be evidence of the close cognitive connection between future time reference and the inherent cognitive (or even metaphysical) uncertainty of the future, but this is not a necessary reason to conflate modality and temporality when there is good reason, for the purpose of semantic modelling, to think that they are, or can be, separate in linguistically expressed cognition. The English putative future meaning of *will* + V is not just a conventionalised expression of high probability. The meaning of *Mary is probably writing the report* is not the same as *Mary will be writing the report (now)* and is unlikely to be used in the same pragmatic contexts, i.e. have the same pragmatic potential effects.

Furthermore, the meaning of *will* + V seems to have a different internal conceptual structure from *must, may* and *might* + V. French linguists have noted that the equivalent French form, the *futur putatif*, presupposes a future point of view at which the present event is predicted to be verified (Damourette and Pichon 1911–36, Sthioul 1998, Saussure and Morency 2012, Saussure 2013). These authors speak of an 'imaginary' future 'perspective' or 'viewpoint'. They also insist that this viewpoint is allocentric, in the sense that it is distinct from the current speaker in the actual world, a claim that DST frames in a slightly different way. The general approach of these authors, however, is adopted here.

The conceptualisations we have to consider arise in the context of certain kinds of prompts for deictic times, illustrated in the following:

(12)   Henry will visit Calais this Thursday

(13)   Henry will be visiting Calais this Thursday

(14)   Henry will be visiting Calais (now).

A simple present version of (14) *Henry will visit Calais now* in the putative sense is not valid because of the regularities governing non-statives.

Sentence (12) is the 'regular future' and (13) is a 'regular future' use of *will* combined with the progressive form (presencing operator) that is represented by the insertion of a new viewpoint in a new R' located at some time in the future relative to S – this is shown in Figure 5.7 below. What do we have for (14), where we have a future progressive form understood as referring to the present *now*? We can simply put the *visit* verb vector at high probability on the *m*-axis. But this is not explanatory. We want to see if the frame-shift model of DST can account for this use of future form with auxiliary *will*.

Figures 5.7 and 5.8 show the different conceptual structures, in DST terms, of the different potential meanings of *Henry will be visiting Calais*, Figure 5.7

5.4 The putative future: a reference frame solution

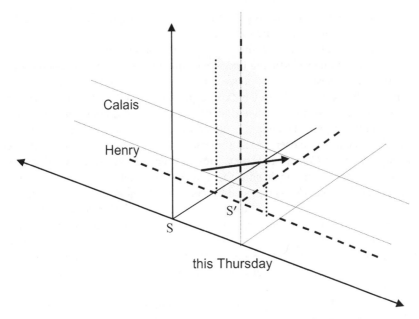

Figure 5.7 Example (7) Henry will be visiting Calais this Thursday [non-putative]

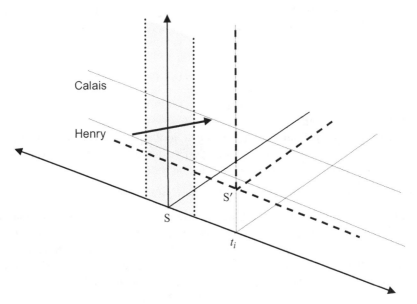

Figure 5.8 Example (14) Henry will be visiting Calais (now) [putative]

modelling the non-putative 'regular future' meaning, Figure 5.8 modelling the putative use of regular future to refer to a present event.

Unlike the future tense form *will* + V, the progressive form introduces the viewing frame (presencing operator), which involves an embedded set of coordinates, transposing a represented or 'imaginary' S to the future time point in the base coordinates indicated pragmatically by *will* at the deictic time *t* represented by *this Thursday*. (To simplify the account, the PRF has been omitted here.) The cognitive effect seems to be one in which the utterer of this sentence is 'closer' to the projected future event than in the sentence *Henry will visit Calais next week*. Next we have to consider whether this modelling approach can handle the conceptual effects triggered when the adverbial *this Thursday* is replaced by, for example, *now*, *at this moment*, etc. or by some equivalent but unexpressed pragmatic indicator, as in (14).

The frame-shifting principles of DST seem to be able to accommodate this putative future use of the future. The putative future is the reverse of (8). Whereas in the latter we have a present tense form referring to a future time, in (14) we have a future tense referring to present time, always relative of course to the speaker's *now*.

Broadly following the proposals of Sthioul (1998) and Saussure and Morency (2012), we have in Figure 5.8 a metarepresented point of view at some $t_i > t_0$. This additionally represented point of view does not have to be an unspecified allocentric other in the sense of a different individual S_j. It is a cognitive avatar of S, S′, and can still be regarded as 'allocentric' in that sense. This is also consistent with examples from contexts in which it is pragmatically manifest that the speaker herself will be in a position to verify, in the future, a current situation. In Figure 5.8 from the viewpoint of S′ the vector *visit* is in the past and it is true (located in the plane $m = 0$); from the viewpoint of S *visit* is in the present. The use of coordinate systems as reference frames makes it possible to model this property of the putative future construction.

Just as 'regular future' tense forms can, in the putative construction, refer to S's present time, so 'regular future perfect' tense forms can refer to events in S's past time. Such a reading is available for (15):

(15)   Henry will have visited/will have been visiting Calais.

In the account proposed by Sthioul and by Saussure and Morency, sentence (15) may be said to take a view of a past event from a future vantage point at which it will be verified. The sentence is of course ambiguous, since it can also be read in a simply temporal sense as referring to a time t_i in S's future but before some future vantage point (i.e. $t_j > t_i$). One may indeed say that it is doubly ambiguous, since for some speakers this temporal reading of (15) may be either simply temporal or putative, again in the sense of Sthioul, Saussure

5.4 The putative future: a reference frame solution

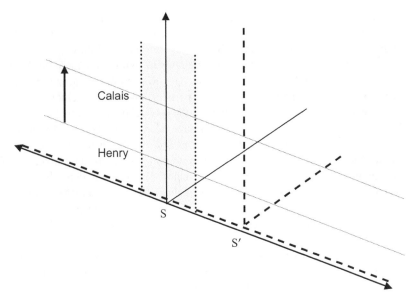

Figure 5.9 Example (15) Henry will have visited Calais [putative reading]

and Morency. If this is so, the ambiguity indicates not that future time is inherently modal but that there are two possible conceptualisations of the future, one predictive and the other something short of predictive.

The geometric DST model for these sentences will have the same structure as in Figure 5.8 except that the vector representing *visit* will be located at some time prior to S's *now*, as in Figure 5.9, which illustrates a model for the non-progressive in order to simplify, and for a reading that is putative. (The embedded axis system has its origin at $d = 0$, in contrast with Figure 5.8, which illustrates a progressive tense form, i.e. presencing operator.)

In this model for the putative reading of (15) the embedded axes are still in the future relative to S, since, in the account of the putative meaning that we are adopting, verification is still in the future relative too S. What is different between (15) and (14) is the location of the event in the past relative to S, i.e. at some $t_i < t_0$. The diagram can show how the *have been* component of the construction, referring to the past relative to S, is conceptually represented simultaneously with the *will* component associated with future reference relative to S.[5]

---

[5] A similar basic structure models the progressive tense version of (15) but with some additional complexities that we shall not attempt to deal with here.

In this chapter I have paid attention to only a small number of temporal expressions, in particular those that refer to the future, that is, the deictic future relative to some speaker S: simple present for future, progressive present for future, periphrastic present *going to*, with brief mention of the 'regular' *will* future. The overall aim has been to demonstrate how a geometrical approach formulated in DST can, perhaps surprisingly, elucidate temporal relations and in particular bring out the role of shifting reference frames. This approach separates the conceptual dimensions of modality and time, and one of the ancillary aims has been to characterise the distinctions between the different constructional means of referring to the future in terms of shifting point of view rather than modal scales. This is not to deny that modal effects may arise, but the present model does not treat them as inherently conflated with time as linguistically expressed. If and when they arise, they arise through contextual factors interacting with the conceptual structures built from reference frames. The solution proposed here to the problem of giving an account of present tense forms referring to past and future times consists of using translation transformations, which are part and parcel of coordinate geometry. The next chapter investigates a further type of reference frame transformation.

6 Counterfactual reflections

> The analysis of counterfactual conditionals is no fussy little grammatical exercise.
>
> Nelson Goodman, The problem of counterfactual conditionals, *Journal of Philosophy*, 1947

It is a remarkable property of human cognition that we can entertain states of affairs and events that did not happen, are not happening, or might not happen. It is equally remarkable that a speaking subject can imagine standing in a position other than the one in which he or she is standing and speaking. In both cases we speak of adopting an alternative point of view. The relevance of this ability has already been discussed in Chapter 2 in connection with physical space and the means we have in languages to relate ourselves spatially to objects in our environment. One of the geometric operations that have been found pertinent, as we saw in Chapter 2, for understanding the language-related conceptualisation of spatial relations is the transformation of reference frames called reflection. This chapter will explore the reflective transformation of axes as a way of understanding analytically several grammatical constructions whose job seems to be to enable people to communicate about their relationship to their environment, physical, social and imaginary.

One of the key advantages of the geometrical approach using base and embedded coordinate systems is that the discourse referents (whose positions are labelled on the d-axis) can, for the relevant cases, carry through their coordinate positions from one coordinate system (conceptual 'world') to another. This property of our approach captures something of the approach of mental space theory described by Fauconnier (1994, Chapter 4) and nicely made already by Lyons:

We can also carry out the psychological process of trans-world identification across real and imaginary worlds of various kinds. We can identify ourselves and others in our dreams; we can create hypothetical situations involving real persons and then talk about these situations in much the same way as we talk about things that are actually happening or have actually happened. (Lyons 1977: 791)

Nonetheless, given the way the geometry of DST is defined, important questions arise concerning certain types of anaphoric relations within the abstract space, in particular the grammatical transitivity relation of S to himself or herself – the reflexive relation. Our main concern, however, will be with the means provided by languages for making communications about the relationship between the S and 'worlds' (entities, states of affairs, events) that do not, in some sense, exist or have not happened.

6.1 Counterfactuality

Counterfactuals have been intensively studied and puzzled over by philosophers of language and science in the last half century, going back at least to David Hume and J. S. Mill. Philosophers are preoccupied with the conditions under which *if*-sentences in English, which do not assert truth, would in fact be true. Nelson Goodman (1947) famously discusses the example 'if the match had been scratched, it would have lighted'. The problem that can arise with such a sentence emerges if one focuses on the causal relation between the *if*-clause and the main clause. For example, to convince oneself that scratching the match would cause its lighting, other conditions have to be assumed: e.g. that the match is dry, that there is sufficient oxygen in the appropriate region of space, etc. The question is how many other things are needed to guarantee the truth of the sentence and how they are all to be computed. The work of David Lewis (especially Lewis 1973) sought to resolve such questions. Lewis's proposals build on the important notion of possible worlds and 'closeness' of different possible worlds to actuality. A much more general way to state the puzzle might be: counterfactuals are by definition not true, so how can they be used, when truth and factuality are required? An even more general way of stating the puzzle might be: given the empirical fact that humans do use counterfactual utterances, amongst which counterfactual *if*-sentences are perhaps prototypical, how, why, and for what purposes, do humans use them? This has been the approach of psychologists (e.g. Johnson-Laird 1983, Byrne 2007). It has also been the approach more recently of cognitive science and cognitive linguistics. Fauconnier and Turner (2002) propose a theory of the way counterfactual thinking works, in language and outside of language, as well as demonstrating the countless manifestations of counterfactual thinking and communicating in science, philosophy and everyday life. Here, however, I focus on a more narrowly linguistic question: how can we analytically describe the fact that counterfactual conceptualisations – in effect thinking that something is *not* the case – spring from a sentence like Goodman's 'if the match had been scratched, it would have lighted' which has no explicit negative word like *not* in it? This may seem a less important question than the philosophical ones, and maybe that is the case. Nonetheless, language is the sine qua non of philosophy and it is rational and even important to ask how it works.

Moreover, there are plainly ways in which the approaches converge in common concern.

Counterfactual thinking is in fact activated and communicated by many different linguistic forms but counterfactual *if*-sentences are of particular interest because they are relatively complex and variable linguistic structures that facilitate not only thinking and communicating about things and events that do not exist or have not taken place, or do not yet exist or have not yet taken place, they are also, as philosophers have been above all aware, instruments for thinking about causal relations. In the realm of social psychology they are, in their contexts, a means to setting up an intersubjectively shared mental space, on which joint attention is focused, for the pursuit of practical reasoning about the past as well as about future action. Fauconnier and Turner (2002: 221, for example) describe such uses minutely in terms of Blending Theory. What Blending Theory seeks to tell us about is the integration of different representations of the world prompted by particular sentences, including counterfactual *if*-sentences, e.g. 'if President Clinton were the *Titanic*, the iceberg would sink'. But it does not tell us how the specific grammatical construction, the *if*-sentence that can give rise to the blending of different spaces can have the form it has. The account I will outline below shows that the seemingly paradoxical form of counterfactual conditionals follows naturally from the fundamental premises in DST concerning the fundamental deictic space. In the Clinton–Titanic example, the blending of factuality and counterfactuality comes from the conceptual frames associated not only with the grammatical frame but also with the conceptual frames associated with lexical items *Clinton*, *Titanic*, *iceberg* and *sink*. Blending Theory demonstrates over many cases how this works. DST shows something in a way more meagre: the basic structure of counterfactual conceptualisation itself and its dependence on abstract geometric transformations.

6.2   *If*-sentences and counterfactual conceptions

Grammatical constructions have meaning. *If*-sentences (conditional sentences) in general are sometimes said to be 'counterfactual'. This is true to the extent that there is a range of ways in which it is possible for a proposition to be 'counterfactual', ranging from assertions of probability, through possibility, to negation of some other proposition. Conditional sentences, that is two-clause sentences consisting of an *if*-clause (the antecedent clause or traditionally protasis) and a consequent clause (traditionally apodosis, syntactically the main clause), yield mental representations that lie somewhere along the m-axis. They may be thought of as having the potential to express degrees of irreality, ranging from the not-quite-certain about the near future to that which is counter to fact. In speaking of these degrees, I am not thinking of absolute truth but the reality as intended to be communicated by a speaker to a hearer, given the conventional structures of the language they are using and the range of conceptualisations that can conventionally be associated with these structures. This is not to say that particular forms, say of the English conditional constructions, are uniquely associated with particular conceptualisations. It is well known that the particular meanings that conditional sentences can give rise to depend on a number of factors – the meaning of *if*, the tense of the verb, the semantics of the particular verb, expected intentions of the speaker and other contextual factors. The tenses of verbs in conditional sentences do not simply denote times; they are often only indirectly related to times to which a speaker may wish to refer. These points concerning the meaning of tenses in conditional constructions have been well made by, among other scholars, Fauconnier (1994), Dancygier (1998), Dancygier and Sweetser (2005) and Byrne (2007).[1]

The past perfect can be used in a conditional sentence to refer to the speaker's present, past or future. One can say, for instance, *if John had been here now, he would have been/would be astonished* but one can also say *if John had been here last week, he would have been astonished*. And one can say *if John had been coming tomorrow, we would all have been very pleased*. A particular time adverbial can then determine how the time reference relative to the interlocutors is interpreted. The use of *were*, and the simple past tense, also allows variation in time reference, since we can say *if John were here now* and *if John were here next week* – although one cannot use this form to refer to the past *(?if John were here last week)*. The simple past tense form likewise allows varying time reference: *if John came yesterday*, *if John arrived at this moment*, and *if John turned up next week* are simple past tense forms referring to different times in the speaker's reference frame.

The degree of certainty or uncertainty itself is not wholly predictable from the tense form, though there is a tendency for past tense and past perfect tense forms, denoting relatively more 'remote' past times, to correlate with more 'remote' epistemic interpretations of the verb meaning in the antecedent clause of conditional constructions. Thus, for example, in making sense of *If John knew the answer, we'd be pleased*, the hearer has to judge whether the speaker intends a counterfactual to be understood or whether John may

---

[1] Some authors (e.g. Byrne 2007: 30–1) use the term 'subjunctive mood' to cover a range of tense forms that may have counterfactual or hypothetical interpretations in certain conditional sentence contexts, including for example the past perfect form (e.g. *if only he had arrived* in time). In terms of English verb morphology the only subjunctive verb *forms* it may be reasonable to call subjunctive are *be* and third person forms without *-s* inflexion: in e.g. archaic 'if that *be* the case', 'it is preferable that he *come* tomorrow', 'if John *were* the winner tomorrow', 'if John *were* to win', 'if John *were* coming' (see Dancygier and Sweetser 2005: 60–1).

6.2 *If*-sentences and counterfactual conceptions

perhaps know the answer. However, in saying *if John had known the answer, we'd have been pleased*, it is more difficult – though not impossible – to imagine a context in which, the speaker intends to communicate anything other than a counterfactual, i.e. that John did (or perhaps does) not know the answer. Even in the case of the subjunctive form *were*, contextual expectations of the interlocutors determine the epistemic interpretation. It is clear that *if pigs had wings, they'd fly* is counterfactual, given our knowledge of pigs. But in *if that man were a doctor, he'd know what to do*, the interpretation depends on what is known or not known about *that man*. Furthermore, if a past perfect conditional, for example, is embedded inside a context that itself is understood as counterfactual, then the embedded space is not necessarily read as counterfactual: *If John had gone to the party he would have met Sarah. If he had found her attractive, he would have flirted with her*. The second sentence can be understood as a possibility rather than an assertion that he did not flirt with her – the scenario of John at the party is set up in the first sentence and in the second sentence a possible happening in that scenario is mooted. A similar case would be *John came home at midnight. If he had been to the party, someone saw him there*. Taken as a whole the two sentences are counterfactual, are in a counterfactual 'space', but the second is not counterfactual with respect to the first (cf. analyses of such contexts by Dancygier and Sweetser 2005: 73–6).

However, it is important to observe that the varying tense forms in *if*-constructions do not allow *any* interpretation but steer the construing hearer in limited directions. Dancygier and Sweetser (2005: 76–7) note that English speakers, despite the crucial role of context, nonetheless have a strong sense that decontextualised past perfect – e.g. *if John had gone to the party* presupposes the counterfactual status of the propositional content of the sentence, an unuttered sentence *John did not go to the party*. A counterfactual reading may be said to be the default reading for the past perfect tense form in an *if*-construction. In the case of the simple past tense form, a default epistemic distancing effect is also found. For example, *if John came to the party, he would meet Sarah* is naturally understood as a non-committal assertion of a possible event in the future. Degrees of past-ness apparently, in the scope of *if*, get interpreted in terms of degree of epistemic certainty, unless contextual factors apply. The correlation of temporal distance in the past with epistemic distance looks like a natural cognitive association.

The use of the 'distance' metaphor in cognitive accounts is different from Lewis's, though perhaps ultimately connected cognitively. Lewis (1973), following Stalnaker (1968), was working with the theory of possible worlds and speaks of the 'closest possible world' in the sense of 'most similar possible world' to the actual world except that the antecedent of the *if*-sentence is true in it. This notion is then held to deal with the problem of when counterfactual sentences are true: they are true if antecedent and consequent are true in such a world. This theory and its subsequent developments are an extraordinarily rich achievement. Its general approach is, however, different in nature and purpose from the cognitive-linguistic approach. The philosophers' problem with counterfactuals arises from seeking to translate them into truth-conditional bi-valued propositions and further treating the relation between antecedent and consequent parts as a causal relation that has to be matched with a plausible account of causation. The question we are posing here is simpler. Counterfactual expressions in human languages give rise to conceptualisations: what kind of conceptualisations are they? What are their properties? What do they tell us about the human mind?[2]

If there is an overlap between the possible-worlds approach in philosophy and the mental-space approach in cognitive semantics it is not in terms of true or false or degrees of similarity. Rather, the scalar notions developed in cognitive linguistics concern cognitive scale, specifically a scale of epistemic judgement concerning the speaker's assessment of degrees of likelihood or real-ness. For linguists investigating conditional sentences the notion irrealis concerns kinds of conceptualisation not 'possible worlds' in the sense of truth-conditional semantics. The term 'distance' (e.g. Fleischmann 1989) is a spatial metaphor for talking about degrees of epistemic scalarity. Linguists working with 'mental spaces' (Fauconnier 1994) are also working with a quasi-spatial scale according to which some conceptualisations, phenomenologically, 'feel closer', some 'feel more remote' (Sweetser 1990, 1996, Langacker 1991, Werth 1997a, 1997b, 1999, Dancygier 1998, 2002, Dancygier and Sweetser 2005; see also Lyons 1977: 718–19, 796 n.1). An important assumption is that types of conditional sentence can be arranged, at least intuitively, in a scalar fashion in terms of their epistemic distance from the speaker. Counterfactual conditionals could, then, be defined as the *furthest* 'possible world' ('mental space' in cognitivist terminology) relative to the 'position' of the speaker. An alternative, but not inconsistent, definition of counterfactuality in cognitive semantics is 'forced incompatibility between spaces' (Fauconnier 1994: 109, Fauconnier and Turner 2002: 230), which can be understood as factual inconsistency, in the speaker's consciousness, between one mental space and another. What also seems to be essential to counterfactual conceptualisation is the way speakers (and thinkers) entertain two incompatible conceptualisations apparently simultaneously (cf. also

---

[2] I am not attempting here to cover the question of causation, which is central to the philosophical investigation of counterfactuals.

Byrne 2007: 34–40 on 'dual-possibility ideas').³ Blending theory (Fauconnier and Turner 2002) gives an account of simultaneous conceptualisation of incompatible spaces in terms of blending theory. The model I propose below takes account of these insights, with the aim of showing how the geometric properties of the deictic space as set up in DST can integrate the sense of contrariness and the simultaneity that is incorporated in the lexico-grammatical details of conditional sentences interpreted as counterfactuals.

In the exploration below, it is assumed that *if* is a cognitive operator that combines with the tense of the verb in the antecedent clause, the semantics of the verb used, interlocutors' expectations and other contextual factors. Without speculating too far into the semantics of *if*, we can say that is inherently modalising: any expression in its scope is going to be some 'distance' along the m-axis. Such linguistic events give rise to, and express, the conceptualisation of things, processes and states of affairs that are non-actual in S's deictic space. We shall consider conceptualisations of the unreal at different points along the m-axis, ranging from degrees of possibility to the main object of concern, counterfactuality. The overarching concern with respect to the latter is why certain grammatical forms that have no negative marker give rise to negative – in the sense of counterfactual – conceptualisations.

6.3   Tense in the modal mirror

In Chapter 2 I discussed the relationship between time and modality. There is clearly a very close relationship cognitively, though it seems to me inadvisable to argue for the reduction of tense concepts to modal ones. Indeed, the fundamental structure of the deictic space that DST postulates has time and modality as independent cognitive dimensions. Nonetheless, it is also implicit in the deictic space framework that there are correspondences between its three axes (and underlying that framework a correspondence with spatial experience). These relationships are made explicit in Chapter 2, Figure 2.8. In looking at the tense forms of verbs in *if*-clauses there is, despite context-influenced variations, a tendency for temporal 'distance' encoded in English past- and present-tense forms to correlate with epistemic 'distance'. This apparent fact is already inherent in the fundamental DST model. Moreover, to focus just on the t- and m-axis correspondence, this correspondence is a geometric reflection. Consider Figure 6.1, which is a partial view of the usual DST diagram, showing the plane defined in t and m – 'looking down on it', so

---

³ The precise way in which the sense of simultaneity emerges is not clear neurologically. Are two representations active over the same time-span? Or does attention oscillate between two incompatible representations in turn?

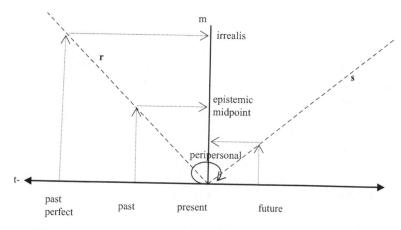

Figure 6.1 Reflection of time onto modality

to speak.[4] The line **r** is a mirror line that maps points on the t-axis onto corresponding points on the m-axis; the line **s** is also a mirror line. The distances are scalar and relative to the origin – the point 0 that is S's immediate consciousness of time and certainty. The tenses correspond to progressively distanced modal meanings, at least in prototypical and decontextualised cases of *if*-sentences.

Figure 6.1 is not intended as a DSM of a particular construction; rather, it is a schematic model of the conceptual (non-linguistic or pre-linguistic) relationship between temporal and epistemic experience. Linguistic forms such as tense affixes schematically, conventionally and by default denote subjectively represented times relative to S, and to one another; these are labelled on the temporal axis. The m-axis is also a non- or pre-linguistic scale – a cognitive gradation of 'realness'. Both t and m represent 'distance' from S, and both reflect also onto the attentional distance d-axis, all three being grounded in spatial perception. The crucial point for the present discussion is the mapping from t onto m: relative temporal distance is mapped onto relative epistemic distance, all in conceptual pre-linguistic space.

In exploring how this cognitive reflection shows up in basic decontextualised *if*-constructions I refer to the following:

---

[4] Linguistic forms used to denote future times are possible in conditionals, as in *if John will come next week, he will help us*. Here there is no epistemic distancing effect; *will* is modal in the sense of expressing will, i.e. volition, the semantic origin of the future-marking auxiliary. In restricted constructions and contexts the future-referring sense of *will* does have an epistemic effect – the 'putative' sense discussed in Chapter 5 above.

6.3 Tense in the modal mirror

(1)   a   If John goes to the party, then he will meet Sarah
        b   If John is going to the party, then he will meet Sarah
        c   If John knows the answer, he is writing it down at this moment/he will write it down

(2)   If John went to the party, then he would meet Sarah

(3)   If John had gone to the party, then he would have met Sarah.

Past perfect maps to the irrealis extreme on the m-axis, as illustrated in sentence (3), which presupposes the counterfactuality of the proposition in the *if*-clause: John did not go to the party. Simple past maps to an epistemic midpoint on the m-axis – neither wholly certain not wholly uncertain – as suggested by sentence (2): John may or may not go to the party. As it happens, these past tense mappings can be modelled geometrically as a reflection through the theoretical mirror line labelled r. The main point here is the default modal reflection from the tense form. The intended time reference in sentences like (2) and (3), when they are uttered, is another question.

Do future tense forms also reflect onto the epistemic axis? This is only partially the case but is worth noting. The future tense form expressed in English by the auxiliary *will* can be used with modal meaning, as in *that will be the postman*. This construction is somewhat restricted but its existence is nonetheless evidence of the general mapping of time concepts onto epistemic ones. In this 'putative' use of the future tense form, it appears that S communicates relatively high epistemic certainty in respect of a process or state, and that this is possible because future time reference by way of future tense morphology is mapped onto epistemic distance. Contrary to what is sometimes argued, despite the fact that future events are inherently uncertain, speakers of languages nonetheless take one another to be confident in referring to future events (see Dancygier 1998: 45–6), though a variety of tense forms reflect the background uncertainty. Geometrically, it turns out that this modal use of future tense forms is consistent with the fundamental design of the deictic space. The t-axis consists in fact of two half-lines, the one being a reflection of the other through the deictic point of origin. If we then postulate a point on $+t$ that is future relative to S and can be denoted by the *will* form, then we can see that it maps by geometric reflection onto the modal axis. Taking future tense forms, in particular the English *will* form, as denoting confident prediction of the future, it is plausible to see it as mapping reflectively onto a relatively 'close' region on the epistemic axis.[5] The mirror

---

[5] There is no reason to think that future tense is symmetrical with simple past in terms of 'distance' and Figure 6.1 does not show this.

line **r** in Figure 6.1 is a geometric way of modelling this kind of conceptual 'backshifting' of the future tense forms.

Continuing to interpret the implications of Figure 6.1, we are left with questions concerning the reflection mapping of present tense forms. The point at the origin – i.e. S's *now* and point of epistemic 'realness' – reflects onto itself. For present progressive tense (*ing* forms) one should include all those points falling within the peripersonal (temporal and epistemic) region. In the case of statives such as *know*, simple present can refer to either time zero or to the future, as indicated in (1c): *if John knows the answer* can be followed by *he is writing it down at this moment* or by *he will write it down*. In the case of non-statives the only tense form in English that can refer to *now* in *if*-sentences is the present progressive *ing* form (called 'presencing' operator in Chapter 4), which extends over peripersonal time. It reflects therefore onto what we might call 'peripersonal modality', and can be understood as high epistemic certainty. Sentence (1b) seems to express more subjective epistemic certainty than (2), which has the more distal simple past tense, and (1b) is indeed readable as 'assuming that John is going to the party...'. The modal reflex of the progressive form might be the maximal degree of epistemic certainty possible under the scope of the semantics of *if*.

This is not quite the end of the story. Sentence (1b) is most likely to be understood as referring to the relative future, but the present progressive in *if*-clauses can also refer to the present ongoing event, e.g. *if John is walking to the party, he's getting wet in this rain* refers to the speaker's peripersonal present time. One cannot use the simple present for this meaning: *if John walks to the party, he is getting wet in this rain*. Outside of context, in *if*-sentences, present progressive for non-statives and simple present for statives can refer to either ongoing *now* or to some future time. What is of interest is that non-statives can be in simple present tense form in *if*-sentences, provided they refer to the future. Can we account for this? As argued in Chapters 4, simple present is connected to 'instants' but also to temporally unbounded, or 'timeless', categorical concepts, contrasting with the 'presence' of events conceptualised via the progressive tense forms. If the conceptual contrast with *ing* meanings that I am suggesting is plausible, then it may be that, under the *if*-operator, the more abstract simple present denotes an abstract certainty but not the 'present' concrete certainty of factual experience expressed by the *ing* forms. What might the epistemic reflexes of simple present be? The simple present in (1a) *If John goes to the party* seems to indicate a relatively high degree of epistemic certainty in the future: not so epistemically 'distanced' as the simple past in (2) and not so epistemically certain or 'close' as the present progressive referring either to present or to future in (1b). The fact remains that simple present with non-statives under *if* can only refer to the future, as is the case in ordinary declarative sentences. In addition, abstract category conceptualisation associated with simple present are compatible with the abstraction implicit in the semantics of *if*.

Finally, one might wonder why the regular *will*-future is not possible in the *if*-clause? The answer has to do with natural cognitive consistency and the semantics of this future tense form. Such forms, as noted above, are in fact taken in context to indicate confident prediction of future events: it may be that this is a degree of confidence incompatible with the abstract hypotheticality inherent in the meaning of *if*. The English *will* form is, unlike for example the Romance future tenses, based on the concepts of volition and intentionality. Consequently, the primary meaning of *if John will go to the party* seems to be focused on John's willingness rather than on the speaker's epistemic assessment of the event's likelihood.

Once granted that past tense forms define scaled regions on the m-axis, these regions, marking epistemic distance relative to S, are available for any time reference, however such time reference is achieved linguistically or pragmatically. This tense reflection phenomenon does seem, in languages generally, to establish a conventional default pattern. This does not mean that tense forms map automatically, when they are used in practice in particular sentences, onto the epistemic m-axis. What I take it to mean is that the cognitive experience of 'temporal distance' is available to be reflected onto the cognitive experience of 'modal distance'. This is a non-linguistic process that provides a resource for linguistic structure. It is a stable mapping that remains in background cognitive structure. Linguistic tense forms are influenced by it and can draw upon it, but they are not univocally determined by it.

6.4 The geometry of *if*-sentences

In this section I turn to the main concern of this chapter. How can we model, using the DST framework, the epistemic distancing effects associated with the different degrees and kinds of epistemic 'distance' associated with different tense forms in the protasis of *if*-constructions? Unlike the preceding section, this section is concerned now with the DSMs of particular sentences rather than the cognitive mapping of tense for meanings to the epistemic scale.

The conjunction *if* is a cognitive operator that transforms the basic coordinate system, S's deictic space R, by translating a copy R' to various points on S's m-axis. How far it shifts depends on the tense forms, together with verb meaning and contextual factors. The second set of axes is a 'new reality' space, similar to a Fauconnier mental space (cf. also Dancygier and Sweetser 2005), except that 'distance', 'direction' and deictic centring are already built into the fundamental structure of DST. Within the new set of axes, R', propositions that are dependent on *if* are represented. They are thus governed by *if* with respect to S's initial coordinate system, but simultaneously 'real' within the new system. That is, all representations activated by conditional sentences are relative to $0'$, the origin of the shifted axes, and the locus of S's conceptualising consciousness. In the new space, anaphors find their antecedents – 'trans-world identity' – across the base deictic space and the translated copy. The discourse entities referred to in *if*-sentences may be real in R, in which case they are labelled on the d-axis in R. They may also, however, be non-real, in which case they are labelled only on the d'-axis in R'. This would be the case, for example, in a sentence such as *If John goes to a party, he will meet people*; the same can be true for de dicto meanings in general. The time axis is aligned in both R and R'. While t represents conceptual time in DST, it is supervenient on metaphysical time and cannot be out of joint in the sense that there are two separate time realities that do not correspond. The actual time references that are intended and understood on the basis of tense forms in both protasis and apodosis of *if*-sentences are represented within the coordinate system of R'.

Taking examples (1a) to (3), simplified and repeated below, as representing something of the graded epistemic possibilities of conditional sentences, Figure 6.1 models two examples that give rise to embedded coordinate systems whose origins are at different points on the epistemic m-axis: (1) is relatively close to, (2) relatively farther from the base system of S. What the DSM represents is conceptual structures evoked by what have become conventional meanings signalled by particular tense forms in the *if*-construction together with situational factors.

(1)   If John goes to the party, then he will meet Sarah

(2)   If John went to the party, then he would meet Sarah

(3)   If John had gone to the party, then he would have met Sarah.

In Figure 6.2a below, the coordinates for *John*, *Sarah* and *party* are labelled in S's base coordinate system, but the vectors representing *go* and *see* are located in the translated coordinate system determined by the verb form, together with contextual and pragmatic factors. Because of the coordinate system, the real labelled entities transfer automatically from R to R'. Figure 6.1 represents the copied and translated coordinate system R' for sentence (1) and its relative position on the m-axis in S's base system R. In the diagram this is an approximate impressionistic system that is intended to correspond with the relative epistemic 'closeness' of events represented under the *if* operator by the simple present – as discussed in the previous section. The verbs in

6.4 The geometry of *if*-sentences

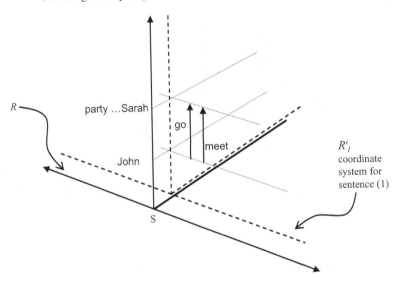

Figure 6.2a Conditional sentence (1) present tense

conditional sentences have the event represented in the protasis temporally prior to that represented in the apodosis (consequent clause). However, both clauses are subsumed under the semantic effects of *if*. The events in the two clauses are part of separate mental representation, a quasi-reality or 'world' in which entities that are real in S's base reality are represented as relating to one another in quasi-real ways (modelled as vectors) with real-time relationships to one another. As also discussed in the previous section, simple present *if*-clauses refer to times later than S's *now*. The exact intended temporal 'location' is a matter of contextual pragmatics, so any diagram can only position the event vectors arbitrarily, while maintaining temporal relationships.[6]

Figure 6.2b differs from Figure 6.2a only with respect to the epistemic positioning of R'_2 within S's base reality space: its origin on the *m*-axis is epistemically more 'distant' than R'_1 in Figure 6.2a. Time reference is still to the future. The diagram shows the origin of R'_2 at the epistemic midpoint. This is of course not a precise metric and contextualised reactions to (2) may vary.

---

[6] The fact that *if* sentences have two conceptually connected clauses poses difficulties for constructing DSMs. Strictly speaking, the framework so far has assumed one clause (verbal construction) per DSM. The DSM for *if*-sentences, therefore, really combines two DSMs. The implications become apparent when the two clauses have one or more different participants. I have not attempted to go further into this matter but have adopted the expedient of double labelling: two different participant entities (here *party*, then *Sarah*) are given the same coordinates ordered in correspondence with the order of the clauses in which they occur. In both clauses *party* and *Sarah* are both relatively distal on the *d*-axis.

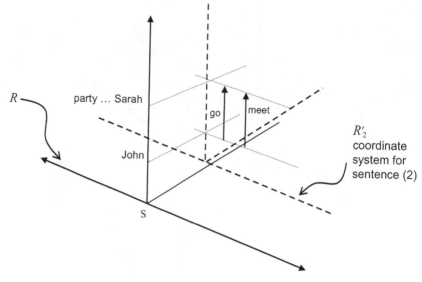

Figure 6.2b Conditional sentence (2) past tense

However, the theoretical positioning at the midpoint is supported by the difference between *If John goes to the party* and *If John went to the party*, where speakers usually report that the latter feels more 'distanced' than the former. In addition, one may note that a paraphrase of the past tense version, *If John went to the party*, can be *John might go to the party and John might not go to the party*, where the speaker's judgement as to possibility is divided. Figure 6.2b, then, is an attempt to model one of the salient cognitive effects of the tense form in the first clause of (2), viz. the simple past, modalised under *if*, in the way discussed in the preceding Section 6.3.[7] It is of course possible to use (2) in reference to a past event relative to S, depending on contextual influences. In such a case, the vectors *go* and *meet* in the R'_2 coordinates would be positioned on the past time axis $-t$. Whether a past-time interpretation affects the sense of epistemic distance is a matter for consideration but perhaps cannot be determined theoretically.

The essential point is that epistemically distanced mental representations occur in response to tense forms combined with pragmatic contextual factors.

---

[7] If the verb is stative the same effect occurs, e.g. *If John was present at the party tomorrow, he would meet Sarah*. The past subjunctive form *were* may have an even more distancing effect, but this may vary from speaker to speaker, and certainly varies with context. See Dancygier and Sweetser (2005: 61–2). Note also that stative forms in the past tense seem to allow a counterfactual interpretation more readily than non-stative verbs.

6.4 The geometry of *if*-sentences

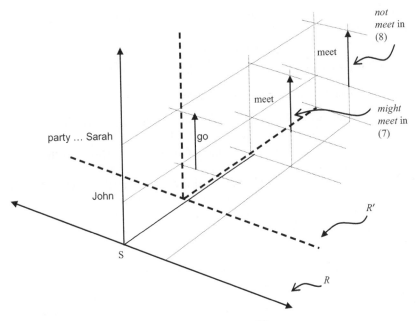

Figure 6.3 Modalised apodosis: sentence (7)

In the DST account, which in many respects follows Dancygier and Sweetser (2005), R' space can glide along the epistemic axis. It may be indeed that epistemic adjustment is in actual processing a matter of variable on-line adjustment.

Within R' the main clause (apodosis) of *if*-sentences can be modalised. This includes negation, producing complete conceptual counterfactuality. This is modelled by the positioning of the event vector on m'. The *if*-clause itself (protasis) cannot normally be modalised, but can of course be negated, as in (6):

(4)   ?If John may/might go to the party, he will see Sarah

(5)   *If John might/may went to the party, he would meet Sarah.

(6)   If John did not go to the party, he would meet Sarah.

The reason for the oddity of such sentences is presumably that *if* is already epistemically modal; conversely, such examples are evidence that *if* is indeed an epistemic operator. The auxiliary *will*, which serves in English to form the future tense is, however, possible in the protasis, as noted earlier and the reason for this is presumably that *will* is not epistemic.

To return to the main clause of *if*-sentences, modalisation and negation in the main clause, illustrated in examples (7) and (8) below, can be modelled in a natural way in the framework developed above.

(7)   If John went to the party, he might meet Sarah

(8)   If John went to the party, he would not meet Sarah.

Figure 6.3 outlines a DSM for (7). The *meet* vector is located in R' at the midpoint of the m'-axis and at some time t_i later than t_j, the time point of *go*. The sentence (8) is as modelled in Figure 6.3: the *meet* vector is at the distal (irrealis) limit of m' in R'.

Modalisation in the main clause is entirely relative to R', which in turn is relative to R. So modalisation in the main clause is, from S's viewpoint, a second layer of modalisation – modalisation within modalisation. Further conceptual nesting can occur. For instance, in (9) the negation operator is itself modalised:

(9)   If John went to the party, he might not see Sarah.

The obvious and perhaps only way to model such a case in DST is to add a further epistemic space R'', with origin at m' in R', such that the event *meet* is modelled by a vector located at m''. We might speak here of first-order (under *if*) and second-order (under *might*) conceptual models.

If-sentences like (1) and (2) – and there are many similar possibilities – can be modelled in DST geometry by translation transformations that set up an irrealis space that is inconsistent with respect to the initial space that S takes as reality. These transformations are geometric translations that glide along the m-axis to a point of epistemic distance that satisfies S's epistemic processing, given the *if* operator, tense forms and contextual factors. These epistemically translated coordinate systems have an important property in DST that we will come across in later chapters. They are not aligned with the base system R. Specifically, the modal axis m' in R' is not aligned with m in R; that is, points on the m-axis are not aligned perpendicularly with equivalent points on the m'-axis. The d- and t-axes are aligned, however.[8] This is as one might expect. The spaces R and R' are modally distinct: by definition the world of *if* is separate from what is the real world for S. For counterfactual sentences like (3), however, the case is significantly different.

---

[8] At least, this is so for DSMs of *if*-sentences. As will be seen in later chapters, axis systems can have their origin at distal points on the d-axis, e.g. for states of mind attributed to other minds labelled on d (see Chapter 7). Time, however, as already noted above, is always 'aligned' across R spaces.

6.5 Through the looking glass: counterfactual *if*-sentences

The significant difference between the *if*-sentences examined so far and sentences of the type illustrated below (including (3), which is repeated) is that all of the latter presuppose the falsity (from S's viewpoint) of the expressed assertions:

(3)   If John had gone to the party, he would have met Sarah

(10) If John had gone to the party, he would not have met Sarah

(11) If John had not gone to the party, he would have met Sarah

(12) If John had not gone to the party, he would not have met Sarah.

Sentence (3) presupposes that John did not go to the party and he did not meet Sarah. In an imaginary world where he did, however, he met Sarah. In each clause the polarity of the expressed meaning is reversed in the presupposed clause. This is the case for (10), (11) and (12). This effect arises even when there is no context. The default time reference out of context appears to be past relative S but, as noted earlier, can be shifted to ongoing present and to future relative to S. In the case of the *if*-sentences with present and simple past tenses, discussed in the previous section, the reversed presupposition is heavily dependent on contextual assumptions: e.g. *if John was a doctor* [and we know he isn't], *he could prescribe a pill*. It is possible, however, and such a conceptualisation would be modelled in the same way as we shall now do for the construction exemplified in (3), which appears automatically interpreted counterfactually, and to which I restrict the term 'counterfactual'.

What sort of DSM should be constructed to model a sentence such as (3)? A first approximation would be to assume that there is an embedded *if*-space, R', as before, as shown in Figure 6.4.

In this first approximation we note that in the counterfactual *if*-construction, the protasis needs to be true in one space but simultaneously false – it being understood that by 'true' and 'false' is meant S's estimation of realis and irrealis polarities. This can be achieved by locating the origin $0'$ of R' at the epistemic limit of the m-axis in R. Thus Figure 6.4 would give us a configuration in which, from S's point of view, John is at the party in R' (*if John had been at the party*), but not at it in the presupposition (or better, in S's base epistemic state). Otherwise put: the negated *if*-clause is simultaneously negative in R (S's real world) and positive in R' (in the *if*-space). However, for any sentence with a negated protasis, as in (11) and (12), $0'$ would have to be located at the epistemic limit m' in R', and this would fail to model the kind of epistemic double vision that counterfactual *if*-sentences

174   Counterfactual reflections

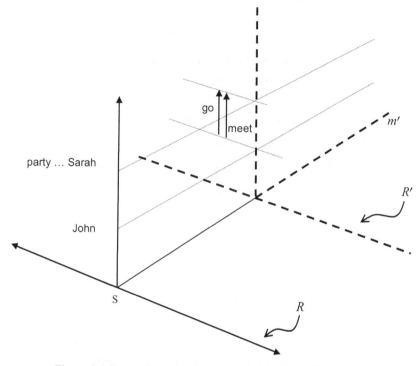

Figure 6.4 Counterfactual sentence (3): first approximation

induce. Furthermore, the entire modal axis would theoretically have to be understood as 'outside' the reality space R of S. While such a configuration is appropriate for some constructions (cf. figures in preceding section 6.4), and modal non-alignment is appropriate as we have seen for certain types of *if*-sentence – those communicating some degree of epistemic distancing short of counterfactuality – it is inadequate for counterfactual *if*-sentences.

It seems that the two representations (the expressed one in the sentence and the presupposed one) involved in understanding sentences like (3) are simultaneous. Phenomenologically, they seem to be experienced as simultaneous. Moreover, polarity reversal is systematic in examples like (3), (10), (11) and (12). What we need in order to model these properties of counterfactual sentences is in fact a reflection transformation, as shown for (3) and (12) in Figure 6.5. Strictly speaking this configuration is a glide reflection: a reflection in the mirror perpendicular to m and passing though its limit point, combined with a translation of the reflected axes such that $0'$ coincides with the limit of m and the limit of m' coincides with 0. The reflection of the space R' interlocks with the base space R.

6.5 Counterfactual *if*-sentences

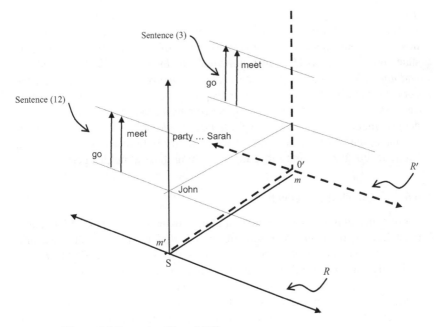

Figure 6.5 Examples (3) and (12)

Figure 6.5 shows the polarised correspondences apparent in (3) to (12). In sentence (3) the vector representing *go* in the protasis, as well as the vector representing *meet* in the apodosis, lie in the realis plane in R' – and simultaneously in the irrealis plane of R, the plane lying at the end of the m-axis in the base space R. Thus in the diagram S knows that John did not go to the party but raises at the same time a mental representation according to which he did go. Conversely, for sentence (12), Figure 6.3 models the *go* and the *meet* vector as lying in the irrealis plane of R' and simultaneously in the realis plane of R'. So S knows John did go but activates a simultaneous representation of his not going. In other words, for sentences (3) and (12), what Figure 6.5 seeks to model are counterfactual states of mind communicated by those sentences.

Counterfactual sentences of this type can have epistemic modals in the apodosis: e.g. *If John had gone to the party, he might have seen Sarah*. The model is simply as for (3), with the *meet* vector positioned at the epistemic midpoint in the *if*-space R'. The *meet* vector then has the same position in both R and R'. This is what is needed, since the meaning we have for the apodosis in both spaces is one of possibility.

This geometrical account thus enables us to model the conceptual effects produced by combining the semantics of *if* with the semantics of the most distal tense (past perfect). We do not have to introduce any more machinery for this: the reflection transformation is given by geometry and is needed for modelling other linguistic constructions. Some of these are related to mental-state verbs such as wishing, imagining and supposing, as in *the boy wished he had not stolen the apple, if only I were a millionaire!, what if she had left the door open that night?, let us suppose that the boson had been found fifty years ago*. As will be seen in Chapter 10, the counterfactual mirror appears also in deontic modal constructions – e.g. *the boy should have apologised for stealing the apple* – where the sense of the auxiliary only makes sense against a simultaneous background reflecting something that did not happen.

6.6 Concluding reflections

A mind engaging in counterfactual thinking, including by means of language, must have two mental representations: one that is epistemically real and simultaneously one that is counter to it – this is what the DSMs model explicitly; in other models, even in mental space and conceptual blending theories, which are the closest to the present account, I believe this dual mental vision is not explicitly modelled as such. It also seems to me that counterfactual conditionals proper are of a different kind, not just a different degree of epistemic distance. Nonetheless, the introduction of 'distance' by Dancygier and Sweetser (2005) has been pivotal to the development of the theory of counterfactuals in natural human language, pointing toward the crucial role of spatial conceptualisation in linguistic meaning.

The exploration of geometric modelling in this chapter, based on the postulate of the three-dimensional deictic space, seems able to bring together a number of the frequently mentioned features of conditional and counterfactual grammatical constructions. The idea of a geometrical transformation on the basic deictic space gives us a way of modelling a secondary imaginary (irrealis or 'counterfactual' in a broad sense) space. Because this secondary space is a copy of the first, it 'inherits' its structure because it is positioned within it in such a way that referent entities are given coordinates that give them 'trans-world identity'. The phenomenon of polarity occurs in a number of linguistic constructions but is most dramatic in counterfactual *if*-sentences, and the geometric reflection modelling handles this in a natural way as a consequence of general geometric principles. Most importantly, the reflection transformation captures the intuition that counterfactual and factual occur simultaneously.

Counterfactuals proper are special. Other conditionals that are epistemically distanced are 'parallel' worlds that do not map isometrically onto the whole of R. The parallel worlds of distanced conditionals are best characterised as 'alternative realities' – they are 'positive' in the sense that a *potential*

6.6 Concluding reflections

(possible) event or relation is imagined, whereas in counterfactuals proper, the reality is *counter*, it is polarity-reversed. Some specific relations between entities are changed that hold in the subject speaker's reality are changed in his simultaneously imagined reality. This seems to me to be cognitively different from entertaining alternative realities, as in conditional sentences that are not strictly counterfactual ones; it is not surprising that models of counterfactuals in DST turn out to have their own special kind of isometry. It also makes the present account different from (though still indebted to) mental-space models (Fauconnier and Turner 2002) and from 'epistemic distance' models (Dancygier and Sweetser 2005), as well as from Lewis's 'closest possible world' theory (Lewis 1973). It is also perhaps what makes them useful to the human mind. Because they retain referent identity and the universal time-frame they make reasoning, retrospection and prospection possible. Curiously, they are tools of reflection in a broader sense. They are essential to highly abstract thinking yet rooted in embodied spatial experience and the cognitive phenomenon of deixis. Finally, however, if the reflection model of counterfactuals has any plausibility, we might notice that they are actually *self*-reflective: thinking in counterfactuals is a self-contained conceptual activity. This makes them different again from the structures that we shall examine in the next chapter, which can be related to a large degree to the imagining of the minds of others.

7 Reference frames and other minds

> ... a person does not, as I had imagined, stand motionless and clear before our eyes ... but is a shadow which we can never penetrate, of which there can be no such thing as direct knowledge, with respect to which we form countless beliefs, based upon words and sometimes, actions ...
>
> Marcel Proust, *The Guermantes Way*

Each human individual has to separate his or her own understanding of the world – his or her reality – from that of another individual mind. In other words, we are looking at the manifestation in linguistic structure of what cognitive psychologists call 'theory of mind' (whether the account is a modularist one, e.g. Baron-Cohen 1995, or a developmental one as in Tomasello 1999). When one person's mind represents (its guess at) the contents of another mind, it is metarepresenting that other mind (see papers in Sperber 2000). Metarepresentations are cut loose from the cognisor's mind – though still its product – and projected onto another mind. Such metarepresentations are assessed epistemically by the cognisor – they may be regarded as real (true), probable, possible, counterfactual, etc. – that is to say, in DST terms, that S can position such metarepresentations at any point along his or her m-axis. Moreover, one epistemically assessed metarepresentation may be nested inside another, with a different epistemic assessment: possibility may be embedded in probability, and so forth.

Can DST represent such complexities in a conceptually motivated way, and how? Before looking at the modelling of beliefs and other minds, it is necessary to look at the geometrical structure. One fundamental point that arises from what has already been said above is that an individual mind can and frequently does form a metarepresentation of the representations assumed to reside in the minds of other individuals. Now parts of these metarepresented other worlds might exist also in the 'real' world for an individual S, but the overall state of affairs may be incompatible with what S holds to be real. We shall see how this property of being partially sealed off or decoupled emerges in the DST modelling. Before looking at embedded metarepresented belief reports, I look at the relationship between the m-axis in R and the m'-axis in R'.

7.1 Epistemic reference frames

The m-axis in DST is taken to be fundamental, since I take it as fundamental that human cognition assesses realities, possibilities, probabilities, impossibilities, and so forth. There are theoretically many possible epistemic reference frames, i.e. planes along the modal m-axis. But human language enables us to concatenate or subordinate one to another different epistemic worlds and assumptions. In a number of places in the semantics literature examples like the following are discussed (see Fauconnier 1994):

(1)   John probably has children and it's possible his children are bald.

What is the puzzle here? Well, the possessive pronoun in *his children* presupposes the actual existence of *the children*, but this is odd because we read his *children* as having the same referent as *children* in the matrix clause and the modalised verb in that clause indicates their existence is only probable.

Fauconnier (1994) proposes a solution based on the theory of mental spaces. Essentially, Fauconnier sets up three labelled spaces, a reality space, a probability space and a possibility space, with referents r, r' and r'' in each respectively, treated as images of a space-builder function (linguistically, in this example, the adverbs *probably* and *possibly*). The present account is consistent with this approach, and in some degree inspired by it. But the present approach dispenses with labels, and seeks a *motivated* representation in terms of the fundamental coordinate system set up earlier. Fauconnier's solution arises naturally from this format. The space-building function is in effect the transformation of axes; the counterpart mappings are the coordinate points in the space. This is illustrated in Figure 7.1.

The new set of axes keeps the same t- and d-axes as R but its m-axis does not correspond to the base m-axis in R – because its origin is defined at the probability point on the m-axis. That is to say, for S, anything in R' is merely probable. Within the new R', processes and states can be 'probably possible', 'probably probable', 'probably wrong', etc. Here, *John* is real for S, but *children* are in the probability space R' defined on m in R. The property *bald* is attributed relatively to the probability space R', i.e. it is a plane at midpoint (possibility) on m'. The attribute *bald* is real in R and can be attributed in all epistemic worlds.

As we shall see also to be the case for belief spaces, there is a non-coincidence between m and m'. The geometric correspondence between a point on the m'-axis and one on the m-axis has no conceptual interpretation. In fact, the strict geometrical expectation does not help us – except contrarily, in the sense that the non-correspondence is exactly as the conceptualisation. That is to say, the embedded axes R' are, as it were, conceptually

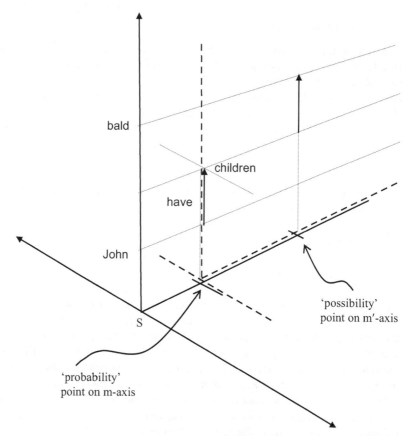

Figure 7.1 Possibility within probability: (1) John probably has children and it's possible his children are bald

'decoupled', or even 'sealed off', from the epistemic dimension of R. We need therefore to stipulate this non-correspondence as part of the idiosyncratic structure of the deictic space in human conceptualisation. Metarepresented happenings and states of affairs that a subject S reports as being in the mind of another subject are decoupled from the world that S 'knows' to be real.

7.2   *That*-ness and other-ness

This section is concerned with the complementising predicates that take *that* clauses. I want to suggest that the word *that* in such cases is more than a mere 'complementiser' – it has a motivated semantic content related to the

7.2 *That*-ness and other-ness

semantics of the predicates that take it. This is not a totally new idea (cf. Bolinger 1977, Wierzbicka 1988, Langacker 1991) but the approach outlined below takes into account that *that* is fundamentally deictic and therefore invites investigation in terms of the spatial structures modelled in DST. More specifically, *that* constructions, or rather the conceptualisations associated with them, can be insightfully modelled in terms of frames of reference, attentional distance (on the *d*-axis) and modal distance (on the *m*-axis).

7.2.1 That

Taking a 'naïve' approach to the relationship between verbs and clause complements, there are two points to make. First, the morpheme *that* which occurs in such constructions, is not a conjunction but is a part of the verb frame; second, in English, it has a meaning and its meaning is the same as or closely related to the distal deictic, or demonstrative, pronoun *that*. As was explained in Chapter 2, proximal and distal deictic pronouns are essential to the general quasi-spatial structure of DST's conceptual modelling, and the present account of the meaning of *that* constructions emerges naturally from that structure. The morpheme *that* is not treated as a purely syntactic and meaningless element (conjunction or complementiser). Rather *that* is treated in DST as always a relatively distal element relative to S on S's *d*-axis, an element that S regards as having certain existence ($m = 0$). Further, *that* is treated as being the impersonal geometric origin of a coordinate system – an impersonal world that can be located where S pleases in S's reality space, including in another mind embedded in that space. In effect, *that* constructions set up a new set of coordinates, a new reality space embedded in S's base space. A *that* space is a kind of abstract entity for S that is deictically anchored in S's world but can be 'projected' into some other world, e.g. that of another discourse entity also embedded in S's world.

What is important is that *that* is a distancing device. (English can of course omit this *that*, as in *Bill imagines Mary will write the report*.) The category of verbs that associate with *that* followed by a clause includes many that can have a noun phrase in the same slot, e.g. *Mary believed the report, Mary believed that, that's what Mary believed*, where *that* is the distal deictic demonstrative pronoun.[1] That is, both events and objects are entities. Clauses can be discourse entities with coordinate on the *d*-axis and be equivalent to the second argument of certain verbs with characteristic related meanings.

---

[1] This perspective is somewhat reminiscent of a philosophical argument of Quine's that there is no distinction between objects and events: 'Physical objects, conceived ... four-dimensionally in space-time, are not to be distinguished from events ... or processes. Each comprises simply the content, however heterogeneous, of some portion of space-time ...' Quine (1960: 171).

What needs to be added to this is the *distal* quality of such entities: they are in some sense more 'mentally distant' from S than clauses associated with *to* and the suffix *-ing*. This is the case whether or not *that* appears before the embedded clause complement: the embedded clause is still distal on the discourse entity axis d, relative to S. In addition, S may regard the embedded clause with varying degrees of mental distance on the epistemic m-axis – S may, for example, regard what is represented in the embedded clause as true (real for S), as possible (S is uncertain one way or the other), or as false (contrary to fact or unreal for S). In this sense we shall be exploring the conceptual structure of what philosophers call 'propositional attitude', implicit in verbs such as *believe that*, *hope that*, *deny that*. The complementiser *that* also objectivises: the content it introduces in its clause complement is not necessarily seen from the point of view of another mind but rather as a whole object from the point of view of the subject S. It is as if S points to 'that' for inspection, holding up a whole event or proposition. It may also be represented by S as such an object 'seen' in the mind of another subject S' – whether S concurs with S' or not.

For modelling purposes, a *that* space has two essential features: it is a discourse referent in S's base reality space, distal to S and thus objectified; and it points to reality space that in itself has no conceptualising origin (i.e. has no S) other than S's and any embedded reality space (the space of another mind) that S's sentence embeds it in. The geometric origin of a *that* space has no S (self, subject, speaker); the origin is the impersonal demonstrative *that* itself. This means that in a DSM for a *that* construction, a *that* space is positioned by a vector anchoring it to *that*, a discourse entity in S's world R. The important point is that *that* is always the anchor in S's base world for a space that can be positioned anywhere in the DSM, including in some embedded mental space belonging to another entity (e.g. John). A *that* space is a metarepresentation that has always an anchor in R but at the same time positions a space (coordinate system in DST terms) somewhere along the base m-axis in R – that is to say, it always has an epistemic valuation from S's point of view, reflected in the semantics of the governing verb (*know*, *believe*, *realises*, *imagines*, etc.). Remember that vectors can have zero length, so it is possible for the deictic centre of the *that* space to be anchored at $m = 0$, that is the plane in which S thinks things are certain, real, existent. This allows us to model both factive and non-factive predicates.

7.2.2 that *constructions*

A subtle survey of several types of complementisers is given by Langacker (1991: 438–63), whose diagrams provide many analogues with the DST approach. Langacker does not investigate subclasses of *that*-clause verbs.

In fact, there seem to be interesting differences among them. Verbs taking *that* as complement have properties that can be modelled directly and naturally in DST. As has been widely observed, *that*-complement verbs are predominantly about knowledge and awareness: they involve different kinds of epistemic mind-states. Such verbs seem to form subgroups along two (at least) variables: type of epistemic state denoted by the semantics of the verb (*know, guess, imagine, ...*) and presence or absence of the ability to trigger existential (factive) presuppositions. For present purposes we will make use of the distinction between factive and non-factive classes of *that*-verbs, though we shall adopt a cognitivist approach to this distinction. Using the usual negation test for presupposition, *John knows [realises, recognises, sees, admits, ...] that Mary wrote the report* does trigger a presupposition, or in cognitive terms, a proposition judged true, i.e. judged real or known by the situated speaker. On the other hand, *John [believes, thinks, imagines, holds, claims, argues, reasons, deduces, concludes, hopes, fears, suspects, ...] that Mary wrote the report* does not trigger a presupposition in that sense. In the latter case, the list in parentheses predominantly includes verbs that denote epistemic states and some that denote more affective ones; the semantic structure of this set is of potential interest but will not be explored here.[2]

Langacker's (1991: 438–63) overview of complementisers places various types (*that, to, ing,* zero) on a scale of 'objectivity' or 'distancing' from the speaker, a scale that he also sees in terms of graded similarity to noun-like thing-ness. Such issues will be considered further in Chapter 8, but it is worth noting here that the feeling of 'objectivity', the 'distanced' quality of a *that* clause, is reflected in the geometrical properties of the DSM.

7.3   Other minds as reference frames

The use of *that* constructions appears when we want to communicate about what we believe to be the contents of a mind or utterance, whether or not we ourselves believe that content to be true or real. Epistemic judgements about the contents of other people's minds do not in fact have to be linguistic – in all likelihood there is a non-linguistic ability, perhaps found

---

[2] There are many adjectival predicates that take *that* clauses (e.g. *be surprised, shocked, horrified, appalled, pleased, delighted, ...*). Interestingly this set is factive. There is a large set of affect verbs that are factive with the complementiser *the fact that*. There is also a class of verbs whose semantics make the complement clause refer to future events or states; the common meanings of these verbs are broadly directive or deontic in character and involve the English quasi-subjunctive seen in sentences like 'John (requires, demands, stipulates, insists, ...) that Mary write the report.'

in the right hemisphere, that is the basis of theory of mind (Baron-Cohen 1995, McGilchrist 2010), the ability of the normal person to understand that another person might have other conceptions and intentions than they do. Though such an ability need not require language, humans like to communicate to one another their thoughts about one another's thoughts. They also can talk about themselves in the same way, saying, for example, not just *Mary fancies James* but *I think that Mary fancies James*. The latter differs from *John thinks that Mary fancies James* only in that *I* refers reflexively to the speaker. In such instances, it is almost as if *je est un autre*, but not quite. Whatever Rimbaud meant by that, we can see clearly in DST how S can project herself as a distanced objectivised discourse entity or, alternatively, can conceive of an other as an S. The focus here, however, will be on two kinds of sentence that enable speakers to report on what they take to be the contents of the mind of others, as reflected in the following sentences:

(2) a John knows that Mary wrote the report
 b John knows that Mary might have written the report

(3) a John might know that Mary wrote the report
 b John might know that Mary might have written the report

(4) John does not know that Mary wrote the report

(5) a John believes that Mary wrote the report
 b John believes that Mary might have written the report

(6) a John might believe that Mary wrote the report
 b John might believe that Mary might have written the report

(7) a John does not believe that Mary wrote the report
 b John disbelieves that Mary wrote the report.

While the subject of *that*-verbs can be any kind of noun phrase or pronoun, including of course the first-person pronoun *I*, the locus of reported subjectivity, what is of particular interest is that *that* constructions are used to report on the self's perception of the contents of *other* minds. This is what the present section is concerned with. The essential idea is that other minds are embedded in the deictic space of S, the speaker, self and subject of the DSM diagrams we have been developing throughout this book. Geometrically these other minds are embedded reference frames, in fact coordinate systems within the base reality space R, whose origin is S. The origin of such embedded reference frames can, in principle, be positioned anywhere in R, and thus have various kinds of attentional, modal and temporal relationship to S. Their origin is an other, labelled O, but is of course an other's subjectivity and self, which can be labelled S'.

7.3 Other minds as reference frames

7.3.1 Knowing that

When S uses *know* they communicate that they take the world of O to be the same as theirs. When a speaker S uses *believe* they communicate that they take O's world to be other – in the sense that S does not commit to believing it to be true. It may be that S has a mental representation of the world that does, or does not match that of O (with respect, of course, to the contents of the *that* clause), or S may be uncertain about it, etc. We need DST to be able to model the differences between *know that* conceptualisations and *believe that* conceptualisations. Attempting such modelling reveals some of the complexity, just some, of communicating in language about other people's supposed mental representations.

In Figure 7.2, and all the following DSMs, the discourse referents are taken as real by S and thus are located at $m = 0$ in R, S's basic space or world. Even when S is representing (or rather, metarepresenting) the mind of some other being, O, the labels for the discourse entities are assumed by S to be real, even if in the mind of O: this follows from the coordinate system, marked by dotted

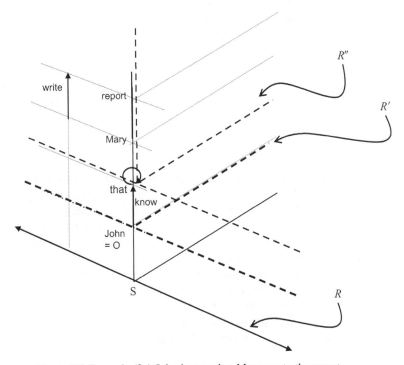

Figure 7.2 Example (2a) John knows that Mary wrote the report

lines in all DSMs. Of course, it is possible for some S to represent the mind of O as containing entities whose existence S does not epistemically endorse but considers only to be figments of O's mind – they may be merely possible as far as S is concerned, or completely non-existent. In such cases the DSMs will have their discourse referent labels in O's reference frame, R'. For the sake of simplicity such cases will not be looked at here.

In Figure 7.2, then, there are three discourse referents, *John*, *Mary*, *the report*, and they are all in the reality plane ($m = 0$) for the speaker S. They have relative discourse distances from S. The complementising verb *know* asserts that S considers real a certain mental state of John's. This mental state is a metarepresentation: what S believes John believes (and perhaps S believes other people believe it too). To say that someone knows something is to make a claim that this someone has the same epistemic stance with regard to the complement clause as oneself. We have to indicate a metarepresented knowledge state connected to *John*, and we do this by the embedded reality space R', of the same form as S's base reality space R – speakers assume that other minds have essentially the same abstract deictic space, albeit anchored on their own viewpoint. Geometrically, it makes sense to think of the new space as a repositioned copy of the S's base coordinate; it is a new reference frame, similar in form to S's, but 'other'. This 'other' space belongs to the discourse 'other', namely, *John*. The verb *know* has the effect of positioning this new space R' relative to *John*, in such a way that the origin (deictic centre) of R' is related to *John* (who is of course already positioned deictically relative to S's base world R). This transposed deictic centre attributed by S to *John* is labelled O: it is the deictic centre of an other mind in S's conceptualisation of the world.

Let's look more closely at the details of Figure 7.2. The verb *knows* is shown both in R and in R': it is asserted by S as a real fact about John and is simultaneously a part of John's own cognitive state. The verb *know* has two jobs, in addition to its core meaning denoting a state of mind. On the one hand, it positions R' at S's point of certainty (the plane positioned at $m = 0$ in R). On the other, it represents what S thinks is John's epistemic stance with respect to R'', which is the space embedded as a result of the *that* complement – a *that*-ness space, if you will. Thus Figure 7.2 shows two vectors connected with *know*: one vector (starting at *John*) points to its complement *that*, while the other (the peculiar zero vector depicted here as a circular arrow) locates the complement of *know*, i.e. the *that* space, at S's point of epistemic certainty (again at S's $m = 0$). This has the effect of aligning R, R' and R'', but the key point is that the vector anchors R'' in the base world R. This is precisely because of the semantics of the verb *know*; a verb such as *believe* gives a different configuration, as will be seen. The origins of R' and R'' are at different points on S's base d-axis – relatively distanced from S: this is as it should be, since the more distal positions on the d-axis correspond to the conceptual distance from S.

So John has a belief space R', epistemically aligned with S's base space, in which John holds that R''. *Write* is epistemically true not only for John in the *that* space R'' but also for John in his own mental world, and furthermore also for S from S's epistemic standpoint. In mainstream semantics it is usually just said that *know* is one of those 'factive' verbs that trigger a presupposition; here, I am presenting a cognitive account and attempting to tease out the cognitive complexities in terms, essentially, of reference frames, i.e. the embedded coordinate systems and their relationship to one another. Restating what has just been said in slightly more geometrical terms, we can say the following.

The embedding, and particular positioning, of a new reality reference frame in S's world arises from the particular semantics of the *know that* construction – and, as will be seen, different verbs give rise to different positionings. Geometrically, we can think of a verb like *know* as a special kind of position vector that has two effects. On the one hand, *know that* positions R' and also establishes a discourse referent coordinate for *that*, distal to S – in effect, objectivising the event represented in R'. On the other hand, it positions the origin of R' on S's m-axis at the point of epistemic certainty, truth or reality – the verb *believe* positions R' differently on S's m-axis, as will be seen. So the space R' is represented both as a relatively distal object and as the contents of another mind – and vice versa.

The epistemic stance of O, then, toward an event or represented event of Mary writing a report, is, as far as sentence (2a) is concerned, the same as S's. Geometrically, the two reality spaces R and R' lie in the same plane – the reality plane – relative to the m-axis; they also share the same time points. The difference between them is in their relative 'closeness' or 'distance' from the discourse entities *Mary* and *report*: S is 'further away' than O is. This is appropriate, since the effect of the complement clause is to denote the contents of John's mind, which are correspondingly more 'distant' from O (cf. also Langacker's remark about '*that*'s distancing effect' – 1991: 447).[3]

The configurations can get more complex. Here is a relatively simple complexity. The verb *write*, in the *that* clause, might be modalised – for example, (2b) *John knows that Mary might have written the report*. In this case the vector for *write* will have coordinates at $m'' = 1$ (midpoint) in the *that* space R''. Put differently, (2b) presupposes *Mary might have written the report*. How is this to be modelled in DST terms?

---

[3] Langacker is here also writing about *that*-less complement clauses, e.g. *John knows Mary wrote the report* which are doubtless more common in English. Citing Haiman (1983, 1985), Langacker points to the iconic role of *that* in phonologically/graphically distancing the complement clause from the matrix verb. There is no reason to suppose DST cannot model this: there would simply be no discourse referent *that* and the *know* vector would simply point to the origin of the R'' coordinate system.

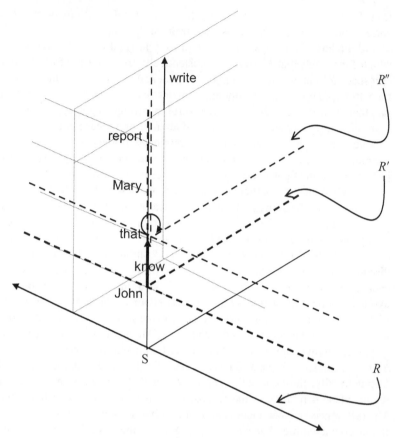

Figure 7.3 Example (2b) John knows that Mary might have written the report

Figure 7.3 proposes one solution. As in Figure 7.2 for (2a), the zero vector anchors R' in S's base world R, but the position of the vector for *write* is now located on the *m*-axis is at its midpoint, corresponding with the English modal verb *might*. Because the zero vector anchors R' (and within it R'') in S's base space R, what John knows might have happened is also represented as what S knows might have happened. Although the important point is that the (zero) vector anchors R' in R, as in (2a) the coordinate systems of S and of the discourse referent *John* coincide, except for the position of the embedded spaces on the *d*-axis. This does not mean epistemic distance but a kind of attentional backgrounding or something similar to deictic *this* and *that*. The distancing implied by the modal auxiliary *might* is, however, epistemic and this is modelled by using the *m*-axis, as explained in Chapter 2.

7.3 Other minds as reference frames

It ought to be noted, finally, that the DSM for *know* claims only to reveal the fundamental cognitive–deictic structures associated with *know* and similar 'factive' verbs, such as *realises, accepts, recognises*. Such verbs share the cognitive–deictic scaffolding we have been discussing, but is not the aim of DST to represent semantic differences between, for example, knowing and realising. It is quite another story, however, for *know*, etc. and *believe*. But before discussing the modelling of *believe* predicates, it is reasonable to ask how we should model sentences like (3a) and (3b), where it is the matrix verb that is modalised.

Perhaps a speaker is conversing with a colleague. She wants to keep the authorship of the report quiet and voices a concern: 'John might know that Mary wrote the report.' In other words, we have a context in which S is uncertain whether John knows or does not know what S and his colleague know. Again, however, the sentence carries a presupposition – that Mary did indeed (from S's point of view, in S's world) write the report. Looking at this in conceptual terms, we can say that S is not sure that John holds a view that would be aligned with S's own base world, only that *if* John did hold such a view, then it would be so aligned. This appears to me to be not the same as the formal semantic understanding of presupposition. It holds certain challenges for a DST account. Figure 7.4 is the DSM for the conceptual structures associated with such sentences as (3a).

As before, *John*, *Mary* and the *report* are all real existing referents in S's world R and there is also a discourse referent, an object labelled *that* which is also really existing for S whose d-coordinate runs through any world embedded in R. The space R'' is embedded in John's reality space R' but is also anchored in S's reality space R. Or, or more precisely, *that* is a discourse referent that positions R'' relative to *that*. Thus, R'' is within *John*'s mind but also anchored to an objective referent in S's mind, or alternatively, what S takes to be real and true (the *that* space R'') is 'projected' into John's mind (i.e. R'). Various kinds of reality spaces R are combined in this complex configuration.

There is a little more that needs to be said about diagrams like that in Figure 7.4 proposed for sentences like (3a). The aim of this book is to pursue an exploration of the frames of reference and vector model that DST puts forward. Now, it is the logic of these simple – but I think cognitively highly relevant – principles that leads to the proposed analysis and model of sentences like (3a). Slightly different diagrams can be proposed,[4] but this seems to me to be the least problematic among possible alternatives. It is

---

[4] For example, R' could remain in the plane $m = 0$ in the base frame R, which would make it automatically 'presupposed' by S. However, this produces confusing results for a modalised *that* clause like (3b), since the 'might have written' vector will then appear in the $m' = 0$ plane of R', i.e. as presupposed as certain for John, contrary to what is needed. Rather than simply stipulate non-correspondence, I've taken it to be more perspicuous to adopt the approach described, using an 'anchoring' or 'projecting' vector from *that* in R to *that* in R'.

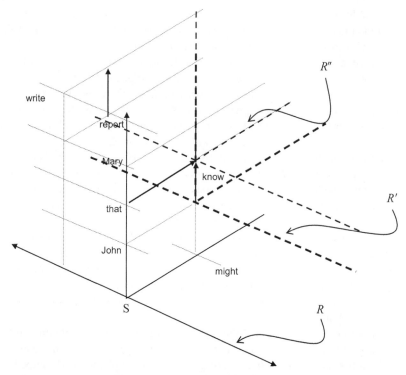

Figure 7.4 Example (3a) John might know that Mary wrote the report

exploratory, as I have said, and the point of the exploration is to see what insights and implications might be gleaned from various possible models.

In Figure 7.4, *that* is not represented by a zero vector (unlike Figures 7.2 and 7.3). In other words it does not allow us to neatly model the 'presupposed' contents of the *that* clause as occurring simultaneously (so to speak) in R and in R', as is the case for unmodalised *know* above. Rather, it shows the contents of the *that* clause as true/real ($m' = 0$) with respect to the contents of John's mind, i.e. R'), but epistemically uncertain (midpoint on m) in R, i.e. S's mind. But for *John might know that Mary wrote the report* there is a truth-conditional presupposition 'Mary wrote the report', and in DST terms this should mean that the *write* vector is shown at $m = 0$ in R. Is there a serious problem here for the DST model, in so far as it does not seem to be quite in line with the truth-conditional account? Well, only if DST is simply a translation of truth-conditional accounts, which it is not. There are two further points. First, the contents of the *that* clause (as is in fact the case for the zero vector too) are still anchored to the discourse entity *that*, which is real/true for S in R.

7.3 Other minds as reference frames

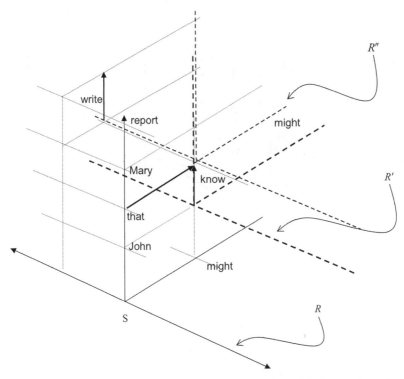

Figure 7.5 Example (3b) John might know that Mary might have written the report

Second, it seems, intuitively at least, that there is indeed a degree of cognitive distance between *that* and the contents it points to: from S's point of view it is more 'distant' though for John it is more 'close'. I am suggesting, then, that in fact it is the vector not the geometrical alignment that models the 'presupposition'. And I am also suggesting that in (3a) and (3b) there is a conceptual distancing, that the 'presupposition' is weaker, harder to recover or more conceptually backgrounded. This is not the same as just applying the negation test, which is basically a logical test, but an intuitive hypothesis about the conceptual backgrounding effect that cognitive approaches to linguistic structure are interested in.[5] If this is indeed the case

---

[5] While a logical negation test yields a presupposition with modalisation of the matrix verb and modalisation of the embedded verb, cognitive processing might make it harder to discover or notice. Only experimentation could test this, and experimentation is certainly not out of the question. Notice that this kind of discussion shows very clearly the distinction between logic-based approaches to language and cognition-based approaches.

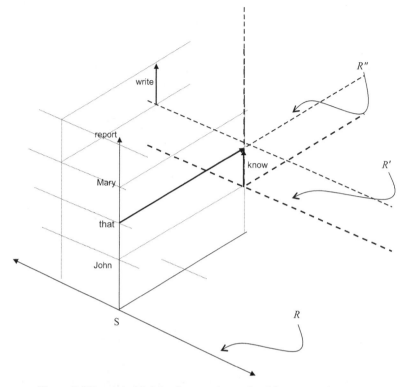

Figure 7.6 Example (4) John does not know that Mary wrote the report (= it is not the case that John knows that Mary wrote the report)

phenomenologically, then that is captured automatically in the geometrical models. The backgroundedness of the 'presupposition' is even more marked in (3b), modelled in Figure 7.5, and in (4), modelled in Figure 7.6.

Looking at Figure 7.5, in which what we have is the following set of reference frames and relations. Relative to R'' Mary might have written the report. Relative to R', i.e. John's reality space, John is certain that this might be the case (i.e. that Mary might have written the report). Relative to S's reality space, S is also certain this might be the case – S has positioned this (by using the factive verb *know*) in John's mind relative to a discourse referent, namely, *that*, which S considers certain, real and existing (it is positioned in R at $m = 0$ for S). The coordinate for *that* runs through to R' and R'', of course. In R'' the midpoint on the m''-axis is aligned with the midpoint on R'' – as it should be in John's reality space: what he knows is a *that* space in which Mary *might* have written the report. This *that* space (i.e. R'') is, however, S's 'projection' – its reference point is on the d-axis in

7.3 Other minds as reference frames

S's reference frame, the reality space R – in other words, it has objective existence for S, whoever else's mind S may think it is also in.

There are situations in which S's epistemic judgement as to what the other might hold in mind can be negative – in the sense here that S can metarepresent a reality that is in S's own mind but not in O's. However, language operates in a curious way in this case of negation and all negations. In saying 'that idea is not in John's mind', S already evokes a conceptualisation 'that idea is in John's mind'. This kind of conceptual structure in modelled in Figure 7.6: the *that* space is projected into O's mind by S (and also anchored in S's), but the whole is located at the distal (i.e. counterfactual) extremity of the m-axis. The structure representing O's mind and its thought *Mary wrote the report* is thus off S's epistemic scale: what is counterfactual is not a part of what is or may be real. This does not mean the discourse entities (John, Mary, report) do not remain real for S; they are labelled on the base d-axis and their coordinates 'persist' throughout whatever R spaces may be embedded therein. Figure 7.6 shows clearly that the embedded reality spaces representing other minds amount not only to a parallel world but to a world that is separate from S's.

7.3.2 Believing that

While DST cannot tell us everything about what it means to 'believe', what it can show is the fundamental difference between verbs like *know* and verbs like *believe*, *think*, *imagine*, *hold*, *claim*, *argue*, *reason*, *deduce*, *conclude*, *hope*, *fear*, *suspect*, etc. Thus, turning to (5a) and (5b), the effect of the semantics of *believe* is that no presupposition appears in the interpretation. That is to say, in DST terms, the embedded proposition does not coincide with the mental representations that constitute S's reality space.

The interpretation of (5a) seems to be that S's sentence means, *inter alia*, that the event is real for *John* but not real for S. The question is: how can this be modelled in DST? The geometrical approach can clarify the conceptualisations conventionally associated with belief-constructions in general. We often speak of 'distancing' ourselves from other people's thoughts, their view of the world and so forth. S's 'distancing' of the epistemic reference frame of O is a key element in the modelling.

There is another interpretation of (5a) and (5b) according to which S is expressing the idea that John believes what S also believes to be true – in for example a context where S and an interlocutor actually want John to believe what they believe, or know, to be true/real (cf. 'John believes that the earth is flat' uttered by one member of the Flat Earth Society to another, when considering John for membership). In such an interpretation *believe* is semantically close to *know*, and the DSM presumably has a similar structure to

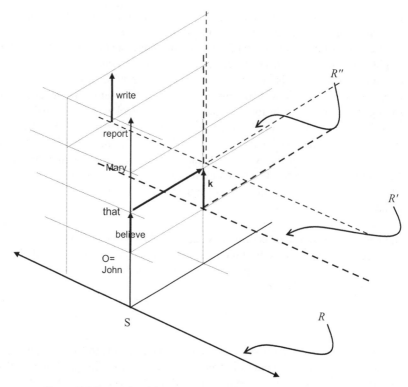

Figure 7.7 Example (5a) John believes that Mary wrote the report

Figures 7.2 and 7.3. What then would be the difference between *John knows that Mary wrote the report* and *John believes that Mary wrote the report*? While the cognitive–deictic structure as modelled in DST is the same, there is additional meaning that differentiates *know* and *believe*, what we might for now call an affective element in the semantics of *believe*. In what follows I will not say more about this aspect of the meaning of (5a) and (5b) and (6a) and (6b) but concentrate on the interpretation in which these sentences express that S's epistemic stance is $m > 0$, i.e. a stance according to which *Mary wrote the report* is viewed by S as epistemically probably, neutrally possible, improbable, etc.

Figure 7.7 reveals a number of elements in the basic conceptual structure of *believe*, when used to report other minds, that is, to make a truth-claim about another mind. In this case the other mind, i.e. John's mind labelled O, is metarepresented as a frame of reference R', represented by S at $t = 0$, the time of S's speaking/conceptualising. Here S locates the origin of R' at the epistemic midpoint – so all the contents of R' (i.e. John's mind) at that moment

7.3 Other minds as reference frames

are represented as epistemically uncertain for S. And what is in this frame of reference, John's mind, according to S, is the representation R'', the contents of which John regards as true – that is why R'' is positioned at $m' = 0$ in R'. Within R'' is the metarepresentation *Mary wrote the report*, arising from the *that* clause. And within that reference frame R'' the *write* vector can be epistemically positioned at varying points along the m''-axis. But its epistemic status is twofold. On the one hand, for John it is at $m' = 0$ (i.e. true) relative to his frame of reference R'. On the other hand, the whole of R', John's current state of mind, together with its projected contents R'', is positioned by S (i.e. in R) as epistemically uncertain (i.e. at the midpoint on the m-axis). Note that in this model R'' is understood to be 'sealed off' from R: it is relative to R' and its internal coordinates within R' are relative to R' and do not 'pass through' to R. But although it is 'sealed off' in this sense, it is still anchored in S's world, R. Any *that* space is initially represented by S in R. However, recalling what has been said about *that*-ness, the coordinate systems representing *that*-complements have to be located somewhere; on their own they have no epistemic status and have to be epistemically positioned. In Figure 7.7 the *that* clause is located in John's mind R' and within that space at John's epistemic plane of certainty (i.e. $m' = 0$). But because S positions the whole of John's mind R' as uncertain, its contents, here the *that* space representing Mary's activity, are also uncertain for S.

Other minds conceptualised by *believe* expressions like (5a) are viewed as closed. This also implies that within his own frame of epistemic reference John 'knows' that Mary wrote the report and to make the point for this example I have inserted a 'shadow' vector k. In a sense, therefore, S is 'distancing' itself from the O's subjective 'knowledge' (or what S claims it to be). Of course, S may report an other mind as holding a *that*-representation which expresses that Mary's activity is uncertain (i.e. at epistemic midpoint $m'' = 1$). This is the case in sentence (5b). Using the same approach, we can model (5b) in the same configuration of R spaces: the *that* space can position the *write* vector at any epistemic point on the m''-axis, as shown in Figure 7.8.

There is a detail in Figures 7.7 and 7.8 that we have not yet commented on. It is important to note the position of the *believe* vector, in contrast to the following Figures 7.9 and 7.10, the models for the meanings of (6a) and (6b). These are sentences by means of which S reports on the possibility that John holds a belief, a belief expressed in the *that* clause. After all, S may not be sure whether John holds the belief in question – an indication, incidentally, that the *that* clause conceptualisation is felt to be independent of John's mind as well as, paradoxically, in it. In Figures 7.7 and 7.8, modelling (5a) and (5b), the *believe* vector occurs in R at $m = 0$, i.e. S is certain that John does hold the belief content represented in R'', the *that* space. In Figures 7.9 and 7.10, not only is John's entire mental space R' epistemically uncertain for S (it is at

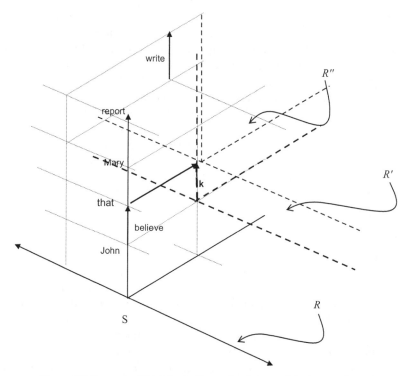

Figure 7.8 Example (5b) John believes that Mary might have written the report

m = midpoint in R), S positions John's believing (the *believe* vector) also at $m = 0$, S's point of epistemic uncertainty. That is, S is uncertain whether John does in fact have this particular belief about Mary in the first place. And the diagram has to be understood as meaning that S is unsure whether John holds *that Mary wrote the report* and unsure, if he does hold it, whether *that* is true. For John, however, the *believe* vector is at the point of certainty $m' = 0$ and subsumes the 'shadow' **k** vector mentioned earlier. If John does believe that Mary wrote the report, subjectively that is knowledge **k** for John. But for S, what John holds true of Mary, real about her, is neither true nor untrue, real nor unreal. The cognitive structure for (6a), and any *that* constructions, is complex: the space R'' that holds *Mary wrote the report* is still anchored in S's world R and positioned in John's mind R' by S, as the vectors connected to the S's discourse entity *that* indicate.

A DSM for (6b), in which the verb *write* in the *that* clause (represented as the *write* vector in R'') would be along the same lines, in parallel with Figures 7.4 and 7.5. But the reader need be taxed no further on this point.

7.3 Other minds as reference frames

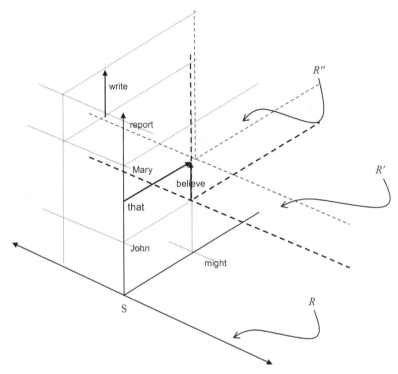

Figure 7.9 Example (6a) John might believe that Mary wrote the report

In fact, S can adopt any number of epistemic attitudes towards the conceptualisation of the embedded *that* clause represented in R'', not directly but by adopting an epistemic position with regard to the space R' that holds R''. This is the case whatever the epistemic stance of O – i.e. wherever on the m'-axis the event, as viewed by O, is located. Take the counterfactual stance that S might take vis-à-vis O:

> John believes Mary wrote the report! He's probably right/maybe he's right/ it's not very likely/he's wrong, etc.

> John believes Mary might have written the report! He's probably right/ maybe he's right/it's not very likely/he's wrong, etc.

But no particular epistemic stance is actually *necessitated* by the *believe* construction, by contrast with *know*. Pragmatic factors of considerable variability affect the reading. For instance, S might hold O in especially high regard as an authority in the relevant circumstances, in which case, far from regarding John as wrong, S might intend to indicate strengthened epistemic likelihood. The upshot of this seems to be that there are non-linguistic

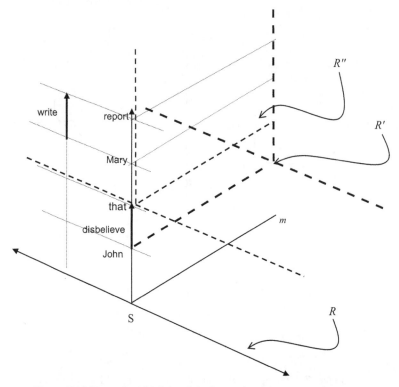

Figure 7.10 Example (7b) John disbelieves that Mary wrote the report

judgements of epistemic likelihood that are not coded in language structure. One final point. It should not be overlooked that O, the other, distal to S, is also a speaker, subject and self, whom we could label S′. Linguistic structure makes it possible to represent and report on other minds that are represented as similar – geometrically they are copies of S's base coordinate stem and frame of reference. They are nonetheless in some sense both distal and sealed off. What is uncovered through this geometric analysing of core language structure is the ability to metarepresent other minds (i.e. theory of mind). The existence and use of *that* constructions, I am suggesting, rests on the non-linguistic ability to adopt the mental perspective of another human.

7.3.3 Not believing and disbelieving

In this section I want to enter slightly more speculative terrain. What is the difference between not believing that something is the case and disbelieving it? And how can the different conceptualisations associated with each be

7.3 Other minds as reference frames

modelled? The point here is to see how much the DST model can deal with. Consider (7a) and (7b) repeated below:

(7a)   John does not believe that Mary wrote the report

(7b)   John disbelieves that Mary wrote the report.

In the case of (7a) the model would be like that of Figure 7.9 for (6a), but the coordinate systems R' and R'' would be positioned at the extreme end – the counterfactual position – of S's epistemic scale m in the base space R. This might be compared with the model for the analogous sentence containing *know* instead of *believe*, viz. (4) *John does not know that Mary wrote the report*, for which the DSM is given in Figure 7.6. The DSM for (7a), which is not given here, would be similar except that the label *believe* would replace *know*. Unlike Figure 7.6 for (4), the DSM for (7a) would not *necessitate* a vector that anchored R'' in R'. The conceptualisation of the *does not believe that* sentence (7a) *could* have the same anchoring vector, if, given pragmatic factors, (7a) were understood as communicating that S did believe what S did not. What differentiates the DSMs for (4) and (7a), then, is the fact that anchoring vectors are *not* necessitated for (7a). I am not clear whether (7a) can imply or be compatible with a midpoint epistemic stance ('it might be the case that Mary wrote the report') on the part of S.

Turning now to (7b), the effect of *disbelieve* seems to be different from that of *not believe* in (7a). What S seems to have in mind is that John does hold a conceptualisation, but it is the reverse of S's. In other terms, in using *disbelieve* (7b) implies S is committed to the truth that Mary wrote the report; its negation also seems to imply this. Whether this is tantamount to formal logical presupposition is something I shall not pursue in detail. As we shall see, a particular kind of geometrical arrangement of reference frames works for *disbelieve*. It does not work for a classic presupposing verb such as *know*, though again I leave this for others to investigate. It is the negative element in *disbelieve* that seems to make a mirror-image model appropriate, as proposed in Figure 7.10.

The point to notice is that R', the space (coordinate system, reference frame) representing John's mind is the reflection of S's reality space R. Looking at the modality axes m and m', what is true/real for O (*John*) is false/unreal for S and vice versa. Further, R'', the *that* space, is a reflection of R' but is aligned with R, because it is what S epistemically judges true/real – that is the interpretation I am suggesting for (7a). We can see, then, that in Figure 7.10 *Mary wrote the report* is true for S (at $m = 0$) but at the counterfactual end of R', the mirror image of R, representing John's mental 'disbelief space'. Notice that, as before, we have to seal off R' (and its R''

contents) from R. This means we can interpret the *disbelieve* vector as relative only to R (S is asserting that it is truly the case that John disbelieves *that*) and as not also located in R' (at the counterfactual point for John). It would be convenient geometrically if there were such a carry-through – but as we have seen consistently, the sealing off appears to be a necessary stipulation for DST geometry and is, as it perhaps more conveniently happens, cognitively plausible to think of other minds as 'closed' to S in this limited sense.

But what of the other end of the m-axis in R', the position where a representation is true for O but false for S? It would be uneconomic to propose this mirror-image configuration just to account for *disbelieve* but the geometry allows us to explore conceptual space. If *disbelieve* is a kind of conceptual act whereby O denies what S holds true or real, then one might expect a converse conceptualisation: O believes what S does not hold true or real. This is quite a coherent concept but is not in fact lexicalised in English: there is no English word that necessarily has that meaning. It is, however, likely that in certain contexts (e.g. an atheist speaking of a believer) that John believes *p* automatically 'presupposes' that what O holds true and real is not true and real. But this is not the general semantic structure of the verb *believe*.

7.4 Connections and disconnections across parallel worlds

7.4.1 Anaphora across belief spaces

How it is possible to refer in language to events, processes, states, etc. that are represented in other people's minds, entities the speaker does not belief to exist, or about whose existence the speaker is doubtful, is a question that has intrigued and provided endless puzzles for philosophers when they attempt to describe the linguistic structures that give rise to them and do so in terms of symbolic logic. How does one separate or connect entities in one's own mind from or to those in the mind of another person? Elegant solutions have been offered in Fauconnier's (1994) theory of mental spaces as well as in Kamp's Discourse Representation Theory (see above Chapter 1, Section 1.2.1). It is not my aim to propose DST as a rival, but it can do at least some of the same work, and in a more explicitly cognitivist framework.

I will consider one limited case. Asher (1987) offers an account of (8) in DRT terms (i.e. in terms of Kamp's Discourse Representation Theory):

(8) Hob believes that a witch has blighted Bob's mare, and Nob believes she has killed Cob's cow.

7.4 Parallel worlds

This is a slightly modified version of a celebrated example discussed by Geach (1967).[6] Geach pointed out that despite the fact that *a witch* is an indefinite NP and does not presuppose the existence of any witch, the pronoun *she* in (8) nonetheless refers to this non-existent entity and does so across a clause boundary. This latter point was the nub of the problem for Geach and others since, and is due to an assertion of Quine's that anaphoric reference cannot reach across clause boundaries (the boundary being in effect the conjunction *and*). Actually, *a witch* can be heard as referring to an actual witch, that is have a de re meaning, but we will proceed with the de dicto reading that interested Geach. The fact that (8) makes sense led Geach to propose the notion of intentional identity: Hob and Nob are intending the same thing, whether it exists or not:

We have intentional identity when a number of people, or one person on different occasions, have attitudes with a common focus, whether or not there actually is something at that focus. (Geach 1967: 627)

Interestingly, Geach explains this by using the etymological meaning of *intend*: stretch (tend) a bow to shoot an arrow at some common target. They are all pointing their arrows in the same way. This intriguingly matches the way we represent vectors, as deictic pointers. The notion of joint attention is also suggested. I believe the solution to the puzzle less problematic in the holistic, visual and geometrical DST approach than it appears to have been if one goes down the sequential symbolic logic route. The key is cognitive frames of reference, the labelling of discourse entities in these different frames of reference and the approach to de re and de dicto meanings discussed briefly in Chapter 2, Section 2.4.2 above. More fundamentally, treating the logical–linguistic sequencing structure as yielding a holistic abstract *space* is what makes it possible to grasp how meaning operates when human minds process (8).

How, then, does DST handle (8)? Essentially, (8) is viewed as a construction that in some fashion coordinates the representations of two states of mind (so much is similar to the classic logical accounts including Asher's). The main problem is how to model *and*. What does it mean for two clauses to be joined – or separated – by this conjunction? I shall not attempt a general theory of coordinate clauses, and what Figure 7.11c models below is based only on sentences strictly analogous to (8), which is a combination of

---

[6] Geach's sentence was: 'Hob thinks a witch has blighted Bob's mare, and Nob wonders whether she (the same witch) killed Cob's sow'. (8) is Asher's simplifying sentence into which I've inserted *that*. But the analysis I propose will work just as well with Geach's original sentence, since the 'wonder whether' construction has the same essential and distancing properties as *believe that* construction, and cows are as good as sows for our purposes.

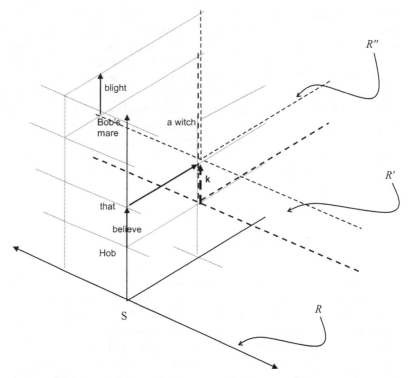

Figure 7.11a Example (8) Hob believes that a witch has blighted Bob's mare

Figure 7.11a (representing the first clause of (8)) and Figure 7.11b (representing the second clause of (8)).

Figure 7.11a models what S reports to be in Cob's mind, as communicated by the first clause of (8). Cob's mind is R'. In it is R'', which is the *that* space. This space is still anchored in S's reality, R. However, as we have noted before, not all discourse entities have to be labelled as existing in the same space – that is, some may be seen by S as existing only in the mind of an other. And as has emerged in all the earlier parts of this chapter, the mind of an other (R' and its contents R'') are sealed off: their coordinate points do not persist through into the coordinates of S's base reality R. So here in Figure 7.11a, on the reading of (8) we are interested in, *a witch* is not a really existing entity in R so is not labelled on the d-axis; it is labelled instead only on the d'-axis in R'. And of course, because of the semantics of *believe*, S does not align Hob's reality space with S's own but positions the whole of it as merely possible at the neutral midpoint on S's own d-axis.

7.4 Parallel worlds

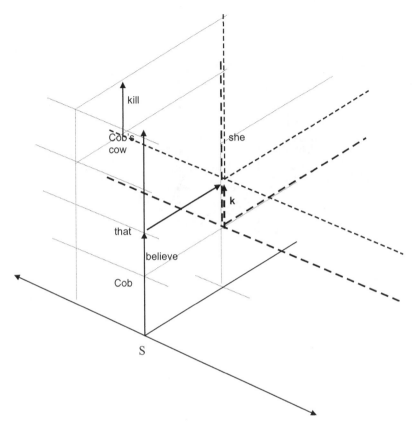

Figure 7.11b Part of sentence (8) Nob believes she has killed Cob's cow

Note that the coordinates for *a witch* do pass through to R'', the *that* space positioned in R' by S. The *that* space concerning a witch killing Cob's cow is still anchored at a *that* discourse entity on the *d*-axis in S's reality R. It is possible to omit that positioning anchor, in which case the diagram would model a conceptualisation in which S regards the proposition contained in R'' as a wholly private one, private to Cob. However, as for similar sentences in earlier sections, I am assuming that (8) is read as reporting on a 'public' proposition, or at least one assumed by S to be 'real' in his world, albeit projected into the mind of O, i.e. Cob in this instance. However, this is not the main issue at hand.

Consider now Figure 7.11b, which models Cob's view of the world. Note that this is still the world according to S: the configuration is, rather boringly, the same as Figure 7.11a with different discourse entity labels.

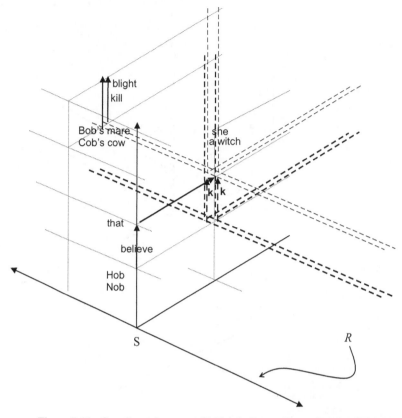

Figure 7.11c Coordinated spaces: (8) Hob believes that a witch has blighted Bob's mare, and Nob believes she has killed Cob's cow

The question now is how the conjunction *and* combines Figures 7.11a and 7.11b. For sentence (8), and sentences like it, Figure 7.11c is a possibility. This drawing approximates the hypothesis that the *and* in (8) operates a kind of conceptual superimposition of the second part of the sentence on the first. Far from being some sort of syntactic barrier, *and* should be seen, on this account, as a kind of operator: first the DSM for the first part of the sentence is constructed and held in working memory, then the second part is constructed and combined with the first.

The structure of (8) yields a configuration that gets us some way towards a model that captures two important facts about the understanding of this sentence: the sealing off of *a witch* from the reality space of S, and the relating of *she* to the appropriate antecedent *a witch*. That is, it shows *a witch* and *she* as occupying the same bit of conceptual space and distancing it from S's own view of the world – which is what we need. However, there remains an obvious problem. How should we handle the relationship between the other discourse entities, namely, *Hob* and *Nob*, *Bob* and *Cob*, and the two vectors *blight* and *kill*? The answer that suggests itself is that two entities denoted by a non-pronominal are allowed to occupy the same position in conceptual space. However, where a pronoun, here *she*, and a nominally denoted entity occupy the same position, the former must conceptually merge with the latter in some way. This idea of 'conceptual merging' needs further exploration that can't be undertaken here, but note that it is likely to be part of the phenomenon of 'blending' as investigated by Fauconnier and Turner (2002). The proposal does nonetheless have the advantage that anaphora resolution combines with the appropriate epistemic positioning of an entity in the mind of an other. It seems also to me that a holistic spatial conceptual configuration must emerge at some level of consciousness as a result of the sequential processing of (8).

7.4.2 Theory of mind, language and communication

Imagine the following scenario. A and B work in the same office. There is a bowl of sweetmeats for office workers to help themselves to. A is particularly fond of a certain brand of chocolate and knows that B is also. One day A puts the chocolate inside a box on her desk, but notices that B was watching her action. Then B goes out on an errand and while he's away, A moves the chocolate to her desk drawer. She expects that B will, wrongly, think that the chocolate is in the box.

Humans and probably primates (see Whiten 1991) are adept at such mental interactions. Humans certainly are sensitive to others' intentions, as Baron-Cohen, Tomasello and others have emphasised. Moreover; humans can *report linguistically* on (what they claim to be) the thoughts of others by means of complex sentences. And they do so with varying kinds of caution. Thus A may say to C: 'B thinks that the boss is coming to check the project.' A is not committing herself to agreeing with B's assertion, because she uses the verb *thinks*. Note also A's utterance is ambiguous. What exactly is A communicating about her own cognitive representation of the state of affairs, and in particular, what is she communicating about her judgement as to the truth, likelihood or counterfactuality of B's assertion? It could be she agrees with it, is unsure of it, or believes it to be improbable, or even counter to fact. These things have been much discussed in the formal semantics literature and it is not my aim to add further discussion within that paradigm (truth-conditional or possible-world semantics). It is rather to show how DST, in a natural and straightforward fashion, might model these cognitive and linguistic–communicative phenomena.

The theory of the theory of mind rests on psychological and neuroscientific evidence that human brains normally attempt to make inferences about the intentions and thoughts of people with whom they interact. It is also worth noting that work on theory of mind generally makes the assumption that people have the ability to *communicate linguistically* their inferences about the thoughts and intentions of others. Baron-Cohen *et al.* (1994) give brain-imaging evidence that words associated with mind, such as *think*, *imagine* or *believe*, activate specific areas in the right hemisphere. In McGilchrist's words 'the right hemisphere is critical for making attributions of the content, emotional or otherwise, of another's mind' (McGilchrist 2010: 57). In well-known work Baron-Cohen and his co-researchers (1985, 1995) argued that individuals on the autistic spectrum suffer from a deficit in this ability: roughly, they do not infer the intentions of other people and more generally do not grasp that other people's minds have different content – that is, autistic people may not have a theory of mind, they are 'mind-blind'. What needs to be spelled out further is what it is about the grammatical structures linked with verbs such as *think*, *imagine* and *believe* that makes it possible to communicate about other minds.

Let us return to the office scenario. This is in fact a version of the 'Sally Anne test' (Wimmer, and Perner 1983, Leslie and Frith 1988). A normal observer S of the office episode might say to a colleague something like (9a) below, while an autistic observer might say something like (9b). The parts in square brackets here are not actually uttered by the speaker S.

(9)   a B thinks that the chocolate is in the box [but I know it is not]
          b B thinks the chocolate is in the drawer [and it is].

We are interested here in the communication of thoughts concerning the mind of an other. Such thoughts, as we have seen, involve the grammar of *that* constructions. The question now is how the cognitive DSMs I have proposed for *that* constructions relate to the theory of mind and related research into the mind's ability to think about the thoughts of other people. The first of the two diagrams below (Figure 7.12a) is structurally similar to Figure 7.7 (for *John believes that Mary wrote the report*).[7]

According to this model for a normal representation of the situation after A has hidden the chocolate, the speaker S's cognitive state includes knowledge that B holds a counterfactual belief about its location. Hence, S metarepresents B's current cognitive state as positioned at the end of her

---

[7] Recall that prepositional predicates, e.g. *in*, are diagrammed as position vectors, so that the location is the tail of the vector, the locandum is at its tip.

7.4 Parallel worlds

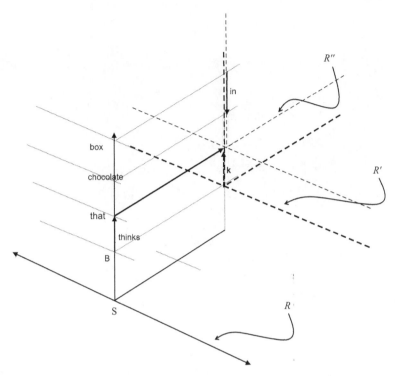

Figure 7.12a Normal representation of other mind communication

epistemic axis (R'). The representation of the chocolate's counterfactual location is in S's mind and is treated as an abstract propositional entity with a referent label *that* on S's *d*-axis – but projected by S into B's counterfactual mind state. This is what, according to S, B really thinks – so the *think* vector is epistemically certain for S (at S's $m = 0$) and has the coordinate for B at its tail and the coordinate for *that* at its tip. The diagram seeks to capture these components of S's cognitive state, which is also the one communicated in the language used by S in (9a). The diagram is not of course a model of the sequential structure of the utterance; it must be taken rather as a holistic model of the conceptual structure motivating S's lexical and grammatical choices in (9a) and, potentially, of the conceptual structure constructed by her interlocutor. If we assume now that the speaker reporting on B has no theory of mind (perhaps she is autistic),[8] we should expect a somewhat different DSM.

---

[8] In this section, in in referring to autism I am thinking of a theoretical extreme case and making no attempt to speak in any way of real individual cases, which vary enormously.

208   Reference frames and other minds

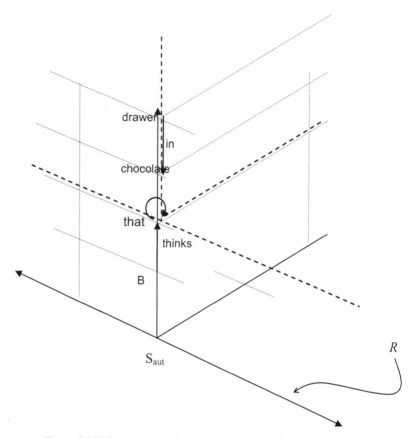

Figure 7.12b Possible autistic representation of other mind communication

There is more than one possibility for such a model. One is that in communicating about B in our invented situation, an autistic subject, here S_{aut}, does have a model of what B thinks – but it is simply located at $m = 0$ on S_{aut}'s m-axis. This might be tantamount to saying that an autistic subject like S_{aut} might simply not have the m-axis at all, that is, have no epistemic judgement, at least in the judging of what other people think and expressing that judgement linguistically. And this approach would be to suggest a somewhat different account of the theory of the theory of mind itself. Another approach, however, more in line with the prevailing idea that the autistic mind does not represent another mind, is outlined in Figure 7.12b.

In this second approach, the model simply removes any representation of the other's mind, the mind of B, which is equivalent to saying that autistic S_{aut}

7.4 Parallel worlds

has no theory of mind. What is left is S_{aut}'s base reality space R. In this space S_{aut} knows the chocolate is now in the drawer.

Assuming S_{aut} does have an m-axis, the *that* representation (*the chocolate is in the drawer*) will be epistemically certain – that is an S_{aut} will not communicate *B might think the chocolate is in the drawer, B probably thinks the chocolate is in the drawer, B does not think the chocolate is in the drawer*, etc. There is always a zero vector locating the *that* space in the mind of the speaker, in the speaker's own reality. This perhaps is an argument for the complete absence of the epistemic m-axis in the autistic mind. The configuration proposed in Figure 7.12b is similar to that proposed for the semantics of the verb *know* – see Figure 7.3 above. In a sense, an autistic speaker who says *B thinks that the chocolate is in the drawer*, in the situation described, is presupposing what is presupposed in saying *B knows that the chocolate is in the drawer*, with the crucial difference that the other mind coordinates (R') are not represented. It is possible that autistic subjects cannot correctly handle the difference between *know* and *think*; empirical investigation is needed.

The main question in this chapter and in Chapters 5 and 6 has been: how exactly do embedded axis systems relate to the base system in which they are inserted? In the case of embedded reference frames that are inserted and translated with respect to the m-axis, it seems that we have to conclude that the embedded m'-axis in R' cannot be projected onto that of the base R in any cognitively meaningful way. This is not neat geometrically but it corresponds conceptually to the cognitive phenomenon under consideration – the ability to communicate linguistically about mental representations judged to be discrepant with the real world as judged from the point of view of S.

This holds for the metarepresentations entertained by S that are attributed by S to another mind. The crucial point is that S makes an epistemic judgement regarding the representation of S' (i.e. the metarepresentation of S): for S, R' can be real (true in R), be a mere possibility (S is unsure or non-committal as to the truth of R' in R) or contrary to R. Now in the latter case, the DSM I have proposed is special, in the sense that it 'projects' beyond S's reality space R it is simply not in the world as understood as real by S. It is in this sense that it is contrary or counter to fact. But there is another kind of 'counter to fact' configuration that involves a different geometric transformation, and which I refer to as 'counterfactual'. This is particular kind of conceptual space with its own properties and whose job is not to metarepresent *other* minds but to metarepresent an alternative state of S's *own* mind. We shall look at this in detail in Chapter 8. The next chapter, however, takes up the question of the meaning of complement constructions, in particular those associating with verbs that require the use of *to* and *ing* – or allow alternation between both, with differing conceptual effects.

8 Mental distance and complement clauses

Languages enable us to embed structures recursively. While this is a syntactic notion, we have seen something of its cognitive *raison d'être* in Chapter 7, and we shall in effect be exploring it further in the present chapter. The key idea will be that particular verbs are associated with particular morphosyntactic operations and that these in turn are related to the kind of conceptualisation that has been referred to as 'perspective' in cognitive linguistics and throughout the present book. The term 'perspective' is useful in so far as it implies 'viewpoint' but viewpoint is not the main consideration here. To be slightly more precise, by 'perspective' I mean 'mental distance' and 'direction'. The spatial metaphors seem unavoidable in trying to describe the cognitive effects produced by the complement clause types that this chapter looks at. The effects are describable in terms of a sense of immediacy or closeness of the events denoted in the complement clauses and in terms of the subject's orientation toward or disposition toward some event in the complement clause. These effects come along with complement constructions that are distinguished by the morphemes *to* (as in *she wants* to *go*), and *ing* on the embedded verb (as in *he considered going*), *that* (as in *they thought that she was clever*) or by the so-called bare infinitive (as in *he hoped that she would go*).

The present chapter and the following chapter seek to model the conceptual effects connected with *to*, *ing* and zero complements. Some verbs have semantic properties that naturally select only one of these three possibilities. Some can select two or three, and can also select *that* constructions. We shall consider the semantic affiliations between particular verbs and the particular construction that select as complements these alternations have varying conceptual distinctions. The overarching conceptual basis seems to have to do with cognitive 'distance'.

8.1 Verb meanings and clausal complements

There have been a number of efforts to categorise verbs in terms of their syntactic properties in the linguistics literature, and semantic effects were noted even in theoretical approaches, such as the generative approach, that did

8.1 Verb meanings and clausal complements

not favour a meaning-based analysis. Generative grammar formalised the fact that predicates, in particular verbs, each carry a specification (known as a 'subcategorisation frame') for the number and type of syntactic argument they combine with. Certain verbs appear in more than one kind of subcategorisation frame, yielding different semantic effects: this is sometimes called 'diathesis (rearrangement) alternation', an example of which is the alternation *the lads loaded hay onto the cart* vs. *the lads loaded the cart with hay* (see above, Chapter 3, Section 3.3.6). In a primarily conceptual approach Fillmore relates the syntactic properties of verbs (their arguments, including clausal arguments) to their meaning (Fillmore 1982b). Levin (1993) classifies verbs in terms of the syntactic operations (diathesis alternations) that are associated with them and finds semantic correspondences, following a number of previous proposals that a verb's syntactic behaviour is predictable from its meaning. But her investigation concerns only the verb-argument structure internal to the clause; it does not explore the correlation of *to*-type, *ing*-type and *that*-type clausal complements with verb semantics. We shall broadly follow Fillmore's conceptual framework. Most importantly, diathesis alternations are extended to include clausal complementation, and alternating clause complement types will be treated as conceptualising alternations – more specifically, *perspective* alternations, manifest in subtly but importantly different conceptual frames of reference. These phenomena are well suited for treatment in DST; I shall not, however, explore the clause-internal argument structure in this framework.

Verbs that take clausal complements, of the kind noted above, raise particular challenges of description. In generative grammar the question of the *meaning* of such clausal complements in relation to the meanings of their verbs tended to be obscured by a fascination with syntactic description. Generative grammar also subclassified predicates in terms of syntactic transformations of invisible, i.e. postulated syntactic structures underlying the sentences one heard or uttered.[1] This led to a classification of verbs in terms of the type of transformation that was postulated, viz. raising to subject, raising to object and control verbs in which a main clause predicate was held to 'control' a subject or object in the complement clause (see Rosenbaum 1967 and Postal 1974 for the earliest idea and the overview in Borsley 1996). I shall use this classification in what follows, but only to a relatively small extent, since my main concern is with the conceptual effects of the 'surface' grammatical structures.

In Table 8.1, *ing* constructions are shown as appearing in 'control' ('equi NP') sentences. What is the justification for this and why do they not appear

---

[1] For a clear summary of the syntactic considerations see Van Valin (2001: 49–55).

Table 8.1 Syntactic description crossed with zero, to, ing and that constructions (sample)

	Zero	ing	to	that
raising to subject verbs (matrix coding)			seem, appear, be likely, turn out, happen, prove	seem, appear, be likely, turn out, happen, prove
			Mary seems to have written the report	*It seems that Mary has written the report*
raising to object verbs (matrix coding)			know, believe, imagine, consider, assume, presume, suppose, expect, intend, want, get, allow, permit, forbid	know, think, believe, consider, assume, presume, suppose, pretend, find, expect, intend, discover, imagine
			John believes Mary to be writing the report	*John believes that Mary is writing the report*
subject-control (equi NP, subject)		like, enjoy, hate, detest intend, try, remember, regret, forget, imagine, consider,	want, like, hate, long, intend, try, remember, forget, learn, fail, refuse, decline, consent, pretend, manage, plan, decide, expect, hope, presume, promise, vow, agree	remember, forget, imagine, expect, hope, presume, promise, vow, agree, deny, pretend, decide
		start, begin, commence, cease, finish, stop, keep, continue, get	start, begin, commence, cease, continue, get	
		avoid, deny		
object-control (equi NP, object)	see, watch, hear, feel, sense	*John imagined writing the report; John tried writing the report; John got writing*	*John pretended to cry; John tried t; John got to write the report*	*John vowed that he would write the report*
		see, watch, hear, feel, sense	want, get, cause, force, persuade, urge, order, ask, implore, command, press, pressurise, exhort, beg, incite, push, remind, advise, warn, encourage, induce, forbid, allow, permit	see, hear, feel, sense
		find, discover, imagine, remember, want, get		find, discover, imagine persuade, urge, order, implore, command, exhort, remind, advise, warn
	make, have, let, help	prevent, stop, keep	help	
	John saw Mary write the report	*John saw Mary writing the report; he got her writing the report;*	*John persuaded Mary to write the report; John got Mary to write the report*	*John saw that Mary had written the report; John persuaded Mary that she [should] write the report*
	John made Mary write the report	*John prevented Mary writing the report*		

8.1 Verb meanings and clausal complements

in the raising (subject and object) structures? The main way of distinguishing 'raising' from 'control' sentences is as follows. Consider the following two sentences:

(1) a John asked Mary to write the report
 b John imagined Mary to be writing the report.[2]

Sentences (1a) and (1b) appear to be syntactically similar. Both have embedded *to* clauses in which there is no grammatical subject: *Mary* is not in the *to* clause as subject but in the matrix clause as grammatical object. However, it is generally claimed that there is a crucial difference between the two sentences, and sentences like them: in (1a) *Mary* is a semantic argument of *ask* in the matrix clause as well as in the *to* clause, while in (1b) *Mary* is not a semantic argument of *imagine*. That is, while in (1a) it is the case both that *John asked Mary* and that *Mary wrote the report*, in (1b) it is not the case that *John imagined Mary*. In the latter, *Mary* is said to have been 'raised' from an embedded clause that we can see in the sentence *John imagined that Mary wrote the report*, where *Mary* belongs semantically with *wrote the report*. Sentences of the (1a) type are generally known as 'control' (or 'equi NP') structures, since the verb in the matrix clause determines what is the missing subject of the embedded clause.

Let us now turn to the *ing* construction, which is an acceptable alternative complement for the verb *imagine*:

(2)   John imagined Mary writing the report.

Intuitively, *Mary* seems to be an argument both of *imagine* in the matrix clause and of *writing* in the complement *ing* clause. It is therefore similar to (1a), a control structure on standard accounts. In general *ing* structures fall into the 'control' ('equi NP') category. However, the dichotomy of raising versus control may be too rigid, if we compare (2) with (1b), which is supposed to be a raising structure not a control structure. In a cognitive framework, one might want to argue that in (1b) *Mary* is in fact felt to be an argument of *imagine*, though the effect is not as strong as for the *ing* construction in (2). Many verbs allow both *to* and *ing* constructions, with varied but related conceptual effects. In such cases it is possible to describe the syntactic differences but what is needed is an explanation of why two constructions exist in the first place. The obvious explanation seems to be that there is a difference in meaning, so what is needed is an account of the different conceptual effects associated with each of the two alternate complement constructions.

---

[2] The *ing* here is part of the progressive form required by non-stative *write* and the complement clause is a *to* construction not an *ing* construction.

The underlying theoretical assumption of my approach is that grammatical constructions are associated with meaning, that is, with conceptual effects. This assumption is most consistently and insistently asserted in the work of Bolinger (1977), Dixon (1984, 1991), Wierzbicka (1988) and Langacker (1991). Wierzbicka's enquiry pays attention specifically to fine differences in meaning between complement clauses associated with *to* + verb, those which have *ing* forms of verbs and those which are introduced by the conjunction *that*. Langacker (1991: 449–63) takes up this lead but in conjunction with the 'raising' and 'control' framework. I shall pursue a similar trajectory in this and the next chapter, with the aim of suggesting how the DST framework might contribute to modelling the conceptual import of *to* and *ing* complements.

Table 8.1 summarises the kind of data that will be investigated in this and the next chapter. There have been some insightful attempts to characterise *to*- and *ing*-complementation (Wierzbicka 1988: 23–168, Langacker 1991: 417–63) and some of the ideas discussed in the literature influence the account I shall give. However, the existing accounts are not entirely satisfying. It would be tedious here to describe what I think may be inadequacies in these accounts, when they have made significant conceptual breakthroughs, so I shall proceed to explore what further insights and unifying generalisations, if any, DST has to offer. The verbs that pattern with *to* and *ing* complement clauses are various in their syntactic manifestations. To simplify matters, I shall leave aside verb constructions such as *think of doing*, *decide upon*, *ruminate about*, etc., which involve prepositions. Although these verbs show important shared semantic characteristics, I shall look only at those verbs that appear in constructions where they are followed directly by *ing*, as in *remember doing*, *consider making*, etc. I shall also leave aside non-verbal predicates such as *be afraid to*, *be afraid of doing*.

The goal is to seek to represent each type of construction in a distinct geometrical model. The underlying assumption in doing this is, of course, that the three cognitive dimensions, the relativisation of embedded axis systems and the relative distancing from the origin will take us a long way (though probably not all the way) toward distinguishing between the constructions at issue in terms of motivated conceptualisation.

This table shows the combinations in which certain verbs in the *that, to, ing* and zero constructions occur: for example, *stop* occurs with *ing* but not *to* constructions. Where a verb occurs in one cell but not in another, it is to be understood that that verb is not possible in that cell. As far as possible I have tried to hold the meaning of these verbs constant across the rows. However, certain verbs can occur in constructions where they do not appear in the table (e.g. *learn, consider, see that, hear that*), provided the meaning of those verbs changes slightly. This is in itself a revealing phenomenon. Without going into

8.2 The meaning of *that*, *to*, *ing* and zero

detail, it is for example worth noting the subtle semantic change undergone by the predicates *see*, *hear* and *feel*, depending on whether they are constructed with *ing* or with *that* complements: there is a switch from physical sense perception to some form of mediated knowledge state. In the case of the *that* column, the tense choices are variably constrained by the verb's particular semantics, and, in the case of 'object control' verbs require either the 'subjunctive' form of the present tense or the use of the modal auxiliary *should*. Speaker judgements may be variable with such cases, or may be known but regarded as obsolete.

The table is not exhaustive: it gives sample verbs and illustrations. It also uses syntactic criteria relating to relations between grammatical subject and object and cross-classifies these with four complement types. The aim, however, is to explore the conceptualisation differences between the cells. As can be seen, some verbs appear to allow for choice, and we shall investigate possible meaning differences among these choices of form. Some verbs appear to permit only one type of complement clause: we shall see how the semantics of such verbs determine the complement type.

8.2 The meaning of *that*, *to*, *ing* and zero

8.2.1 that

The meaning of *that* and *that*-ness was considered in some detail in Chapter 7. It is sufficient here to recall that *that* constructions communicate objectivised statements that are located in modalised worlds, and in the minds of other people. The degree to which *that* constructions are epistemically modalised is variable and can produce considerable complexity. This means that they contrast with *to*, *ing* and zero constructions, whose distribution is more restricted. The difference in conceptualisation effects is particularly clear when we consider verbs that admit two or more of the four complement clause possibilities; the alternations are not conceptually equivalent.

8.2.2 to

What does *to* mean in *to*-complement constructions? The proposal here follows what has already been said about the essentially spatial structures that underpin DST. The first step is to treat *to* as related to the semantics of the verb that patterns with it, not to the subordinate verb: it does not have to do with the 'full' infinitive (*to*-infinitive). The second step is to treat it as having meaning, and the third is to treat it as having a meaning, a meaning related to that of the verb that licenses it. The expectation is that this meaning will be highly abstract, as the detailed studies of Wierzbicka (1988) and Langacker (1991) already show. What they do not make particularly clear is why English, and some other languages (e.g. French), use a particular spatial preposition with particular verbs for particular complement clauses.

Taking a 'naïve' approach to the meaning of *to* in so-called *to*-infinitives, consider the following:

(3)   a   Mary can walk to her office in a half an hour
       b   Her office is to the south of the city centre
       c   It is close to the railway station, on the way to the piazza
       d   Mary moved the file from her desk top to the out-tray.

(4)   a   Mary appears to have written the report
       b   Mary intends to write the report
       c   John intends Mary to write the report
       d   John persuaded Mary to write the report.

Let us adopt the strategy of treating the word *to* in (4a) to (4d) as having meaning and as semantically related to (3a) to (3d). This means regarding *to* in the latter examples as akin to prepositions. It is certainly the case that there are differences between the spatial senses in (3), between these in (4) and between the different examples in (4). Following Tyler and Evans (2003) and Evans and Tyler (2004a), I shall assume these are not different senses of *to*, but that *to* has a highly abstract conceptual schema, which is spatially geometric and which is modified by the sentences it occurs in, specifically in the case of examples like (4) by the verb semantics.

In Chapter 1 it was noted that spatial prepositions can be described in part at least, though not entirely, in terms of geometric image schemas. The DST framework suggests that spatial conceptualisation is also relevant to a description of *to* clauses, though of course we are speaking here of abstract mental spaces. Such spaces can still be described in geometric terms. What is needed at this point is to outline the conceptual elements of the image-schema for the preposition *to*.

Tyler and Evans argue that the conceptual schema for *to* contributes concepts of orientation and goal, but not path, trajectory and motion. In this 'core' schema for *to*, the orientation is to a conceptually salient entity, the goal. How can this account of the primary spatial sense of *to* be related to the abstract sense that we find in the *to* constructions we are concerned with in this chapter? Clearly, the elements of orientation and goal are compatible with DST's use of the essentially spatial notion of vectors in reference frames – a vector has orientation or directionality, and coordinates that specify their goal, that is the position of their tip within a coordinate system, or reference frame. The spatial sense of *to*, in addition, implies a starting point – in DST terms a coordinate that specifies the position within a reference frame of the

8.2 The meaning of *that, to, ing* and zero

tail of a vector. In terms of physical space, directionality is not just a geometrical notion but is fundamentally an embodied concept, arising from human front–back orientation and from human movement from one location to another. Directionality is involved not only in physical locomotion but also in cognitive attention: one 'directs' one's attention *to*, or *towards* something. The 'attentional spotlight' of visual focusing is also describable as a vector in a reference frame, with tail anchored at the eye (or perhaps the self in a more general sense) as deictic centre. The notion of intention is even more abstract, but can also be described in terms of directedness and goal-orientation deictically anchored at the self and focused on some goal, an end point of a highly abstract vector that may be constituted by a highly complex entity. The etymology of both *attention* and *intention*, it may be noticed, involves interesting spatial conceptualisation: stretching, reaching or aiming toward some object – a point we have already seen Geach making in connection with intentionality in Chapter 6 above.

A 'goal' is also an embodied concept and involves more than a mere position vector: goals are human constructs – desires and needs. They are also representations of virtual not actual events or states of affairs – more specifically, representations of desired or needed states – coupled with motor planning that provides a schema for the path to the goal. The prototypical example would be the motor grasping schema, but while the grasping schema is about 'present' perceptible goals that are desired or needed, human cognition can also represent non-present goals – non-present in the sense that such goals may be beyond the peripersonal space. Indeed, goals may not even be perceptible, or may be intrinsically *im*perceptible. They are also inherently not present in the sense that the reaching of goals, present or remote, involves a time course projected into the deictic future.

Telicity is relevant to the conceptualisation prompted by *to* in sentences, in combination with particular verbs. In context *to* may emerge as telic or atelic: (3a) and (3d) are telic sentences but (3b) and (3c) are not. The core conceptual schema for *to* is likely to be highly abstract and may not strictly imply that goals are actually reached. Rather, discourse entities, including non-physical ones, may simply come under mental focus, that is, have attention *directed towards* them. In a similar fashion, *to*-complement clauses may represent realised or unrealised events: (4b) and (4c) are not telic, at least not telic in the sense that such a sentence does not convey communicate that Mary has, so far as the speaker knows, reached her goal of writing the report, though it is clearly conveyed that she had such an intentional 'orientation'. Sentences (4a) and (4d) are telic, at least in the sense that the speaker communicates that Mary has 'reached' the goal of writing the report. (How orientational *to* comes to be associated with the verb 'appear' is considered below.) In all these cases, whether the goal is understood as reached (or realised) has to do with the semantics of the main clause verb.[3]

The most general core schema appears then to be directedness. Even (3b), which says nothing about the orientation of *her office*, i.e. the way it is 'facing', has a directional component in discourse. Compare the following two sentences with one another:

(3)   b   Her office is to the south of the city centre
      b′ The city centre is to the north of her office.

Here the alternation is between attentional foci. In (3b) the office is the focus of attention ('theme' or 'topic') and is located relative to the city centre; in (3b′), it is the other way round. On the *d*-axis, *her office* in (3b) is 'closer' to S and *the city centre* is – conceptually – more distant; the reverse is the case for (3b′). The relation is noted by *to*, a vector representing attentional direction generated by the construction. In (3b), it points from the (attentionally 'closer') office to the city centre, in (3b′) from the (attentionally 'closer') city centre to the office.

How can we think of this schema? The most general point to emerge from the above discussion seems to be: abstract *to* expresses directedness toward a distal goal. This directedness is not just spatial but temporal and epistemic, similarly the distal goal. If we try to relate this back to physical spatial experience and its representation in geometrical form, it is natural to think in terms of vectors, if we allow for unspecified length. The abstract meaning is still vectorial: for vectors do not always have to be about displacement (that is, movement, physical or metaphorical) but may define a position of a point in relation to a reference frame (see Tyler and Evans 2003: 150–2), the 'locational sense' of *to*. Similarly, one might speak of 'disposition' when considering the sense of 'attitudinal' verbs that appear with *to*. A closely related term might be 'tendential' for some uses of the core schema of *to* (e.g. verbs such as *want to, hope to, intend to,...*). The most general schema appears to be purely relational: one thing is related to another by *to*, and *to* allows for attentional selection. To ground this idea further, it is worth noting that what we have outlined in the preceding paragraphs might be pointing us to the general concept of *deixis* itself, which can be seen as an image schema grounded in a cognitive–bodily structure. Deixis is simply pointing; pointing is a combination of gesture and (joint) attention (Tomasello 1999, 2008). I have suggested from the outset of this book that the geometrical concept of vectors is a well-grounded way of abstractly

---

[3] It is also worth noting examples such as *hostile to Mr Smith, to the contrary, to his astonishment, to the cheers of the populace*, which can be argued to be compatible with the present account, though this may require further abstraction of orientation to a mere relation.

modelling linguistic structures in a way that links concepts to embodied cognitive structures (image schemas).

There is, finally, an important consequence of treating the *to* constructions considered below (Section 8.3.3 and Chapter 9, Section 9.1) as consisting of a meaningful preposition-like morpheme *to* associated with the matrix verb followed by a complement clause containing an unmarked lexical verb. The verb appears as a 'bare infinitive', the verb stem without inflections. This will be regarded as associated with instancing conceptualisations of the kind discussed in Chapter 4. Instancing is conceptually connected with completeness, summation and abstraction.

8.2.3 What does ing mean?

Both Wierzbicka (1988: 59–97) and Langacker (1991: 442–5) insist that *ing* has a conceptual component relating to simultaneity. This approach seems to be entirely relevant but I believe it needs to be widened. In Chapter 4 it was proposed to introduce a cognitive 'presencing' operator to account for frequently commented conceptual effects of the so-called progressive tense form of English. The idea of 'presencing' involves more than progression along a time-line. Taking numerous accounts by linguists into consideration, *ing* (or the progressive present tense) communicates a quasi-spatial sense of 'closeness' to an event, or 'being present at' that event, attentional focus on an ongoing event, inclusion within peripersonal space and concurrent with peripersonal time, and consequent occlusion of start and end points. Chapter 4 made proposals for modelling this kind of conceptualisation within DST and the same geometrical schema will be explored when we consider how to model *ing*-complement clauses below.

8.2.4 Zero

There is a traditional distinction between bare infinitives (e.g. *swim*) and *to*-infinitives, also called 'full infinitives' (e.g. *wants to swim*). But this distinction leaves us with the problem of how *to* should be described and categorised. In most theories it is assumed to have no meaning and is called a 'particle' or a 'dummy' or a 'complementiser'. By contrast, as we have seen, it can be argued that *to* does have meaning and is in fact treated as a preposition attached to the grammatical frame of the main verb in a sentence. This seems to lead to a classification that is more economical and that is more coherent with an overall cognitive framework: there is no need to have the meaningless category of 'particle' or for that matter the doubtful distinction between bare and full infinitives. Instead I shall simply refer to the verb stem, that is, the morpheme (or combined morphemes) to which inflectional affixes are added.

An infinitive is thus simply an inflectionless verb stem and it is found in a small number of grammatical constructions. These are often said to include the occurrence of the uninflected verb stem after modal auxiliaries, e.g. *Mary may write the report*, but our main concern will be with the occurrence of the uninflected verb stem in complement clauses following verbs of perception and verbs linked with certain kinds of causation, e.g. *John watched Mary write the report*, *John made Mary write it*. The term 'infinitive', like the similar term 'non-finite', is not pointless, however. Inflectional affixes generally are connected with conceptualisations tying the event denoted by the verb stem to a relative time point, or duration, completeness, reality, irreality, etc. Verb stems are thus 'infinite' in the sense that they are maximally general, totalising and timeless.

This account of 'infinitives' should really make the term 'zero' constructions or complement inappropriate. Its use arises from the notion that in sentences like *John watched Mary write the report* there is no 'complementiser', whether *to* or *that* – whence 'zero' complementiser. In addition, it is odd to call *ing* a complementiser. I shall therefore abandon the idea of 'complementiser', but continue to use the term 'zero complement' or 'zero construction', because the term 'zero' usefully carries the notion that the verb stem in the clause complement is without the conceptual trappings carried by inflections and also contrasts with the meanings attached to *to* and *ing*.

But how exactly is the meaning of zero constructions to be described? Whereas *ing*, we have just argued, is linked with the cognitive presencing operator, zero is linked with the cognitive instancing operator, sketched out in Chapter 4. Instancing conceptualisations, it was argued, are of two kinds. On the one hand, they are conceptualisations of the *now* in peripersonal time and may be point-like. On the other, they are, precisely because they are like points and points geometrically speaking have no dimension, timeless. And, as many linguists and philosophers of language have noted, the English 'simple present' often communicates, depending on the verb type and context, either a timeless generalisation or an infinitesimally small and point-like *now*. What needs to be proposed here is that the uninflected verb stem ('bare infinitive', if you will) in zero constructions is also an instancing conceptualisation.

The sentence *John saw Mary write the report* necessarily implies that the writing is complete, which not the case in the alternate *John saw Mary writing the report*. The zero construction here clearly entails that the event is total and finished. Although the meaning of *write* includes that it is a temporally extended process, and seeing it is therefore also a temporally extended process, the effect of the zero construction is to instantise it, rendering it abstract, timeless and complete. The precise effect depends on the particular lexical verb that is in the zero complement: for example it may be the final point of a process that is conceptually focused under the influence of the instancing operator brought into play by the 'bare' verb stem. In Langacker's example (Langacker 1991: 442–4), *we saw the ship sink*, the zero complement may lead to a conceptualisation in which the instant of final immersion is focal. The general point here is that the zero complement, i.e. the uninflected verb form, has a primarily aspectual effect irrespective of the natural implications of what perception verbs denote. What I am claiming then is that, despite the morphological differences (the inflected structure of the simple present tense form versus the uninflected form of the 'bare infinitive'), they both, on an intuitive examination, appear to yield similar kinds of instancising conceptualisation. Despite the association of the zero construction with sense-perception verbs, there does not seem to me to be any salient effect of 'immediacy' compared with the presencing conceptualisations induced by *ing* constructions, and this is consistent with the occurrence of the zero construction with a group of causation-related verbs. There is perhaps, however, a sense of closeness of contact brought on by the perception verbs. Consistent with this, zero constructions are restricted to the formal syntactic category of object-control: the construction communicates direct contact with the entity denoted by the grammatical object, as well as communicating that this denoted entity is involved in the subordinate clause verb (as agent or undergoer in the above examples).

8.3 Constructions with the verb *seem*

This section takes a look at the so-called 'raising to subject' relationship, examples of which are displayed in Table 8.1. The standard examples of this syntactic relationship show that it differs from the other kinds of complementation illustrated in the table. The construction in question involves not only the complementisers discussed above (in particular, *to* and *that*) but also a particular matrix structure.

The first row of Table 8.1 shows an alternation between *to* and *that* constructions for a small sample of verbs that have to do with two kinds of epistemic meaning. The first kind has to do with mental impressions that S does not credit with total certainty (*seem*, etc.) and the second kind has to do with epistemic state changes (*turn out*, etc. have to do with events that are unexpected). While we might think of these verbs as taking a *that*-complement, it is more accurate to say that they are involved in an *it* + V+ *that* construction. They alternate with a construction in which the subject of the *that* clause is the subject of verbs such as *seem, appear, be likely, turn out, happen, prove*. Sentences such as (5) below alternate with 'raised' subject constructions such as the one illustrated in (6):

(5) It seems that Mary wrote the report

(6) Mary seems to have written the report.

Rather than treating (6) as 'derived from' (5), the two sentences can be regarded as exemplifying separate constructions, each with its own conceptual structure. Obviously they are lexically and thus semantically related to a large degree, but the hard question is how they are *conceptually* different and how to describe this difference. The specific challenge is to describe these separate meanings in terms of the DST geometry.

Continuing with examples (5) and (6), the idea of 'derivation' by a 'transformation' that 'raised' *Mary* from an 'underlying' structure *Mary wrote* to a 'surface' subject position in *Mary seems* was justified by syntacticians on narrowly conceived quasi-logical semantic criteria. It was thought that while *Mary wrote* makes sense, *Mary seems* does not. Consequently, as noted earlier, a syntactical rule was devised to move elements of the two sentences around. The analysis here is that, actually, structures like *John seems to, appears to, happens to*, etc. do make sense.

8.3.1 What is it?

To begin with, we are not concerned with the deictic or anaphoric pronoun *it*. This does not pose any particular conceptual problems different from those of the other pronouns of English. Other uses of *it* also can be subdivided and commonly are by linguists and others. These go by the names of: dummy and anticipatory. 'Dummy' *it* was the term used in versions of generative grammar that used transformations. The sentence *it seems Mary wrote the report* was said to contain a dummy *it*, which was devoid of meaning. Some other accounts treat the *it* here as anticipatory, i.e. anticipatory of the subordinate clause *Mary wrote the report*. The transformational account held also that the subject of the underlying (or 'deep') subordinate clause, here *Mary*, could be raised to give the surface form *Mary seems to have written the report*. The raising transformation was also held in some way to explain that in the latter sentence *Mary* appears as the grammatical subject of the verb *seem*, which was thought to be semantically and/or syntactically odd. So there are two questions to consider here: what is the sense of Mary seeming? And what is the sense of *it*? In cognitive linguistics these questions attract attention and need to be answered to maintain the consistency of the cognitive-linguistics principle that linguistic structure pairs with conceptualisation differences. That is, the conceptual difference between *It seems that Mary wrote the report* and *Mary seems to have written the report* needs to be described and explained.

8.3 Constructions with the verb *seem*

The so-called 'dummy' *it* is closely related to so-called 'dummy' *there* that is found in constructions such as: *there is a problem with this dummy theory*. I shall not here attempt to deal with *there* constructions, but the problems of characterising the meaning of *there*, assuming as we are that it is not semantically empty, are similar to those of *it*.[4] This is especially clear in the descriptive accounts of Bolinger (1977) elaborated and modified by Langacker (1991).

Bolinger (1977) relates *it* to the 'ambience' or 'all encompassing environment', a concept that he also uses to explicate the occurrence of *it* in expressions concerning weather conditions, and further notes:

[*it*] is a 'definite' nominal with almost the greatest possible generality of meaning, limited only in the sense that it is 'neuter' . . . It embraces weather, time, circumstance, whatever is obvious by the nature of reality or the implications of context. (Bolinger 1977: 84–5, cited by Langacker 1991: 365)

There are two key ideas here: ambience and 'what is obvious by the nature of reality'. It is not exactly clear what Bolinger had in mind, and Langacker's summary does not seem to me to advance greatly. But there are, I think, insights here that can be taken up in the DST architecture. Ambience is what is around S, thus peripersonal space. What is obvious in reality, is what is manifest, what is obviously real to S, i.e. cognitively 'close' and also in peripersonal space. This description also corresponds with the notion that *it* means generality. However, since we are speaking of a pronoun *it* is referential, and deictically referential in the sense that it picks out an area of reality 'close' to the speaker within which inferences and epistemic judgements can be made.

Given what has been said above, *it* and *that* contrast conceptually: the former is proximal to S, in fact within peripersonal space, while the latter is relatively distal. It has to be admitted that the appropriate modelling of *it* in DST is not entirely clear, since the geometric formalism deals in points and we have proposed (following Bolinger and Langacker) that *it*, as it occurs in the grammatical constructions we are concerned with, generally denotes an 'abstract location', a diffuse setting or 'maximal generality'. Yet, as Bolinger notes, *it* is definite, as are points. Perhaps we can simply say that modelling *it* as a discourse entity is no different than allowing DST to model mass nouns and abstract nouns as points on the *d*-axis. If so, then where should *it* be located? It is conceptually natural to locate it at the periphery of the peripersonal space of S – after all, *it* has an 'ambient' meaning. Ironically, if not entirely geometrically, this has some advantage, as will be seen.

---

[4] Translation equivalence between English and German bears this out also (Langacker 1991: 351): *es lebt ein Einhorn im Garten* is roughly the same as *there lives a unicorn in the garden* or *there's a unicorn living in the garden*.

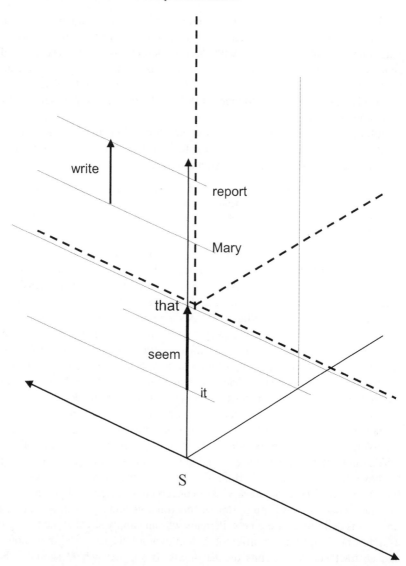

Figure 8.1 *It seems that* construction: (5) It seems that Mary wrote the report

8.3.2 *The* it seems that *construction*

We can model the *it* construction by setting up an embedded set of axes, R', taking it to be a metarepresentation of what S takes to be manifestly the case and which coincides with S's R axes. Figure 8.1 is a DSM for (5).

(5)   It seems that Mary wrote the report.

In line with the arguments outlined in the sections above, *it* and *that* are treated as discourse entities and therefore appear on the *d*-axis.

The reference frame R' is different from other translated reference frames examined in previous chapters: the origin of R' is not labelled as a virtual S', but as a virtual entity *that*. The implication is that R' is depersonalised – a relative distal object attended to by S. What S foregrounds, is *it seems*, where *it* may refer to whatever it is in S's environment that is the source of evidence supporting the use of *seem*. The configuration of the two reference frames makes the *write* vector (which is in the past relative to S's *now*) appear simultaneously in both R and R'. This seems to be required in order to capture the conceptual content of this kind of construction: S is presenting an assertion in the *that* reference frame: the *write* vector is thus located in a plane running through $m = 0$. At the same time the *write* vector is at $m = 0$ also in the base R reference frame – S regards the contents of the *that* clause as real. There is doubtless pragmatic variation in the exact way such constructions are conceptualised in context, and such variation may include degrees of epistemic distancing. However, it should be noted that if S wishes to indicate epistemic detachment, the verb in the complement clause itself would be modalised (*Mary might have written the report*, for example). It seems justifiable to configure the core meaning of the *it seems that* construction as in Figure 8.2 below.

8.3.3   The seem(s) to *construction*

Langacker (1991: 449–57) argues persuasively that in 'raised subject' constructions such as *Mary is likely to write the report* we do not have 'raising' at all but an independent construction that, essentially, thematises Mary as grammatical subject, while the alternate construction *it is likely that Mary will write the report* does not do this. We shall see how this point emerges naturally with the DST approach, but the main focus here is on how to model the semantics of *to*, in particular in connection with the semantics of the verb *seem*. The approach, once again, involves using the deictic space of DST and the transformation of reference frames in a way that is intended to correspond to the three-dimensional deictic conceptualisation associated with and activated by, one particular lexico-grammatical construction. This is an exploratory undertaking and begins with the 'naïve' approach to meaning of the constituents of constructions. Figure 8.2 is a possible DSM for sentence (6).

In line with Section 8.2.2 on *to* above, Figure 8.2 models *to* as part of the verb, rather than as part of a '*to*-infinitive' (*to have written*). The semantics of *seem* and *to* combine. The *seem to* construction is thus modelled as a vector

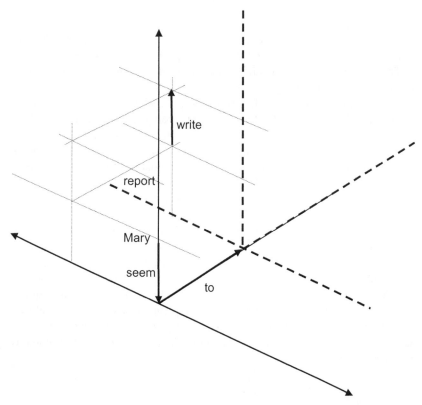

Figure 8.2 *Seem to* construction: (6) Mary seems to have written the report

relating *Mary* to a new reference frame R', depicted by the dashed coordinate system. The new axis system is a translated copy of S's view of the world. It has the position it has in this space, shown in Figure 8.2, for the following reasons.

Its origin is at $d = 0$, since this still represents S's viewpoint: it is still S's view that Mary has written the report, not someone else's belief or a something that is counterfactual for S.

This means that the coordinates for *Mary* and *report*, which are labelled as real discourse entities for S in R, are aligned with those of R' – they exist in both the base world and in the metarepresented *seem*-world. The origin of R' is also temporally aligned: the event of writing happened, from S's point of view, at the same temporal distance in both worlds.

However, on the *m*-axis of R, Figure 8.2 is shifted. In fact, this is an arbitrary choice for the diagram, because R' could, it seems to me, be positioned at varying degrees of epistemic distance from S. The reason for this is that *seem to* is not itself an epistemic modal, as is often assumed, but something more like an evidential. The semantics of *seem* have to do with highly general evidentiality rather than epistemic certainty or uncertainty. Different epistemic judgements on what 'seems' are possible. While the default seems to be some degree of certainty, phonetic emphasis gives an uncertainty meaning: 'Mary *seems* to have written the report (but she might not have done/did not).' In Figure 8.2, for the sake of illustration, the DSM shows a condition in which S expresses communicates the following conceptualisation: there is an evidentiality relation between Mary and her having written the report. At the same time, while communicating that there is evidence, for him, that Mary has written the report, S is modelled in Figure 8.2 as entertaining an epistemic judgement of uncertainty either way as to the reality of the case. This is not linguistically expressed. A hearer may simply be unsure of S's actual epistemic state: 'S says Mary seems to have written it, but does he actually think she did?' I am suggesting that *seem to* is evidential and that evidentiality and epistemic judgement can co-vary.

According to the diagram, then, Mary is foregrounded ('closer') on the *d*-axis, which matches Langacker's analysis, according to which Mary would be his equivalent of foregrounded (viz. trajector). This model also makes clear that the complement clause is a metarepresentation anchored at S. Figure 8.2 is consistent with another interesting fact about the construction that *seem* is associated with:

(7)   Mary seems to me to have written the report.

The PP *to me* indicates that *seem to* constructions involve a mind receiving evidential information from a source, here Mary. How is this modelled in the appropriate DSM? As it happens, it is already implicit in Figure 8.2. The *seems to* vector that set up R' is a vector pointing to the origin of the embedded reference frame, whose coordinate is that of S (i.e. S = 0) on the *d*-axis of R.

8.4   Further notes on seeming

There is, then, intuitively, a meaning distinction of some kind between (5) and (6) and the proposed geometric DSMs are an attempt to probe what this may be. This seems to have to do with two factors. First, in (5) we have *that* relatively more foregrounded attentionally than *Mary*, while in (6) it is *Mary* that is relatively more foregrounded. Foregrounding a *that* space focuses attention on an impersonally anchored space. Second, taking (6) we have a cognitive effect produced not only by the relative foregrounding of a person (*Mary*), but also by locating R' as in the consciousness of S, i.e. in S's base reference frame R. But most significant perhaps is the epistemic distancing effect of *to*, which translates R' to some point, say epistemic midpoint, on the m-axis.

The predicates like *seem to* that take grammatical subjects are not semantically strange and thus in need of explanation via transformational rule. One of the literal meanings of the verb *appear* points towards the concept of visual manifestness. But the others in this group can be understood in a similar way. The construction in question merely treats the grammatical subject either as (i) the source of evidence for the embedded metarepresentation in the complement clause in the case of *seems to* and *appear to*, or as (ii) the source of evidence that S perceived over a time course, in the case of *turn out*, *happen* and *prove to*. The latter would require more complex modelling on the t-axis. Their meaning involves a representation in which there is a prior state that is fulfilled or contradicted in some sense by the current state of the world, as conceived by S. In principle it should be possible to model this kind of representation in DST, but I shall not pursue the issue here.

9 Verbs, complements and their conceptual effects

The last chapter introduced a *to* construction associated with the verb *seem* and similar verbs. But the *to* construction occurs in different forms with other verbs too. In this chapter I look at these constructions and contrast them with the *ing* and zero constructions. The aim, as before, is seek a geometric representation within the terms of DST that is capable of modelling some at least of the conceptual effects of these different constructions.

9.1   *to* constructions and grammatical subjects

As can be seen from Table 8.1 in Chapter 8, *to* constructions appear in four subtypes that have been distinguished on purely (it has been supposed) syntactic criteria. However, following the usual assumption of cognitive approaches to grammar – that grammatical constructions reflect conceptual properties and choices – I shall examine each of the syntactic subtypes of the *to* construction within a cognitive framework, specifically that of DST. I shall deal first with *to* constructions that have been subdivided because of syntactic phenomena relating to their grammatical subjects and second with the *to* constructions that have been separated because of syntactic phenomena relating to grammatical objects.

9.1.1   *Subject-control structure (equi NP, subject):* want, *etc.*

Subject-control verbs are a syntactically defined category in which the grammatical subject of a verb in the matrix clauses is also the subject of another verb in a complement clause. The following is a sample of such verbs, limiting our focus for the moment to verbs combining with the *to* construction (see Table 8.1 in Chapter 8):

> *want, like, hate, long, intend, try, remember, forget, learn, fail, refuse, decline, consent, pretend, manage, plan, decide, expect, hope, presume, promise, vow, agree*
> *start, begin, commence, cease, continue, get.*

Some of these verbs appear not only in the subject control (equi NP, subject) row but also in the raising to object (matrix coding) row: *presume, expect, intend, want, hate, get*. We can say, for example, both *John expected to write the report* (subject control) and *John expected Mary to write the report* (raising to object). The other verbs just listed do not allow this: we can say for example *John believes Mary to write reports* but not *John believes to write reports*. Conversely, we can say *John pretends to write reports* but not *John pretends Mary to write reports*. The reason for the dual possibility for a subset of verbs, compared to the restricted possibility for the other verbs, may have to do with conceptual compatibilities between the semantics of the individual verbs and the meaning of the *to* construction. The meaning of one and the same verb – consider, for example *presume* – also seems to vary depending on whether it is in a construction with an object or a construction with only a subject. It is not possible to investigate all the complexities involved here in a systematic fashion; the principal aim of this section is merely to outline a way in which the meaning of the *to* construction can be modelled in terms of DST.

It can also be seen that the verbs sampled in Table 2 seem to fall into two broad semantic categories. The first – the verbs just listed above – clearly have to do with mental activity, including, volition, intention and (in a broad sense) their negation, directed, so to speak, toward as yet unrealised activities. Another group seems at first glance to do with external rather than internal mental activity and consists of verbs that have been considered aspectual:

start, begin, ?commence, cease, continue, get,*finish, *stop, *keep.

The starred examples, although they are appear to be aspectual, do not, however, occur with the *to* construction. They do occur with the *ing* complement construction. Again, the reasons for such a distribution are conceptual and the motivation for it will be considered briefly below.

Speaking very generally, the conceptual basis for most of these verbs reflects the mental state that English calls planning, which implies desire, volition and intention. All these mental conditions are inherently directional, and deictic, in so far as they imply a deictic anchor oriented towards some target that is non-actual, or irrealis. Such events may be mental representations of unrealised events or of future-oriented events that are inherently unrealised.

Given that the most abstract schema for the preposition *to* is orientation toward a goal that is not necessarily reached, it is scarcely surprising that verbs with the relevant semantics – such as those illustrated above – are apt to associate with the abstract *to* clause complement. This requires we no longer think in terms of infinitive forms. Thus, for example, in

(1)   John expects to write the report,

9.1 *to* constructions and grammatical subjects

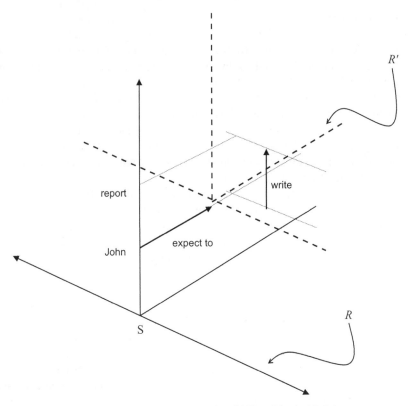

Figure 9.1 Subject-control structure (equi NP, subject): (1) John expects to write the report

we are not dealing with *expect* plus a *to*-infinitive *to write*; rather, we are dealing with a semantic whole, viz. *expect to*, complemented by a 'bare infinitive' construction *to write the report*.

All the verbs concerned involve the representation of another mind, so far as S's viewpoint is concerned. The verb *want*, and the other verbs listed, are in fact making a public claim by S about the contents of another mind – in this case about John. Figure 9.1 attempts to show how the reference frame and vector geometry of DST might model this kind of conceptualisation, as illustrated in (1).

As in the other cases in this chapter and in Chapter 6, S is making a claim about the contents of another person's mind. In the DSM given in Figure 9.1, S conceptualises two entities as real, *John* and *report*. It is also the case that for S John has a certain mental state, R', the contents of which S regards as epistemically uncertain (hence the origin of R' is at S's midpoint on the *m*-axis). This is similar to the configuration for *believe that* (see Chapter 7). Further, the DSM is trying to model some of the semantic structure of the verb *expect*. I shall assume for now that the state of mind denoted in English by the verb *expect* contains an epistemic concept of high probability. An expectation that something will take place is not absolute certainty, but an epistemic assessment of high probability, though not of course conceived as a mathematical probability. An approximate indication of this element of the verb *expect* is meant to be captured in Figure 9.1 by the positioning of the event vector for *write* at a high probability point on the embedded *m*-axis of John's 'expectatory' world, the reference frame designated R'.

Putting this slightly differently and in more detail, from S's point of view, the *expect* vector is a position vector that anchors R' at *John* and represents the claim that John has this mental reference frame at the time of speaking. It is possible also to think of the *expect* vector here as a kind of 'attentional beam' from John to a particular world (R') within which an event *write* is represented as highly probable at some time later than now. In John's expectatory representation of the event of report writing, John positions the event on his *m*-axis at a high probability point. That is a key part of the meaning of *expect*. It should also be noted that the *write* vector has no time extension; it is, in the terms of Chapter 4, a manifestation of a cognitive instancing operator – akin to the simple present, and applied here to a process verb – which means that the writing event is conceptualised as a complete abstracted event. Finally, note that R' and its contents are judged epistemically uncertain in S's base reference frame – S is not committed to the truth of what S claims is contained in John's expectatory mental state.

9.1.2 to *constructions and grammatical objects*

The criterion generally used to distinguish raising-to-object verbs from object-control verbs states that the objects of raising verbs are not semantic arguments of such verbs while those of control verbs are. For instance, *John persuades Mary to write the report* seems to imply that John exerts some force (in this case verbal force) on Mary, whereas in the case of *John expected Mary to write the report* it is not so easy to understand a force to be impacting Mary. This is the essence of the criterion. Consequently, the raising-to-object verbs and the object-control (equi NP, object) verbs appear to be in complementary distribution, and that is how they appear in Table 8.1. However, the fact that both categories allow the *to* construction suggests that the criterion may not be wholly relevant. There could, for instance, be a more abstract conceptual relation between verb and grammatical object that accounts for the

9.1 *to* constructions and grammatical subjects

possibility of noun phrases appearing as apparently inappropriate objects of the matrix clause while 'semantically' being subject of the verb in the *to* clause. What follows is a brief exploration of some aspects of this hypothesis formulated in the DST framework.

I shall be suggesting that any proposal to separate object-control constructions (equi NP, object) from raising constructions on semantic grounds is questionable and proposing instead that a cognitive approach along DST lines can treat all grammatical objects in *to* constructions as being conceptually meaningful. This can be seen as a consequence of treating *to* in such constructions also as meaningful. However, I shall look at the two formal syntactic categories (raising-to-object and control/equi NP constructions) separately, taking so-named object-control (equi-NP relating to object) constructions first, and raising-to-object constructions second.

9.1.2.1 *to constructions: object-control structure (equi NP, object)*

In formal syntactic accounts of verbs like those listed in Table 8.1 for object-control *to* constructions are acknowledged to have two semantic roles. Thus in (2)

(2)   John urged Mary to write the report,

it is intuitively comfortable to understand the grammatical object *Mary* as both the undergoer of the verbal act of urging and the doer of the act of writing. It is even more comfortable for the verbs *cause* and *force*, and also plausible for verbs such as *persuade* and *beg* listed in Table 8.1. In some sense, then, a sentence such as (2) involves a cause-and-effect schema. The main verb is a kind of causal force exerted on an entity and the entity is in turn affected in some fashion – not necessarily within the represented physical world, but within the overall conceptual configuration communicated by S. Thus, in the case of *persuade*, Mary, as conceptualised, does not actually undergo a physical change if someone persuades her to do something, but her conceptual representation does enter into another conceptual reference frame in which she performs an action. In the case of a verb such as *urge*, on the other hand, no real caused result is presupposed in S's conceptual configuration. We start by looking at *urge to* in Figure 9.2.

Given the semantics of *urge*, what is communicated does not presuppose that John was successful – that Mary did indeed write the report. At some time $t < 0$, that is before S's speaking time, S is claiming that John engaged in the verbal act of urging Mary 'toward' an action that he, John, was mentally representing as really taking place at some time later than his verbal act of urging. However, in using the verb *urge*, S is not committing himself to this reality: that is, S's use of *urge* does not entail that S represents that the effect of urging Mary, the actual writing of the report, had taken place at any time up to the time of S's speaking. For this reason, the whole reference frame

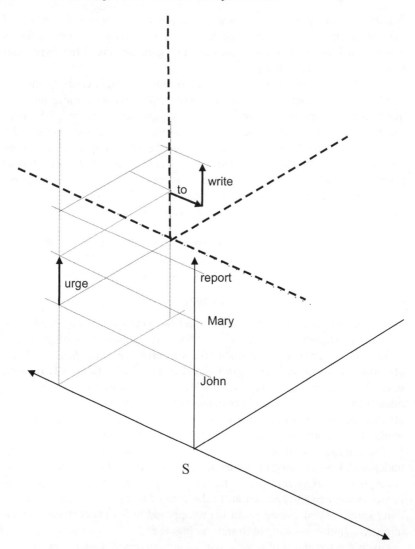

Figure 9.2 Object-control structure (equi NP, object): (2) John urged Mary to write the report

representing John's mental state, the mental state in which he represents the effect of his urging, is located by S at the point of epistemic uncertainty, the midpoint on S's m-axis.

John's urging of Mary is a kind of force applied to Mary, and is modelled by the *urge* vector; in turn this brings about the mental movement of the

9.1 *to* constructions and grammatical subjects

discourse entity *Mary* into the mental space R' that S attributes to *John*.[1] This means that *Mary* has a coordinate on *John*'s axis d', which makes *Mary* the source of the action *write* – not at the time of S's speaking but at some time in the future relative to John's act of urging. The important point here is that *Mary* is both the point of impact of the *urge* vector and the point of departure for the *write* vector – translated into conventional linguistic terms *Mary* has two semantic roles, one as the undergoer of urging and another as the doer of the writing. This seems to be a natural way of representing the semantic intuitions in geometrical, specifically vector format.

Consider again the different reference frames, that is the base reference frame R and the reference frame for John's mental state R'. John represents this future effect as real (i.e. at $m' = 0$). The semantics of *urge* (and *ask*, *request*, *beg* and the like) implies that the urger is confident of the effect (in urging, one intends and envisages a future effect), but the speaker using the verb *urge* (as opposed to persuade, for example) does not make an epistemic commitment to the effect of urging having taken place up to the time of speaking. Other verbs, however, in particular *cause*, *force* and *persuade*, presuppose that the action in the *to*-complement clause is fulfilled. This means that verbs in this category (the cell in Table 8.1 that intersects *to* constructions and object-control verbs) fall into two types according to whether they presuppose or not the reality of the event in the *to* complement. DST can model this distinction. Taking *persuade* as an example, Figure 9.3 is an attempt to model (3).

(3)   John persuaded Mary to write the report.

Comparing the DSM (Figure 9.3) for a *persuade to* construction with Figure 9.2 for an *urge to* construction, the crucial difference concerns the position of the embedded reference frame R' standing for John's mental state at some time $t_i < 0$. Why is it necessary to represent John's mental state, for after all *persuade* is a real action for S that has a real effect in S's frame of reference? The answer is that at time t_i John is envisaging a future time $t_j < 0$ at which Mary will write the report, according to his thinking. This is implicit in the semantics of *persuade*: John engages in verbal acts oriented to causing Mary to perform particular future acts of report writing; as with *urge* this act is modelled

---

[1] The reader may think at first glance that the *urge* vector in Figure 9.2 does not connect with the *to* vector, in contrast with Figure 9.3. What perhaps needs to be made clear for this and similar diagrams is that the coordinates of the Cartesian space are points on the three axes that apply throughout (unless otherwise specified for particular cognitive modelling reasons) the three-dimensional space, as indicated by the faint dotted reference lines. The relevant consequence for the present case and Figures 9.2 and 9.3 is that the *urge* vector, for example is understood as 'impacting' the coordinate for *Mary* in R and in R': the faint reference lines make this clear. Consequently, the *urge* vector still impacts on *Mary*, even though the embedded R' is epistemically distanced by S from the d, t plane in R, the reality plane where S locates the real-for-S action of urging, and where the diagram locates therefore the *urge* vector.

236   Verbs, complements and their conceptual effects

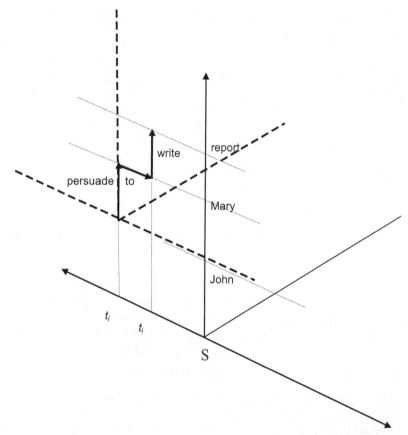

Figure 9.3 Example (3) John persuaded Mary to write the report

as being entertained by John as epistemically certain within his mental world R'. Now this outcome is also represented by S in S's reference frame R as epistemically real, having occurred in the past relative to S's now, the time of speaking. These considerations account for the positioning of R' in Figure 9.3.

The point of modelling (3) in this fashion is that it results in the event of Mary's writing of the report being real and true in both S's and John's frames of reference – capturing the presupposition of (3) to the effect that the persuading was successful.

9.1.2.2 to *constructions: raising to object verbs (matrix coding)*
While it can be made plausible to model the object-control verbs in terms of abstract force, as we have attempted to do in the last section, the so-named raising verbs seem less amenable. Sentences like (4) below, with verbs like

9.1 *to* constructions and grammatical subjects

expect, have caused puzzlement to linguists because it is not obvious that the grammatical object of *expect*, namely *Mary*, can be given the usual sort of meaning, call it 'undergoer', given to grammatical objects.

(4)   John expects Mary to write the report

In other words, the long-standing problem in linguistics is how to account for – how to model – the apparent dual grammatical and semantic function of the grammatical object, in this example, the NP *Mary*. This NP is certainly understood to be, on a conceptual level, the doer of the writing, although *Mary* is not the formal grammatical subject. At the same time *Mary* is certainly the formal grammatical object of *expect*, but is it the conceptual undergoer of the verb *expect*? It was supposed that the NP could not have two semantic roles (Chomsky's 'theta criterion') but it is legitimate to ask why not, in particular if a cognitive approach is adopted in which complex conceptualisations are perfectly admissible. In transformational–generative grammar the puzzle was resolved by postulating two levels of structure, one that made the doer of the writing the grammatical subject of *write* and one that rearranged the words so that such a noun (here *Mary*) was moved 'up' the tree structure to 'become' the grammatical object of *expect*. The approach outlined by Wierzbicka (1988) and Langcker (1991) tries to avoid the 'raising' machinery by giving a conceptualist explanation for the apparently odd appearance of *Mary* as a grammatical object when she is semantically a doer (typically therefore a grammatical subject). I shall do the same but end up with a different analysis based on following through the implications of the DST geometry developed so far.

Figure 9.4, like Figure 9.1 (modelling equi NP/subject control with *to*), has a reference frame R' for the mental state that S attributes to John – in which according to S *John* represents a high-probability event of writing the report. Following what the geometrical approach seems to suggest for the word order of (4), there is a vector representing *expect* oriented to *Mary* and a vector representing *to*, which reflects the prepositional meaning schema discussed in Chapter 8, Section 8.2.2. How do we understand the vector in this case? The *expect* vector can be understood as a kind of force vector causing *Mary*, who is in S's reference frame but outside John's, to quasi-move, in S's overall reference frame R, towards or to what S claims to be John's expectatory mind state, viz. the reference frame R', which has its origin at *John*'s d-coordinate.

This, one may well think, is a strangely literal-minded way of modelling this kind of *to* construction. But it is consistent with the geometrical modelling of other constructions, and is cognitively plausible, I suggest, provided one keeps in mind a *metaphorical* understanding of the usual interpretation of vectors in the elementary mathematics and mechanics. If that is done, then it

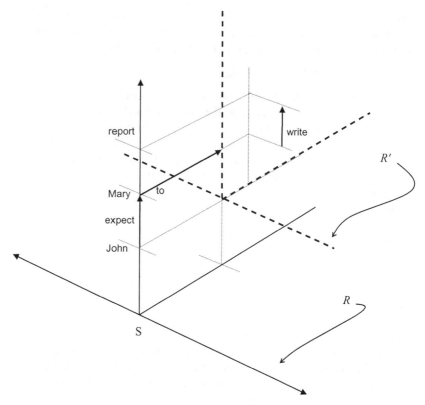

Figure 9.4 Raising to object: (4) John expects Mary to write the report

is possible to outline a self-consistent metaphorical model of the meaning of lexical and grammatical integration that is grounded in embodied conceptualisation. It is worth noting that one conventionalised meaning of *expect* in English does indeed imply a force concept, specifically social force. One communicable meaning of (4) involves deontic force. In such a meaning, S is communicating that John employs some sort of pressure or force on Mary, with a conventional presumption concerning a relevant normative frame (see Chapter 10) on deontic meanings in DST. The same seems to be the case for *want* and *intend*.

Verbs such as *allow*, permit, forbid, which are included in Table 8.1 as raising-to-object verbs on syntactic grounds (by the passivisation test), seem to me to be semantically unclear, if one asks whether their grammatical objects are also their semantic 'object'. If John allows Mary to write the report, does he do something to Mary or not? Many linguists would probably say not, but it is at least arguable that here too there is some kind of social force in play in our mental workings. Such considerations are another reason why it seems to make sense to explore the possibility that the so-named raising-to-object and the object-control categories can be merged in an overarching cognitive perspective.

The drift of the geometric modelling, then, is that other raising-to-object verbs, e.g. *know, believe, consider*, also express a kind of conceptual force, albeit of a highly abstract kind that relates to epistemic 'movement' from one frame of reference to another, from the cognitive point of view of S. So the vector for *expect* (in its simple epistemic meaning or in its social force meaning), and also any vector for *know, believe, consider*, etc., can be thought of as a force vector, standing for particular kind of 'mental contact' (a notion used by Langacker 1991: 91), or better attention, and having the effect of relating Mary *to* the act of writing the report. Note that attention is a directional notion and that it also involves volition and thus directed energy or force of some sort whose target, in this example, is Mary.[2]

9.2 Modelling *ing* constructions

The subject-to-subject raising alternation does not work with the *ing* construction. For instance one cannot convert *It seems that Mary is writing the report* into *Mary seems writing the report* – though *Mary seems to be writing the report* is fine, and we have outlined a DST model for this in the last section. And above it was argued that *ing* sentences that look superficially like raising-to-object sentences are better analysed as object-control sentences in which the grammatical object of the matrix verb has a dual semantic role. In this section, therefore, the cases we consider are *ing* constructions that are analogous to control constructions involving *to* complements (see Table 8.1).

In constructing DSMs for *ing* sentences, it will certainly be the case that some metarepresentation is involved and that this can be modelled by means of an embedded coordinate system – a reference frame. For instance, if we consider sentences like 'Odysseus imagined returning home', 'John considered writing the report', 'Jo remembered locking the door' and the like, some speaker S is making a truth claim both about the kind of mental state another person is in (say, imagining as opposed to remembering) and about what they are representing (say, going home). A further question will be where such a metarepresentational set of coordinates should be located. In the

---

[2] Dixon (1984: 589–90, cited in Wierzbicka 1988: 45) distinguishes two kinds of *to* construction, one relating to the *expect x to* type, and the other to the *know x to* type. Wierzbicka (1988: 45–6) disagrees and considers all cases of *to*-complement clause to be abstractly related, and offers explanations of certain semantic differences in terms of 'distancing' effects and different kinds of knowledge and judgement. I agree that there is a unifying account of *to* construction but suspect that Wierzbicka's analysis of semantic reflects pragmatic factors.

case of *to* constructions they are modelled as positioned at a point of epistemic certainty (e.g. in the case of the verb *persuade*) or uncertainty (e.g. in the case of *urge*) on S's *m*-axis; in the case of *that* constructions, the metarepresentational set of coordinates, they are modelled as occurring variably along S's *m*-axis, depending on S's epistemic judgement (or truth claim) about the state of the world (Chapter 7).

In outlining possible DSMs for *ing* complements the general principle is the same as for *that* and *to* constructions: the morpheme is taken to have a meaning rather than being arbitrary. It has already been argued (Chapter 8, Section 8.2.3), that *ing*, wherever it occurs, performs a cognitive presencing operation. In practice this will mean inserting into DSMs the kind of schema presented in Chapter 4 (Section 4.2.2) in discussing the modelling of the English progressive verb forms. In this chapter the hypothesis is that the same conceptual schema is activated whether we experience a verbal tense form or a present participle or what is traditionally called a 'gerund'. Here we consider the gerund manifestation of *ing* in sentences such as *John liked writing reports*, *John saw Mary writing the report*, *John stopped Mary writing the report*. What is striking from a semantic point of view about this category – the verbs listed in Table 8.1 in the column for *ing* constructions – is that the verb meanings are of three main kinds: emotion (e.g. *like singing*), sense perception (*see Mary writing*), sensory representation (*imagine Mary writing*) and aspectual meanings (starting, continuing, ceasing, etc.). The question is how to make unified sense of these various meanings in *ing* constructions, and of course how to develop a DST-style geometric modelling.

9.2.1   ing *constructions and grammatical subjects: subject-control structure (equi NP)*

There seem to be at least three semantic subgroups of sentences where the semantic agent (the grammatical subject) of the matrix verb is also the semantic agent of the complement clause (see Table 8.1). Some of these are able to take *to* complements as well as *ing* complements. We shall begin by looking at *imagine*:

(5)   John imagined writing the report.

(6)   *John imagined to write the report.

(5) and (6) illustrate the reverse of *expect*, which does take *to* but does not take *ing*. Figure 9.5 modelling (5) may be compared with Figure 9.1, which models *expect*. What they share is the embedding of a mental reference frame and a vector anchoring this embedded coordinate system to *John*, making *John* a secondary represented subject S' in S's base reference system. The crucial difference lies in the way the meaning of the verb in the complement clause is represented.

9.2 Modelling *ing* constructions

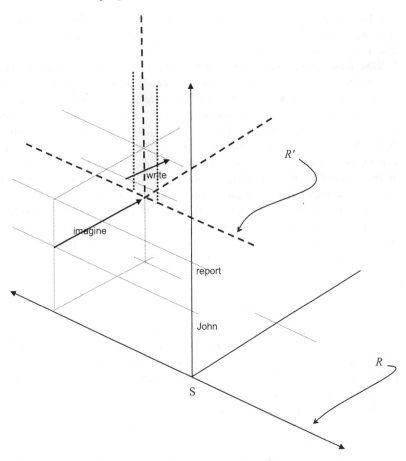

Figure 9.5 Example (5) John imagined writing the report

The *ing* morpheme is modelled here exactly as it was when investigating the present progressive tense form in Chapter 5. Here it is used to model an ongoing process of *writing*. The conceptual frame for *write* has a start and end point but the *ing* morpheme is defined as a presencing operator that brings the subject (here S′, *John*) 'into' the event. John is imagining himself in the process of writing, without regard to his starting or finishing. Thus there is a frame of reference R' set up by the verb *imagine* itself, which also has the same effect; this is the space in which John is imagining something. It is of course S who uses the verb *imagine*, and thus S communicates that from S's point of view there is no epistemic certainty: consequently R' is anchored, as shown, at S's midpoint on his *m*-axis. For S, according to sentence (5) John's imagining is in the past, but for John at that time the imagined representation – the picture, if there is one in John's mind – is in the present.

The DSM makes this clear. The crucial difference between Figure 9.5 modelling (5) and Figure 9.1, then, lies in the content of R', namely the different cognitive schemas for *to* and *ing*.

This now gives some clue to why *expect* and *imagine* take different constructions exclusively. The intrinsic semantics of the verb *expect* involve a mental representation that is oriented to the future, not surprisingly, since *to* entails the path image schema, and path is a conceptual metaphor implicated in the representation of time. The semantic frame for *imagine* is different. While it clearly involves a mental representation that need not, as also for *expect*, be endorsed by the speaker as real or true, it also involves a quasi-sensory, possibly pictorial, representation (as its etymology also would suggest). Sense impressions and pictures are experienced in the present – hence the suitability of the presencing operator *ing*. Of course, I can imagine doing something in the future, but the point is that, phenomenologically, the act of imagining brings an event framed as future into the present: the future is 'present' or 'presented'. This is what the DSM is trying to get at.

9.2.2   ing *constructions and grammatical objects*

In discussing the distribution of verbs in Table 8.1, it was argued that *ing* constructions be treated as 'control' (or 'equi NP') constructions, whether involving a grammatical subject or a grammatical object. The main ground for treating those involving grammatical objects as 'control' constructions rather than raising constructions was that the grammatical objects in question can be experienced as having a dual semantic role. In sentence (8) below, the noun *Mary* has the semantic role of agent for the verb *write*, as it does in (7). In (8) the noun *Mary* is not the affected participant with respect to the matrix verb *imagine* – at least, that claim could be made from a purely syntactic point of view, though from a conceptual–semantic point of view this is not so clear.

(7)   John imagined that Mary is writing the report

(8)   John imagined Mary writing the report

(9)   John imagined Mary to be writing the report.

The difference in meaning between (8) and (7) is striking, as the reader may verify subjectively. It seems that in (7) *Mary is writing the report* is held up as an objective proposition, whereas in (8) we are being told about some mental image in the mind of *John*. Sentence (7) can be modelled along the lines discussed for the verb *believe* (see Chapter 7, Sections 7.2.2 and 7.3.2 and Figures 7.7 and 7.8). Note, however, that *believe* does not behave exactly like *imagine*, in that it does not allow the *ing* construction: **John believes Mary writing the report*. What explains this is that the semantics of *imagine* is specialised for denoting 'present' visual mental images – and thus consistently allows the 'presencing' operator *ing* (see Chapter 4, Section 4.2.2). The verb *imagine* alternates with the *to* construction, as shown in sentence (9), a grammatical fact that can be explained as follows. The *to* construction picks out the non-real element in the semantics of *imagine*, because that construction is conceptually linked with orientation to goals that are not necessarily reached and thus potentially with what is unrealised and indeed unreal. Reading (9) seems not necessarily to evoke visual imagery in the mind of John, but merely some kind of erroneous belief (from S's viewpoint). This is consistent with what has been said about the abstract meanings of *to* in Chapter 8, Section 8.2.2, as well as in section 9.1.2 of the present chapter.

The immediate question is how (8) might be modelled in DST. In particular, any DST model needs to be able to reflect the dual semantic role of the grammatical object. Figure 9.6 has the same ingredients as Figure 9.5, which seeks to model the effect of the *to* construction verb *expect*. These two configurations are similar in so far as *expect* and *imagine* both involve S ascribing mental states to another mind, so they both have an embedded R'. They both also deal with the dual semantic role of *Mary* in (8) and (4), but Figure 9.6 for (8) requires some comment. The *imagine* vector in Figure 9.6 points to the *Mary* coordinate on S's *d*-axis, and the reference lines run through to the imagined *Mary* in *John*'s embedded mental representation R' – so the *imagine* vector is still pointing to *Mary* (on this see also footnote 1 of this chapter). The immediate and 'present' event of Mary writing is what John, according to S, imagines, so the *imagine* vector points direct to the coordinate for *Mary* in both 'worlds' – S's base reference frame and John's embedded reference frame. The *imagine* vector also serves to anchor R' – which is S's claim about the contents of John's mind – at S's real discourse entity *John*.

As noted above, there is a strong sensory component in the semantics of certain verbs that take *ing* such as *imagine, consider, contemplate, think about* and even *remember* and *forget*. It is not therefore surprising that primary perception verbs *see, hear, feel* take *ing* constructions but not *to* constructions. They can, however, take *that* constructions, as in *John saw that Mary was writing the report*. But this alternate results in a specific cognitive distancing effect, one that is clearer in *John heard that Mary was writing the report* and that leads to a shifted though still related meaning in *John felt that Mary was writing the report*.

9.3 Modelling zero constructions

We come now to the bottom left-hand corner of Table 8.1: verbs that allow what I have termed the zero complement, because the complement clause contains a verb in its simple root form, i.e. without affixes and thus often called a 'bare infinitive'. This means that the verb meaning appears without tense or aspect markers and their associated meanings. Although I have used

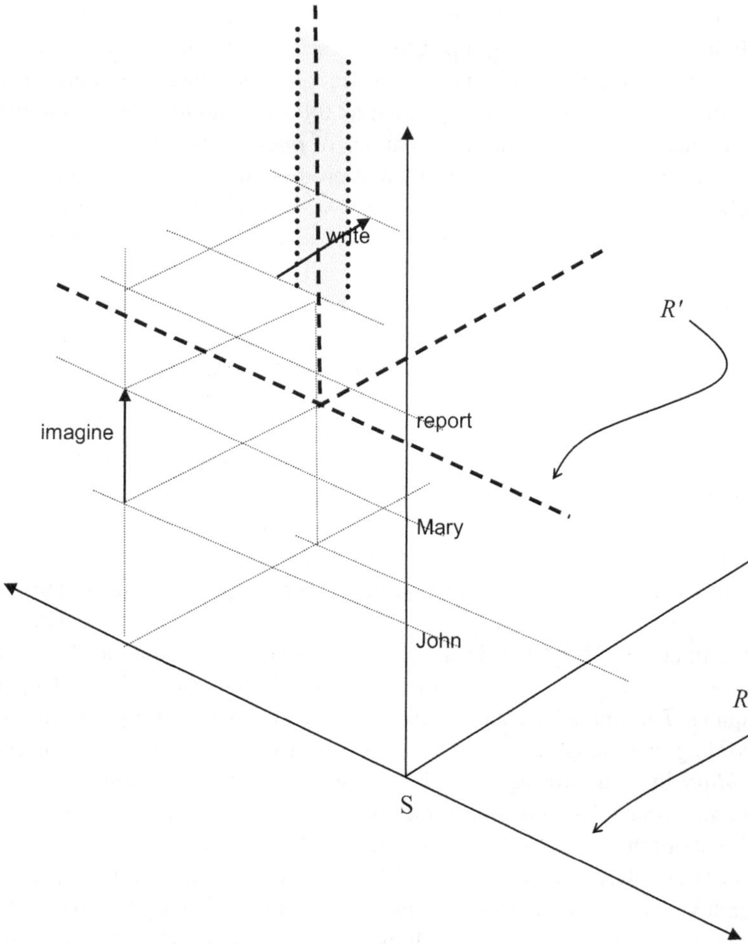

Figure 9.6 Example (8) John imagined Mary writing the report

the term 'zero complement', this is not to be taken as meaning there is a complementiser that is 'really' there but manifest as nothing in the surface morphology of sentences like (10).

The examples we shall look at are:

(10)   a John saw Mary write the report
       b *John saw Mary write the report but she never finished it.

(11)   a John saw Mary writing the report
       b John saw Mary writing the report but she never finished it.

9.3 Modelling zero constructions

The difference in meaning between these two sentences will be apparent. And this fact of intuitive language processing is evidence that the alternate constructions are not arbitrary, and more generally that grammatical constructions communicate abstract conceptual meaning. Within cognitive linguistics both the form of the verb and the overall grammatical shape of the zero complement construction have been treated as having meaningful motivation. Langacker's analysis of the sentences 'we saw the ship sink' and 'we saw the ship sinking' (Langacker 1991: 442–4) has abstract similarities with the DST account I will propose – it uses, for example, the notion of 'viewing frame' that has the effect of limiting and giving 'immediate reality' to some portion of a temporally extended process; Langacker also relates the *ing* construction with the semantics of the English progressive tense form and the zero construction with the simple present. DST differs in using reference frames and in its characterisation of the cognitive effects of the simple present form (and bare infinitive) and progressive tense form (and other *ing* forms).

In (10a) and (11a) we have the two main cognitive operators outlined in Chapters 4 and 5 illustrated. The workings of the presencing operator in the *ing* complement construction have been discussed in detail above. For (11a) a DSM will represent the verb *writing* in the same way as the diagram in Figure 9.6 modelling (8) – except for the labelling of the matrix verb vector as *see* (rather than *imagine*), and apart from the important question of whether or not the DSM should show the *see* vector within an embedded reference frame standing for John's view of the world. But the main issue here is how to establish a DSM for the zero complement construction in (10a).

What we have in (10a) is an effect of, or creates the effect of, the instancing operator, which was introduced in Chapter 4 to explain the conceptual effects of the English 'simple present'. It will be recalled that the instancing conceptualisation is timeless – in the sense that an event is experienced as timeless, or is timeless in the sense that it is generic and non-contingent. Unlike the presencing operator, which is essentially durative and occludes starts and finishes of events, the two conceptualisations associated with the simple present are summative. It is duration and process that are outside the picture, as is represented in the modelling of the simple present as point-like. It is clear from examples like (10b) and (11b) that the instancing conceptualisation is what is required to model a zero construction such as that illustrated in (10a), which presupposes that, in the speaker's perspective, the action of report-writing is completed. This is consistent, it is worth noting, with the high evidentiality and high epistemicity of visual perception in knowledge representation, and thus of the verb *see*. The verb *imagine* may well have an important visual element in its semantic frame, but more important is the fact that it implies low epistemicity: the midpoint of uncertainty or outright counterfactuality. The latter (a counterfactual understanding of an *imagine* sentence) is more likely with a *to* construction (*John imagined Mary to be writing the report*) and especially with a *that* construction (*John imagined that Mary was writing the report*), both of which have distancing effects.

The high epistemic validity given to seeing now needs to be considered in relation to the question whether an embedded reference frame is needed in the case of such verbs. Unlike *imagine, see* can take either *ing* or zero constructions. There may be two reasons for this. First, *imagine* inherently places the imagining subject 'close' to and present at the event, so an *ing* construction (thus presencing operation) is strongly induced. Second, it may, perhaps additionally, just be the case that the zero construction (which means the use of a tenseless simple present) in itself presupposes completeness of the denoted action, as is indeed implied if the simple present activates an instancing conceptualisation with the properties we have proposed.[3]

Taking this second point further, abstract completeness and truth are defined by the dimensionless ('vertical') vector for an action coinciding with time zero for the reference frame of some S (who might be an S', e.g. *John*). Given this definition, it seems that we do need an embedded reference frame. If we have such a frame, it can be positioned at any time t in S's base reference frame, so long as t is appropriately earlier later than $t = 0$. One further point is worth making. It is likely that the semantic frame for *seeing* does not merely denote perceptual modality but also has an epistemic element related to high epistemic confidence. Taken together, the foregoing remarks may be regarded as justifying the model in Figure 9.7, which treats S as attributing a reference frame representing his mental (and perceptual state), in the same way as for the *imagine* sentence (8) modelled in Figure 9.6. This also seems pertinent in the light of what was said earlier about the visual element in the semantics of *imagine*.

According to what Figure 9.7 makes explicit, S is certain that John saw something. At the same time (literally at the same time in the model), this visual experience attributed to John is part of John's mind state, attributed to him. John's seeing is understood by S not only as an external event; it is part of John's subjective experience (so he is also S'). The *see* vector is oriented to the *write* vector whose tail coordinate is *Mary*, that is, *Mary* is the initiator of *write*. For S, Mary's writing is an event in S's past; for John it is an event that is in his present, relative to the reference frame that S attributes to him and which is in the past relative to S. Note that *write* is tenseless: it is not identical with the simple present (which takes tense-marking morphemes) but shares its central semantic property, viz. that of activating the instancing operator.

---

[3] And this is irrespective of the action's temporal position relative to S's now: *John might see Mary write the report next Monday*. But note that, for quite natural reasons, *see* cannot coincide with S's now (?*John is seeing Mary write the report*, ?*John can see Mary write the report*).

9.3 Modelling zero constructions

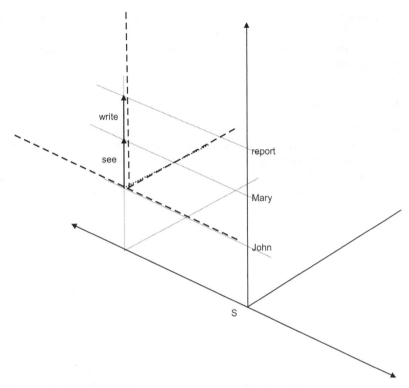

Figure 9.7 Example (10a) John saw Mary write the report

Of course, in the natural world the act of writing is certainly not instantaneous, but as has been said, the instancing conceptualisation is not about instantaneous events but about timelessness, which is about truth and certainty assertion. It is this that gives rise to the epistemic alignment, by S, of John's frame of reference (John's view of the world) with S's own.

The two verb vectors *see* and *write* require further comment. Both are in the base reality world of S, R'. Both *see* and *write* appear as instancing conceptualisations: the past tense form of *see* makes it appear punctual therefore complete, the tenseless *write* vector likewise, as already noted. What is the relationship between these two vectors? The *see* vector cannot be interpreted here as a force vector, in contrast with some cases considered earlier. But it does make sense to take the *see* vector as specifying position, or better, direction, specifically direction of *John*'s attention. Thus in Figure 9.7, *John* is oriented toward the coordinate, labelled in S's base world, as *Mary*. Further, *Mary* is the tail of the *write* vector, and thus is associated with two vectors. This is how DST accounts for the dual semantic role of the matrix grammatical object in the zero construction, as indeed has been seen in earlier sections (9.1 and 9.2) for *to* and *ing* constructions. It looks as if all questions regarding the dual semantic role of grammatical objects can be brought under one theoretical account.

9.4   Overview of alternations and restrictions

Let us return to Table 8.1. It is not my main purpose here to give an exhaustive and conclusive account of the distribution of verbs laid out in this table over the four construction categories zero, *to*, *ing* and *that* intersecting with the formal syntactic categories. Nonetheless the detailed analysis in terms of geometric DSMs does point towards some general observations. The overarching assumption seems to be borne out – that the distribution is not random and arbitrary but semantically, that is conceptually, motivated. However, a number of puzzles remain to be investigated and detailed models need to be worked out for particular verbs. Contextual input is often crucial and potential differences of meaning that appear when particular verbs alternate between complement constructions are not always evident in schematic and uncontextualised forms.

In the course of designing DSMs for the zero, *to*, *ing* and *that* constructions a number of points have emerged. Scanning Table 8.1 (Chapter 8 section 8.1) top–down, it was suggested that the 'raising to subject' category that coincides with the *to* construction and with a certain type of *that* construction be treated as *sui generis*. On the next row, the formal category of 'raising to object', could, it was suggested, be merged with the 'object-control' category. On the basis of the geometric and frame-of-reference investigation it was further suggested that there is an overarching cognitive and conceptual basis for this, provided the metaphorically spatial dimensions of the DST framework are kept in mind. Merging these two categories, we thus need to consider the two rows of Table 8.1.

What is of interest is the distribution of the verbs – that is their specific restriction to one of the zero, *to* and *ing* categories and the possibility of alternation between two or more.

9.4.1   Zero, ing *and* to *verbs: restrictions and alternations*

Zero complements – that is, 'bare infinitive' construction complements – appear not to occur in subject-control sentences. It is not possible to say for example, 'John wanted win the race'. The explanation may be that the zero construction prompts an instancing conceptualisation, which means completeness and thus reality in S's reference frame. Zero complements are therefore unlikely to occur with mental-activity verbs such as intend, want,

9.4 Overview of alternations and restrictions 249

hope, etc. One might think that the verb *remember* might be compatible but 'I remember write the report' is also impossible, perhaps because the semantics of the verb so strongly involve sensory images that they call for the close-up presencing conceptualisation activated by *ing*. The verbs allowing zero complements seem in fact to be restricted to those denoting perceptual contact with the external world or with cause-and-effect relations in the external world – and these occur in object-control structures, to which we now turn.

The sensory verbs *see*, *watch*, etc. alternate between zero and *ing* constructions, making alternate cognitive perspectives available. As predicted by the theory of instancing and presencing operators (Chapter 4), the sensory verbs communicate completeness and abstraction of the event perceived when combined with the zero construction (uninflected verb stem); when combined with the *ing* construction the sensory verbs communicate an event whose beginning and end are not attended to. Again as suggested by the theory of *that* complements in Chapter 7, the sensory verbs appear to have their meaning rendered abstract when they take *that* complements, shifting from sensory to mental perceptions, with the contents of the *that* clause emerging as a cognitively distal and perhaps public proposition endorsed by S. With regard to the combination of the sensory verbs with *ing*, it is worth noting that verbs like *imagine* and *remember* may fall into the same category because their semantics includes an element of visual imaging that is stimulus-independent.

However, other verbs (*make*, *have*, *let*, *help*) taking the zero construction do not denote sensory or imaginal experience, but are related to kinds of causation. The semantically related verbs *prevent* and *stop* are restricted to *ing*, the second of these two occurring also as a subject-control verb. The verbs *make*, *have* and *let* have to do with a generalised concept of causation, removal of countervailing force and enabling force. They occur uniquely with the zero complement and are thus viewed as unified and completed acts, without temporal duration, and are regarded as real. Langacker (1991: 444–5) is concerned with lack of temporal coincidence in 'I made/let/had him clean the garage.' However, DST escapes this problem, since temporal coincidence is not the key conceptual feature that defines the zero construction. Rather the key feature is the aspectual quality of the instancing operator – its summative or perfective aspect.

In the subject-control use of *stop*, e.g. *Mary stopped writing*, we might explain the restriction to *ing* by noting that *ing* is associated with close engagement in an activity (up to a certain point in time), which is implied by the construction. The verb stop is, precisely, about *dis*engagement from a process. When *prevent* and *stop* occur with grammatical objects, as in *John prevented Mary writing the report*, something similar may be at work. Their lexical meaning would suggest that English has specialised these verbs for communicating a concept of forceful blocking seen as immediate engagement in some ongoing action. If this is so, then it would be expected that *prevent* and *stop* take an *ing* complement because immediacy and engagement are properties of the presencing conceptualisation prompted by *ing*, and *ing* further denotes a situation in which a process is continuous relative to an observer (the speaker S). In fact, the conceptualisation varies depending on whether an accomplishment verb or an achievement verb is in the complement clause. With achievement verbs, e.g. *John prevented Mary reaching the summit*, the default understanding seems to be that Mary completed part of the journey but did not get to the top. With an accomplishment verb, however, it seems that the event denoted in the complement does not even begin, e.g. *John prevented Mary writing the report* – so we cannot explain *ing* on the basis of the blocking of an ongoing process. Both cases may perhaps be – somewhat speculatively – accounted for by the fact that immediacy and engagement concepts prompted by *ing* are sometimes purely mental, as seen most clearly in the case of *imagine*. It could then be argued that the restriction of *prevent* to *ing* complements has to do with the fact that in order to forestall an event the event first has to be mentally represented or imagined by the actor.

It is easier to explain the non-occurrence of *prevent* and *stop* in the *to* construction and the zero construction. We can at least say it is naturally expected that *prevent* and *stop* do not take the *to* construction, since in the case of verbs that do take *to* complements (e.g. *require, urge, expect, ...*) there is a goal-oriented force that is intended, desired or predicted by the subject to propel an entity towards (if not *to*) a goal rather than to block it. Similarly, *prevent* would not be expected to take the zero construction, which expresses the completeness and reality of the event denoted by the verb complement. This is corroborated by verbs similar in meaning to *help* (opposite in meaning to *prevent*), which alternates between the *to* and zero constructions but is less happy with *ing*.[4]

This alternation between zero and *to* constructions in the case of *help* (e.g. *John helped Mary write the report* versus *John helped Mary to write the report*) may be connected with subtle conceptual differences that are not always activated. There is no need to expect conceptual motivation of constructions to be always clear or activated. In line with what has already been said about zero and *to* constructions, the zero version of help sentences may favour completeness and truth in S's reference frame, while the *to* version of *help* sentences may favour epistemically or temporally more distal interpretations of the relation between the helping and the action, as well as bringing out the understanding of the action as intended and willed by its agent (the

---

[4] *John helped Mary writing the report* seems awkward but with a PP, *John helped Mary in writing the report* one senses again some concept of engaged closeness.

9.4 Overview of alternations and restrictions

grammatical object of the matrix clause that is also the subject of the complement clause). The restriction against *ing* sentences (one cannot say *John helped Mary writing the report*) may seem surprising, since helping appears to be about involvement, thus presence, in an event, and thus be related to stopping and preventing (which do take *ing*). A plausible explanation lies in the semantic frame of the verb *help* itself. Helping implies efficacity. That is, helping is scarcely helping if it does not result in some effect, and an effected result is not in the conceptualisation induced by *ing*, which inherently occluded end points.

9.4.2 Remarks on to and ing verbs: restrictions and alternations

Table 8.1 shows that *to* complements range across formal syntactic categories of raising to object, subject-control and object-control constructions. Many or all of the lexical verbs sampled here, and appearing in the *to* column in the table, can be argued to have semantic frames that include the concept of goal-orientation and since goals may be not only distant in space but also in time, they are strongly associated with epistemic uncertainty. Many can be seen to communicate some kind of force, whether physical or social, and I propose that there may be also a kind of mental 'force' (e.g. imagination) that 'moves' some entity into a mental reference frame of epistemic uncertainty. That physical-force verbs (e.g. the verb *force* itself) appear alongside social-force verbs (e.g. *order* someone to do something) and verbs of subjective mental effort (imagines someone to have written the report) is not surprising if some overarching abstract force concept is part of their semantics. But for reasons of space it is not possible to probe further here. Verbs in the *to* column in Table 8.1 are probably either goal-oriented in part of their lexical meaning or have a force component, or both.

The *ing* complement verbs in Table 8.1 range only over the traditional subject-control and object-control categories, appear to be less numerous, and to exclude a force component, as well as goal-orientation. We are of course talking here about their schematic lexical meaning, not about specific context effects in utterance. The restriction of certain verbs to either *to* or *ing*, and the possible alternation of certain other verbs between both types of complement, may be argued to reflect the semantic effects already proposed as being associated with *to* and *ing* constructions – goal-orientation and presencing conceptualisations respectively. If we may generalise, the verbs in the *ing* column can all be analysed as involving direct sensory perception or mental images, and/or subject presence at or in some time-extended process. The cases of alternation between *to* and *ing* deserve further brief comment because what looks at first glance like mere syntactic variation can be seen as switch of cognitive 'perspective'. Since each alternating verb has its own complex semantics, I shall briefly consider just two examples – *try* and *remember*. This permits us to touch also on restrictions relating to *that* complementation, since *remember* can take it while *try* cannot.

9.4.3   to *and* ing: *the case of* try

Sentences (12) and (13) below indicate that *try* lexicalises a concept that denotes activity, perhaps mental activity, that is internal to the actor denoted by the grammatical subject. Sentence (14) indicates that *try*, unlike verbs such as *force* and *urge*, does not conceptualise the kind of force that can be applied to another entity.[5]

(12)   John tried to write a novel

(13)   John tried writing a novel

(14)   *John tried Mary to write a novel.

There is also an intuitively perceptible difference in meaning between (12) and (13). Given what has been said about the conceptualisation associated with *to* complements, we should expect goal-orientation to be part of the meaning – and this seems to be the case. It was also noted (see Chapter 8, Section 8.2.2 and this chapter, Section 9.2.2) that *to* does not entail attainment of a goal. This is compatible with the fact that sentences with *try* such as (12) strongly implicate that John put in some effort towards achieving a goal but was not successful. It is not at all clear that the semantic schema for uncontextualised *try* necessarily presupposes failure of effort. This is also the case for sentences like (13), although the possibility of failure seems to be much more weakly implicated. In describing these semantic effects, I am relying on intuition: there seems no way to demonstrate these effects empirically in the current state of knowledge. English speakers may have varying kinds of conceptual experience in response to these forms. Assuming there is indeed a conceptual difference between reading (12) and (13), this is predictable in terms of what has been said already about *to*, and explicated further by the way in which we have described the effects of *ing*.

Viewed in this way the differences between (12) and (13) appear to arise from the selection of either *to* or *ing* (cf. Wierzbicka 1988 and Langacker 1991). However, it should be noted that matters could be the reverse, since *try* is polysemous and encodes two different (etymologically and perhaps

---

[5] We are of course talking here about a construction, the *try to* construction and not other uses of *try*, such as the transitive construction in: *John tried the car, Mary tried his patience, the court tried the thief*.

conceptually related) meanings. One meaning is saliently about subjective effort and goal-directed force, a meaning that only emerges when *try* takes a *to* complement. The other meaning emerges when *try* has a direct object NP in, for example, 'Tom tried the matsutake mushrooms', where effort is not salient but direct personal sensory experience is. This is reminiscent of the presencing operator, and it is therefore not surprising that the 'trying out' meaning of *try* selects, or is selected by, the *ing* complement construction.

Another group of verbs that alternate between *to* and *ing* are the 'aspectual verbs' *start, begin, commence, cease, continue*. It may be the case that the selection of *to* or *ing* yields a difference in meaning or 'perspective' that has to do with, on the one hand, subjective engagement with the action denoted in the complement (*ing*) construction or goal-oriented activity on the other (*to* construction). Another way of viewing the matter is to note that these verbs are compatible with either. The restriction of *finish, stop, keep* to the *ing* complement may arise from inherent semantics of the verb: there is no implied goal to orient to in the case of *finish* and *stop* (Wierzbicka 1988: 78). The verb *keep* has a quasi-synonym in *continue*, which does alternate between *to* and *ing*. The fact that *keep* is restricted to *ing* suggests it is specialised in English for focus on engagement in an ongoing activity, and this borne out by the fact that it is not happy with stative complements, e.g. 'she kept liking him' versus 'she continued to like him' (Wierzbicka 1988: 87). *Continue* simply allows either close-up or prospective perspectives, and also combines with both statives and non-statives. The verb *cease* appears to allow *to* as well as *ing*. Wierzbicka (1988: 79–80) claims this is because *cease* inherently means gradual slowing[6] while *stop* does not. If this is the case, then *to* would be compatible because cease leaves the prospect of the end of the denoted activity available, whereas for *stop* that prospect is cut off and the activity's ending is viewed, so to speak, from 'within' that activity (a description of *ing* meanings found in several accounts; (see above, Chapter 4).

9.4.4   to, ing *and* that *alternation: the case of* remember

The English word *remember* changes its meanings – related, of course – in interesting ways, depending on the complement construction it is used with.

(15)   Mary remembered writing the report

(16)   John remembered Mary writing the report

(17)   Mary remembered to write the report

---

[6] *Cease* derives from Latin *cessare*, the freqentative form of *cedere* (meaning, among other things, 'to yield 'or 'give way') and meant something like 'go slow'.

(18)   John remembered that he had written the report

(19)   John remembered that Mary had written the report

(20)   *John remembered Mary to write the report

(21)   *John remembered Mary to have written the report.

These examples show that *remember* involves subject-oriented mental experience, or rather, the speaker's report, or truth claims about, another person's mental experience, an experience of a particular kind. In example (17) the *to* construction has a quite complex effect on the conceptualisation of the sentence, about which more will be said below.[7] Examples (18) and (19) illustrate the distancing effect of *that* complementation: what John remembers is a publicly available proposition recalled by John. Example (20) shows that it does not have the quasi-causal force that *urge* or *persuade* have, for example, though this kind of conceptualisation can be expressed with the verb *remind*. Nor, as (20) and (21) show, can *remember* be used with *to* in a way that sets up an epistemically uncertain reference frame, as in *John imagined Mary to have written the report* (cf. (9) above). This latter restriction is consonant with the fact that *remember* sentences appear generally to imply the truth of the complement clause in S's base frame[8] – that is, from S's point of view what another person remembers is treated as really having taken place. There is scope for reflection on this fact of natural-language semantics.

How will DST model the complexities of *remember* conceptualisations? I shall confine myself here to remarks on the effects of *ing* and *to*.

The DSM for (15) and (16) would be analogous to (5) *John imagined writing the report* (Figure 9.5) and (8) *John imagined Mary writing the report* (Figure 9.6). There are two key differences. One is that in S's reference frame John's mental state, modelled as a reference frame R', coincides with S's: the DSM would locate its origin at $m = 0$ in S's reality space R. The other is that the event (whether John writing or Mary writing) is necessarily in the past relative to both S and to John, whose mental world S is reporting. Both *imagine* and *remember* (for the relevant sense) require *ing*. Their meanings both have to do with subjective presence in an event, and additionally both

---

[7] It is worth noting that many languages do not work in the same ways as English *remember*: for example, French does not have an analogue of the *remember to do something* construction. This is also the case for the quasi-English synonym *recall*: one cannot say 'Mary recalled to write the report.'

[8] On the classic negation test only *John remembered that Mary wrote the report* triggers a presupposition. It seems that the negative *ing* version *John doesn't remember writing the report* could be interpreted pragmatically as epistemically true, neutral or counterfactual; *John remembered to write the report* entails that he wrote it but its negated form does not.

9.4 Overview of alternations and restrictions

denote subjective experiences 'calling up' (imagining) or recalling' (remembering) some experience that is highly likely to be visual. The verb *remember* with *ing* in (15) and (16) is thus a linguistic means of claiming to report another person's experience of episodic memory, memory of events that have been personally experienced by that person. Of course, S can report on their own episodic memory experiences by using *remember*. This episodic conceptualisation differs markedly from what is reported in (18) and (5).

Sentence (17) *Mary remembered to write the report* will need a more complex DSM that I shall only outline here. The *remember to* construction again makes a claim to be reporting a subjective experience of recall on the part of another person, so S's base reference frame R contains Mary's reference frame R', epistemically aligned with S's own frame R. This is the same configuration as for *imagine* and *remember* with *ing* complement, in the sense that S appears to be accepting that Mary's state of mind represents the same reality as S's own. R' is anchored at $t < 0$ in S's frame R, that is, $0'$ represents a point in time that was *now*, the time of the remembering experience, for Mary. In addition, however, Mary's state of mind R' contains a further reference frame R'', anchored at some time $t < 0'$, in which Mary had the mental goal-orientation *to* do the writing of the report at some time t later than $0''$ (the *now* at which Mary forms the goal-oriented stance) and also later than $0'$, the *now* at which Mary experiences remembrance of $0''$. This, in part, is what is packaged in *remember to* sentences in English as their core semantic scaffolding. DST makes it possible to sum up formally the complex relationships involved, in particular the use of embedded reference frames.

Chapters 7, 8 and 9 have been concerned with exploring in geometric terms some of the more abstract meanings of English grammatical constructions – certainly more abstract than the concrete spatial foundations on which the theoretical framework was erected in Chapters 1 and 2. These constructions have turned out to be best modelled in that framework by drawing on the idea of the displacement (translation) transformation of the base axis system. The final chapter concerns an area of meaning that is perhaps the most abstract of all – deontic meaning in the sense of concepts of duty and obligation. In order to set up adequate models in terms of DST we return to counterfactual reflection transformations discussed already in Chapter 6.

10 The deontic dimension

> From these [propositions] we understand not only that the human mind is united to the body, but also what should be understood by the union of mind and body. But no one will be able to understand it adequately, or distinctly, unless he first knows adequately the nature of our body.
> Baruch Spinoza, *The Ethics: Demonstrated in Geometric Order*

Counterfactuality, as we have seen (Chapter 6), plays a major role in the conceptualisations afforded by grammatical and lexical structures. And the geometrical approach has a simple way of modelling this remarkable attribute of the human mind. This chapter is about the role of counterfactuality in an important category of cognitions and linguistic expressions – those that also involve assumptions or claims about, or appeals to, values, norms or forms of authority. If a speaker S says that her friend *ought* to do so-and-so, two things are clear about the use of *ought*: the friend is not doing so-and-so; the speaker is invoking some kind of authority or moral duty. The two elements go hand in hand.

Many accounts have noted these elements of deontic expressions. The present approach shows how these essential elements can be integrated in the geometric modelling method of DST. This is not simply a formal redescription of what has already been said in the semantic literature, including the cognitive-linguistic literature, but carries some fundamental new claims. The most important aspect of the DST account of deontic expressions concerns the epistemic, specifically the counterfactual, dimension. Contrary to some accounts, the epistemic dimension, rather than the deontic dimension, is taken to be fundamental.[1]

10.1 Deontic meanings presuppose epistemic meanings

In cognitive linguistics the most influential analysis of modals comes from Talmy's papers on force dynamics, summarised systematically in Talmy (2000 [1988]).[2] Talmy makes four connected proposals that are relevant here,

---

[1] Parts of this chapter are based on Chilton (2010).
[2] Among other important accounts of deontic meaning, perhaps the richest is Lyons (1977: ch. 17). Kratzer (1981) and Papafragou (2000), who seek to understand the relationship between linguistic modal expressions and the domains of knowledge assumed in making sense of them, are in

10.1 Deontic meanings presuppose epistemic meanings

subsequently developed by Sweetser (1982, 1990). The first is that force-dynamic concepts, arising in embodied experience, are cognitively primary structures that are recruited in a wide range of lexical and grammatical structures. The second is that force-dynamic structures have a psychosocial interpretation as well as a physical one. The third is that sociophysical force-dynamic concepts are associated with deontic modals. And the fourth is that the deontic modals are 'basic' in the sense of being 'metaphorically extended' from an experientially physical source. Sweetser modifies and develops Talmy's force-dynamic account of deontic modals, and places them in a diachronic account of the meanings of English modal auxiliaries, arguing for a historical progression from deontic meanings (more concrete) to epistemic (more abstract) meanings, a claim supported by the work of Traugott (e.g. 1989). Stated in this way the claim is about the historical development of a vocabulary. The more general claim is that the relationship between the deontic and the subsequent epistemic meanings is metaphorical, in the sense of conceptual metaphor theory. The implication (possibly unintended) is that such a relationship is synchronically true also – i.e. that such a relationship is to be found in the semantic description of the synchronic meaning or utterance processing of deontic and epistemic modals. This ought surely to mean that if we seek to describe the conceptual structure associated with epistemic words, deontic concepts are somehow involved, but not vice versa. I do not think that this is the case.

Langacker's account (1991: 273–5) endorses and claims to be a 'refinement' of Sweetser's force-dynamic account of modal meaning. However, despite similarities, his account does not appear to take the metaphorical mapping of force onto epistemic modals to be primary. Rather, it appears to offer a basic model in which time and degree of known-ness is fundamental and is characterised by directionality and distance. Importantly, Langacker seems to give primacy to the idea of the speaker 'as the person responsible for assessing the likelihood of reality evolving in a certain way', while the force-dynamic element appears to be derivative: 'the notion of evolutionary momentum [in the sense of a temporally evolving reality] might well engender the conception of the speaker being carried by the force of evidence along

---

some respects compatible with the present approach, and I have drawn on distinctive insights from Frawley (1992: ch. 9), who maintains a deictic account of deontic modality based on notions of distance and direction thus making his account essentially a vectorial and geometric one. Jackendoff (2002b) outlines the conceptual structure of rights and obligations. While stating that the descriptions of spatial concepts are a foundation for investigating other domains, he makes no direct connection in this case, but rather analyses the conceptual structures of the constructions *X has a right to do Y* and *X has an obligation to do Y*. As Jackendoff acknowledges, there are similarities but also important differences between such concepts and those encoded in the respective modals *must* and *may*.

a deductive path' (Langacker 1991: 274). There may be elements in this account that are in line with what DST assumes about the fundamental role of epistemic assessment. However, as far as I can see, it leaves us without any account of deontic conceptualisation as such.

That metaphorical mapping from more concrete to more abstract concepts underlies semantic change is not at issue here. But there is a different problem, one that was not in the focus of the Talmy–Sweetser framework. The problem is that, so far as the synchronic semantics are concerned, deontic meanings appear to presuppose epistemic ones. The connection is not metaphorical but presuppositional. For example, it is awkward, it seems to me, to say *John must clean up the mess but maybe he won't*,[3] while it is acceptable to say *John should clean up the mess but maybe he won't*. This shows not just that *must* is 'stronger' than *ought* but that both verbs presuppose an epistemic judgement about the action denoted by the complement over which they scope, namely, *clean up the mess*. On introspection, to make a deontically modalised assertion seems to involve having in mind a representation of a world that is not actual, in parallel with one that is actual. In the parallel world desired or required actions are executed by actors who exist in the actual world, and this parallel world can be imagined as possible, probable or necessary, etc. as well as in varying degrees subject to or free from obligation (however that is defined).

To put these points in a slightly different way, if a speaker S says that her friend *ought* to do so-and-so, two things are clear about the use of *ought*: the friend is not doing so-and-so, or not yet; the speaker is invoking some kind of authority, moral duty or practical rule. At least, it would on the most usual understanding be strange to say she ought to be doing such-and-such if she already is, and at least there will be a tacit invocation of a norm, even if that norm is simply what is required in order to accomplish a practical goal. Intuitively, when S says 'she ought to do it', the associated conceptualisation seems to involve some imaginary domain in which she *does* do it simultaneously with a *real* domain in which she does not. Such mental constructs are not confined to deontic expressions. There are two components here that go hand in hand: counterfactuality and normativity. The normative element can indeed be insightfully modelled as a sort of force, as the Talmy–Sweetser approach does, and as the DST approach is also equipped to do. But counterfactuality is essential to what constitutes deontic understanding and counterfactuality, as proposed in Chapter 6, is essentially a reversal, that is, in geometrical terms, a reflection of the epistemic dimension.

---

[3] Unless we treat the first clause as a case of free indirect style, perhaps. Such a reading would be a special case rather than the indicating the default conceptualisation.

10.1 Deontic meanings presuppose epistemic meanings

In the accounts of modality outlined above (those of Talmy, Sweetser and Langacker), the term 'root' modal has been adopted to place dynamic modality (loosely, to do with physical ability) and deontic modality (to do with obligation and permission) in the same category and epistemic modality in a separate one. The present theory does not adopt this division, since the term 'root' is potentially misleading in appearing to already embody a claim of priority, as well as in putting together under the label 'root' two types of modality (dynamic and deontic) that can be distinguished (see Palmer 1986: 102–3). And the epistemic dimension in DST is a fundamental cognitive dimension, not a derivative of supposedly more basic cognition. The present account of deontic modals shows why both the epistemic dimension and the category deontic are required.

The approach that is adopted here starts with conceptual space. Rather than examining the lexical expressions first in terms of their interrelations and their polysemy, the aim is to explore the structure of the conceptual space that gives rise to the need for lexically communicated modal meanings in the first place. This is the approach followed by Winter and Gärdenfors (1995) in their account of modals and also, more broadly, by Gärdenfors (2000). The present chapter, then, seeks to outline the structure of deontic concepts, and to do this it uses a geometrical model of conceptual space that is developed in DST, not only for modal concepts but for other conceptual structures too that are communicated via language.

Like Winter and Gärdenfors (1995), this chapter also derives an important part of its account of deontic modals from Talmy's insights concerning force concepts, but does not use them in exactly the same way. Winter and Gärdenfors remain interested in the nature of the etymological shift from deontic to epistemic, which they claim is gradual rather than metaphorical, and they do not focus on a synchronic conceptual relation between epistemic and deontic domains. The DST approach does not treat epistemic modal concepts as 'extended' from sociophysical concepts in the sense of some kind of conceptual one-way relation in the mind of a speaker. It is not clear in any case what this would mean. Note that as soon as one formulates the problem in terms of concepts in the mind rather than lexical items the blurring of synchronic and diachronic in the Sweetser account becomes obvious. If we ask what is the nature of the epistemic concept that people have in mind when they use a modal like *must* or *should* in an epistemic sense, we need also to ask what it means to say that such an epistemic sense 'derives from' or is 'extended from' a deontic sense. What I want to look at in this chapter is the structure of the deontic conceptualisation people have in mind when they express it in linguistically available code.

The normative part of this conceptualisation can, as already noted, be modelled in terms of a force-dynamic component: loosely, the sense of compulsion or 'oughtness' that is activated by the use of deontic expressions can be explained in terms of a force-dynamic image schema that metaphorically has a source and has impact on an affected participant. In this I do not fully follow Winter and Gärdenfors (1995), however, who reduce the deontic source to 'social power' existing between speakers in a speech act (see Section 10.3 below).

I have suggested above that deontic conceptualisations probably involve contrasting parallel conceptualisations, one in the actual world, one in a counterfactual imposed world. This view is similar in certain ways to Frawley's account of deontic meaning as 'involving two kinds of world' (Frawley 1992: 420). But the DST account is distinct from Frawley's in viewing deontic conceptualisation as dependent on epistemic conceptualisation. More fundamentally it is distinct from all approaches in taking advantage of Fauconnier's theory of mental spaces as a way of understanding the idea that two 'worlds' are somehow involved in deontic conceptualisation. But in turn, DST is different from mental-space theory too. DST incorporates Fauconnier's idea of multiple cognitive spaces and referent mappings across such spaces, but it goes beyond mental spaces in the following ways: DST proposes an essentially deictic grounding for such spaces, it takes the spatial underpinning of many conceptualisations seriously by using geometrical descriptions and it uncovers unsuspected relationships between spaces when viewed geometrically.

The DST framework was not set up specifically to explain deontic modals. An account of deontic modals does, however, come out of it and the reason for this is that, precisely because of its geometrical conception, it intrinsically provides both for multiple related spaces and for directed forces.

10.2 Deontic reflections

The task now is to show how geometrical modelling can be used to model deontic conceptualisations. This might seem a curious notion at first sight but I hope to show that the DST approach can capture a number of interesting dimensions of deontic meanings. In particular, in this section I show how epistemic meanings are part of the conceptual structure of deontic meanings. The essential epistemic component is counterfactuality, and as we have seen, counterfactual conceptions can be modelled as geometric reflections of the base space R. This does not mean the notion of force is irrelevant. Force also an essential ingredient. It is not necessary to introduce any extra machinery to deal with this, as the DST approach already has vectors, which can be interpreted as standing for force directed from a binding source.

10.2.1 First-order reflection of R: obligation

Deontic meanings presuppose epistemic meanings, primarily counterfactual ones. While a clear-cut distinction between epistemic and deontic modality is generally accepted, many of the theoretical semantic accounts make the point that deontic meanings, like directive meanings (for example, Lyons 1977, Langacker 1991, Frawley 1992), involve a non-existent or not-yet-existent or desired or irreal or imposed action or state of affairs. This surely means that the epistemic is interwoven with the deontic in the conceptualisations that arise in the course of producing and understanding deontic utterances. Within the present framework I refer to the mental representations of the epistemically non-real simply as 'counterfactual'. The type of conceptualisation in question involves the combination of factual (for the speaker S) elements as well as (again for S) counterfactual elements. The geometrical format that DST uses is able to precisely characterise this combination.

We will consider examples such as the following:

(1)   a Mary must write the report
        b Mary must not write the report.

The crucial point here is that there is a counterfactual presupposition in (1a) and (1b): that Mary is not at the time of speaking writing (or intending to write) the report, so far as the speaker S knows. At the same time, the sentence leads us (as presumably its utterer wanted) to entertain the notion that in some possible world Mary *does* write it. This is the basis on which I am proposing the idea of the *deontic mirror world*. Mary is not doing the desired action in the speaker's reality but is represented as doing it in the speaker's ideal or desired world. We need a way of modelling this kind of ideal or desired world relative to the speaker's real world and we can do it semi-formally as follows. There is a base axis system R, which is real-for-S, and within R Mary is not writing the report. Simultaneously there is another axis system R', in which Mary 'is' writing it. And hearers of (1) mentally represent both R and R' because of the conventional meaning of the English modal *must*.[4]

In describing deontic meanings semanticists sometimes talk of the 'imposing' of one world on another by some individual or institution with authority or power (see Lyons 1977: 827). Frawley (1992) speaks of the 'imposition' of

---

[4] I don't mean of course that they have a little picture in their minds that resembles the figure we are about to contemplate. Rather the geometrical diagrams are abstract models of cognitive states. But it is important to note that these abstract models, unlike some others found in cognitive linguistics, are *motivated* in two senses. First, they use standard elementary mathematical ideas from geometry, viz. coordinates and vectors. Second, I suggest that geometric coordinates and vectors are themselves *motivated* because they are rooted in bodily experience of direction, distance and force (see Chilton 2005).

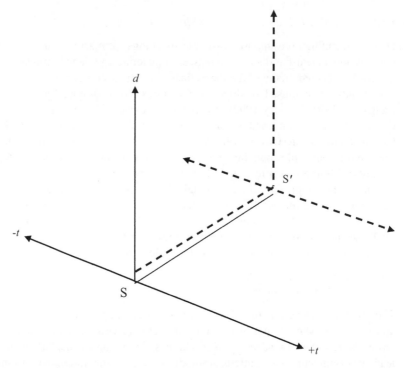

Figure 10.1 Base axis system and reflected copy

an 'expressed world' on a 'reference world'. Langacker's account appears to emphasise a similar epistemic structure (Langacker 1991: 269–81). Force-dynamic accounts use a similar metaphor, but do not speak of this simultaneous dual structure with its differentiated epistemic components.

As we have already noted, there is another component of the deontic meaning of *must*, namely the concept of some kind of compulsion, resembling the force-dynamic account proposed by Johnson, Talmy and Sweetser. Since DST draws on the idea of vectors, we already have a force concept in the elementary physical understanding of vectors themselves. We shall explore how force vectors can in a natural motivated fashion be combined with insights about the different kinds of 'reality' involved in deontic meanings.

The model I am proposing for deontic obligation concepts is built on the same reflection geometry that structures counterfactual conditionals, as described in Chapter 6. The configuration required is repeated in Figure 10.1. We can think of this as representing the fundamental counterfactual space that is apparent both in the kinds of conditional sentences we have examined and in deontic conceptualisations. As we have already noted, this kind of

10.2 Deontic reflections

conceptual space superimposes the non-real on the real (for the speaker).[5] The concept of counterfactuality makes no sense unless both a base world and a counter-world are simultaneously held in mind.

The range of obligation concepts facilitated by English modal verbs can be seen to be structured within this double world. In this world the time axis (t) and the referential axis (d) correspond, but the m-axis, the scale running from what is real for S to what is not-real or counterfactual for S, is reversed. So for S' at the origin of this reflected system R' what is counterfactual in R is real, and conversely for the counterfactual in R'. The whole reflected space of R' corresponds to a state of affairs that is desired, or desirable, including ethically desirable. Of course the precise deontic or ethical basis is determined pragmatically, that is, in the interlocutors' shared expectations, which can include merely S's desires or some source external to S.

Figure 10.2 takes the fundamental counterfactual configuration of Figure 10.1 and positions vectors within it corresponding to modalised action verbs, in particular, obligation concepts encoded in the lexemes *must* and *should*. Various tense forms are considered.

For example, we consider *Mary must not write the report* and *Mary should have written the report*. And we are concerned not only with the deontic meanings themselves but also with the effect of certain tenses that exclude deontic meanings but allow epistemic ones. The sentences we shall consider are listed below. They are variations on (1). The sentences are ambiguous between deontic and epistemic meaning but it is the deontic meaning that is to be assumed here: the starred sentences are those for which a deontic reading is not available without further specification of context.

(2) a *Mary must be writing the report
     b *Mary must not be writing the report

(3) a *Mary must have written the report
     b *Mary must not have written the report

(4) a Mary should write the report
     b Mary should not write the report

(5) a Mary should be writing the report
     b Mary should not be writing the report

(6) a Mary should have written the report
     b Mary should not have written the report.

---

[5] We could say that the alternative counterfactual world is 'presupposed' or 'supposed': this could account for the English quasi-deontic modal in 'Mary is supposed to be writing the report.'

264   The deontic dimension

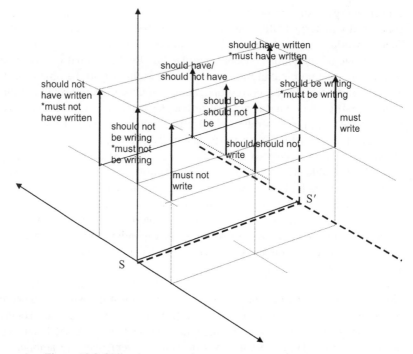

Figure 10.2 Obligation expressions

In Figure 10.2 above the vectors stand for the directed action *write*, the tail starting at the coordinate for the writer (*Mary*) and the tip touching the thing written (the report). The writer, Mary, and the report have their labels in *R*, the coordinate system representing the world as real for the speaker (subject) S, though of course their coordinates run through into *R'*, since *R* and *R'* are aligned. But the vectors are another matter: they are modalised for S, that is, they may be true, untrue, desired, required, possible, probable, untrue, etc.

Let us consider each of the cases in turn as they are positioned in the dual space configured in Figure 10.2.

10.2.1.1 must + V, must not + V

In example (1) the event of Mary writing the report is understood to be at some time $t > 0$. While (1) is perfectly normal, the examples in (2) and (3) are not. This follows from the natural conceptual structure of illocutionary acts that would be being performed, were a speaker to use sentence (1) in some plausible context. In more general terms, it also follows from assumptions about causations. That is to say, you cannot cause something to happen in the

10.2 Deontic reflections

past. Nor can one impose on somebody the obligation to have done something in the past. Of course, we can say things like *John was obliged/had (got) to do X*, but this is a report of an obligation that John was under in the past, not a use of this utterance to bring about an action in the past that has not taken place. Moreover, a sentence like (1) requires that we understand the source of the obligation as being the speaker; it is not understood as a report of an obligation imposed by some other person. The use of *must* in the present tense with second person subject has a pragmatic effect related to that associated with imperative forms understood as commands.

In terms of Figure 10.2, the *must* + V construction gives a conceptual configuration that can be described as follows. In R', the reflected deontic world, S' views reality R from an opposed viewpoint. From this viewpoint, the event of writing referred to in (3a) is 'real' in R', thus positioned at $m' = 0$, with tail coordinates at *Mary* and tip coordinates at the target of the action, *the report*. The location on the t-axis is at some time $t > 0$. And t must be greater than zero for the same reason that it cannot be less than zero – one cannot require or command an action that is already taking place at utterance time $t = 0$. In R, however, it is assumed to be not the case that the writing is taking place. Accordingly, the *write* vector in R is found at the counterfactual end of the epistemic modal axis m, as required. This result, which follows from the mirror-reflection geometry, corresponds to a large part of the epistemic semantics that is incorporated in the deontic semantics of *must*.

Consider now (1b), *Mary must not write the report*. At the distal (counterfactual) end of m' in R', i.e. at what is the opposite of what S demands in the deontic space – is located an act of writing. Simultaneously, there is an assumption that Mary will, contrary to what S requires, write the report in R.[6]

Now (1b) and its geometric representation correspond to a prohibition, and in Section 10.2.2 below we shall use this point in order to build a representation of deontic prohibition and permission modals. A consequence is that the obligation scale and the permission scale are not, *pace* Frawley (1992: 422), totally unrelated to one another; in fact, they seem to be conceptually entangled in certain ways that will be discussed below. For the moment it is important to point out that m' in R' corresponds to commands at $m' = 0$ and prohibitions at m' = counterfactual (irrealis).

---

[6] In practice, she may or may not, but I am proposing here that, so far as the conceptualisation of *must not* is concerned, there is a presumption that she will. Cf. the following: A. *Mary must not write the report.* – B. *Who said she was going to?* This kind of exchange suggests that the utterance used by A always has a strong presumption which has to be challenged if some speaker wants to assert the weaker meaning (that Mary might not be intending to write the report).

10.2.1.2 must be V-ing, must not be V-ing

The construction with present progressive clearly does not behave in the same way as the *must* with bare infinitive. In Figure 10.2, examples (2a and b) and (3a and b) are starred because they are not acceptable on a deontic reading. The sentence *Mary must be writing* can only express S's epistemic assessment of the degree of truth of the assertion that Mary is writing – viz. that it is highly probable that the activity is occurring. For this the relevant DST diagram would not involve reflected axes, merely the base system, which already incorporates an epistemic scale, viz. the m-axis. It is true, of course, that a deontic reading can be imagined for *must (not) be* V-*ing*, but a special context is needed, one which embeds this construction in a future event. Thus for a deontic interpretation of (2) we have to invent a context like: *Mary must (not) be writing the report when the boss comes in.* The progressive in this reading is relativised to some future time (when the boss comes in) relative to S.

If the progressive is given a present time reading relative to S, the only interpretation is epistemic. This is consistent with the fact that deontic modalisation can only apply to events viewed as taking place at some $t > 0$.[7] If we put event e at $t = 0$, then the meaning is limited to the epistemic meaning. Conversely, we cannot make the epistemic reading of *must* + V apply to a future event e at $t > 0$. The strong epistemic modal *must* applies only to events at $t \leq 0$. The reason is noted by many commentators: it is only present or past events about which we make a *strong* epistemic judgement about the probability of their occurrence, while future events are *inherently* epistemically uncertain. This means that weakly predictive *Mary may*$_{epistemic}$ *write the report next week* is acceptable with reference to a *future* event of writing, but strongly predictive *must*$_{epistemic}$ is not acceptable in this construction.

However, we can have a future reference for the following progressive construction with *must* as epistemic: *Mary must*$_{epistemic}$ *(not) be writing the report next week*. Nonetheless, it is arguable that *must* in such expressions relates to an inference carried out by the speaker at a time coinciding with utterance time. The time of the event can, as in (5), be tied to the future by way of the time adverbial. Equally it can be tied to the utterance time in the same manner: e.g. *Mary must (not) be writing the report at the moment*.

Further, it is worth noting that when the natural conceptual constraint prevents the deontic reading of *must* for present time points, the epistemic meaning comes up as a default. This may be a corroboration of DST's assumption that the epistemic scale m is fundamental, not a derivative, at a conceptual level, of deontic meaning.

---

[7] The same applies to commands and similar speech acts. If event e is taking place at the same instant as utterance u, then the preparatory condition for uttering u is not satisfied.

10.2 Deontic reflections

10.2.1.3 must have *V*-ed, must not have *V*-ed

The cases (3) follow the same pattern, *mutatis mutandis*, as (2) and for the same reasons, namely, that the conventional meaning of *must* is linked to illocutionary pragmatic force of a quasi-causal nature and thus cannot apply to a relation between present utterance and an event $e_t \leq 0$. When we turn to *should*, however, the situation is somewhat different.

10.2.1.4 should *V*, should not *V*

In cases (4), it is required that the future event is at $t > 0$, as for *must* in (1). How then does *should* differ in meaning from *must*? Whereas for *must* there is (it seems to me) a counterfactual inference connected with *must* in (1), this is not so obviously the usual interpretation for (4). We can, for instance, have the following:

(4') a Mary should write the report and she will / Mary should write the report but she won't
      b Mary should not write the report and she won't / Mary should not write the report but she will.

The assertion that Mary should (not) write the report can be overridden by its negation by the same speaker. This does not work for *must*. Taking another look at (1), we have:

(1') a Mary must write the report and she will / ? Mary must write the report but she won't
      b Mary must not write the report and she won't / ? Mary must not write the report but she will.

In using *must*, it seems that the prompted conceptualisation (and the speaker by implication) does not admit the possibility of non-compliance or contradiction. The assertion that Mary must (not) write the report is questionably compatible with its contradiction by the same speaker. This rather subtle difference corresponds in part with the difference in meaning between English *must* and *should*.[8]

In addition, a word such as *maybe* is compatible with *should* but not *must*:

(4") a Mary should write the report and maybe she will / but maybe she won't
      b Mary should not write the report and maybe she won't / but maybe she will.

---

[8] The intuitive native-speaker judgements reflected in (1') and (4') seem to require that both clauses are uttered by the same speaker and that the source of the deontic force is also the speaker (as the usual meaning of *must* in such examples does indeed seem to require). However, these sentences seem acceptable if two speakers are involved. Thus: A. *Mary must write the report*. B. *But she won't* and A. *Mary must not write the report*. B. *But she will*.

With *should*, the prompted conceptualisation seems to accept indeterminacy. But not so in the case of *must*:

(1″)   a   Mary must write the report ?and maybe she will / ?but maybe she won't.

With *must*, taking the speaker as deontic source, epistemic uncertainty about the required event is not entertained.

In terms of the DST format, these points suggest that the construction *should/should not* V is located at the medial point on m, as it is in Figure 10.2. In saying 'Mary should write the report', a speaker is not necessarily presupposing that Mary is not writing the report: accordingly, Figure 10.2 does not show any instance of the *should* vector at the counterfactual end of the m-axis in R or at the factual end for *should not*. This situation is, however, slightly different for the present progressive and present perfect constructions.

10.2.1.5 should/should not be V-ing *and* should/should not have V-ed

A further important property that distinguishes *should* from *must* is the difference in the way the two English auxiliaries interact with aspect and tense meaning. Differently from the *must* construction, a deontic reading is available for the progressive aspect (5) construction and for the present perfect construction (6).

First, the present progressive and the present perfect tolerate deontic interpretation. This implies that the deontic semantics here do not, unlike those for *must*, include quasi-causal illocutionary force. They are not used in order to bring about some action; rather, they are used to pass judgement (with respect to some norm) on the doing or not doing of some action, whether or not these actions are viewed as present, past or future.

Second, though we cannot compare directly with *must* (since *must* has no deontic meaning for progressive and perfect constructions), it is worth considering whether there are counterfactuality inferences for *should* in sentences such as (3) and (4):

(3′)   a   Mary should be writing the report and she is / but she isn't
       b   Mary should not be writing the report and she isn't / but she is

(4′)   a   Mary should have written the report and she has / but she hasn't
       b   Mary should not have written the report and she hasn't / but she has.

Although I think the default, on hearing 'Mary should be writing the report', is to assume that she is not (i.e. to add 'but she isn't' is unproblematic), 'Mary should be writing the report and she is' does not seem impossible; certainly it is more acceptable to my mind than (1′).[9] Something similar seems to be the

---

[9] I am using native-speaker intuition as a heuristic here; corpus-based methods could possibly contribute.

case for *should/should not have* V-ed. It is also important to note that *maybe* also seems to be within the limits of acceptability for (3) and (4), as it is for (2). The situation for the progressive and the perfect constructions thus seems to be different from that of the *should* + V construction. They are both compatible with inferences that a reference event has, has not or merely may have taken place, and that such an event is, is not or may be taking place. This is not the case for future events, since only *maybe* can be asserted for future events in R.

10.2.2 Second-order reflection: prohibition and permission

Deontic meanings are inherently more complex than epistemic ones because they imply and interact with epistemic meanings. The geometric approach seems to be an elegant way of capturing the complexities. Though this does not mean the diagrams are simple, they are systematic. In standard accounts of modal expressions, not only for English, the concepts of *prohibition* and *permission* are treated as members of the deontic category. The question that now arises is whether they can be modelled in the kind of integrated theory of modal concepts that DST attempts to build.

Frawley (1992: 422) insists that obligation and permission are on two separate, non-overlapping scales. This seems to be true up to a point but the two scales (corresponding to the two worlds of obligation and permission) are clearly in some sort of relationship one to the other. To make sense of the conceptual relationship, permission needs to be defined by reference to prohibition. Prohibition can be regarded as the issuance of an obligation to *not-p*, e.g. *Mary must not write the report*. In part, prohibition appears in the obligation deontic world that we have been exploring so far. I say in part, because *you must not do p* is significantly different from *you may not do p*. Nonetheless, these two sentences are in practice prohibitions, despite the conceptual difference.

The difference has to do with the fact that obligation is imposed on the real world R while permission is imposed on the obligation world R'. The presumed background to permission-giving is the obligation to not do p. This is why, in Figure 10.3, permission emerges as a mirror transformation R'' of R'. That is to say, the permission world is a reflection, or reversal of R', which brings it into alignment with R. Note that this does not mean it is the same as R, since all transformations are copies of R that remain in the DSM in their own right, as a kind of conceptual 'background'. If we appear to be just going back to where we started, that is what we need, because for one thing R is the world of fact that is resumed, as it were, once the 'imposed' worlds are lifted and for another thing we can see how we got there through ideal mirror worlds and their removal. From now on I shall refer to R' as the 'obligation space'

270   The deontic dimension

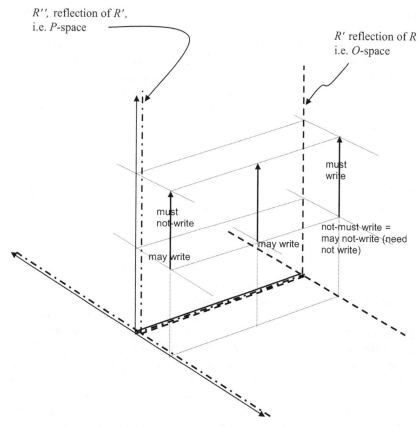

Figure 10.3 Permission and exemption: example (8)

('O-space') and to R'' as the 'permission space' ('P-space'). Every event (vector) that is located in O is an obligation (positive, negative or medial); every event (vector) in P is a permission (positive, negative or medial).

The permission world throws a different light on (1b), which can be understood in two ways, as shown informally in (7) corresponding to (1b) above):

(7) (i) Mary must not-write the report (i.e. it is the case that Mary must not write the report).
     (ii) Mary not-must write the report (i.e. it is not the case that Mary must write the report).

The first of these (i.e. 7(i) corresponding to one reading of (1b) above) can be regarded as prohibition. The second of these can be called 'exemption' (Lyons 1977: 773, 837–40). The sentence 7(ii) is not of course a natural

10.2 Deontic reflections

English sentence but a way of representing a conceptualisation of what would more naturally be expressed as 'Mary need not write the report.'

Turning to the P-space created by the transformation yielding R'', the expressions we are concerned with are the following:

(8) a Mary may write the report
     b Mary may not write the report

   (i) Mary may not-write the report
   (ii) Mary not-may write the report.

Of examples (8), (8a) and (8b(i)) can be epistemic or deontic but (8b(ii)) can only be deontic. The progressive and perfect forms follow a different pattern:

(9) a Mary may be writing the report
     b Mary may not be writing the report

   (i) Mary may not-be-writing the report (if the boss comes in, it doesn't matter)
   (ii) Mary not-may be writing the report (I don't want the boss to see her writing it).

Of these examples (9a) can only be deontic if the event is relative to a time point in the future relative to S. However, as in the examples in (8), (9b(i)) can be both epistemic and deontic, while (9b(ii)) is deontic only.

As for the following, none of them can be deontic:

(10) a Mary may have written the report
     b Mary may not have written the report

   (i) Mary may not-have-written the report
   (ii) Mary not-may have written the report.

None of them can be deontic, for the same natural reason that governs present perfect deontic *must* (though not present perfect deontic *should*) – viz. that permission cannot be given for a past event or non-event to be enacted if it already has been (or has not been).

Let us now see how these meanings are produced by the mirror–mirror transformation that yields the permission space R''. Recall that this transformation sets up a reflection of the obligation space or R'. The result of this can be thought of as the reversal of the m-axis. The setting up of the P-space can be also thought of as the 'lifting' of obligations, whether positive ones (akin to orders) or negative ones (prohibitions). We illustrate this for case (8) in Figure 10.3.

In the P-space, S is the source of a representation which entertains the scenario of Mary writing the report. We understand *may* as setting up such a mental space: a world in which Mary's writing the report is conceptualised as taking place against the presupposed backdrop of the O-space, the space in which S represents the scenario of Mary's *not* writing the report.

Consider the extreme points of m'', the m-axis of R'', the P-space, in Figure 10.3. The point at m'' = counterfactual coincides with m' = 0 (i.e. what S wants to be real in O) and with m = counterfactual (i.e. what S regards as really untrue in the base space R). In other words it is presupposed in R that Mary is not writing the report. In R' (i.e. the O-space) S conceptualises a world in which she *will* write it. In R'' (i.e. the P-space) the entire axis system is in turn reversed (reflected) and the obligation is dissolved. Thus at m'' = counterfactual we have something like *it is counterfactual that Mary must write the report* (or in more logic-oriented terms, *Mary not-must write the report*). This corresponds to (8b(i)) *Mary may not-write the report* or in other words *Mary is permitted not to write the report* – more naturally expressed as 'Mary need not write the report, is not obliged to write the report, is exempt from writing the report, etc.' This position (i.e. m'' = counterfactual) in the DSM thus represents the complex conceptual structure of an 'exemption'. In general, P presupposes O and O presupposes its reflected 'opposite'.

At m'' = 0 we have a different effect. The prohibition *Mary must not write the report* is represented in R' (i.e. in O), where it is located at the counterfactual end of m'. But this point coincides with m'' = 0, what is entertained as real in the second-order reflection R'' (i.e. P). What does this represent? It represents S's conceptualisation of the event of Mary's writing the report occurring in a conceptual space of permission – in which a presupposed prohibition, which in turn presupposes its reflected opposite (i.e. Mary is conceptualised as intending to write the report in R), is being lifted. Just as the O-space entertains what S desires to impose, so the P-space entertains the lifting of the prohibition. In other words, in P the speaker S is representing the occurrence of the event of Mary's writing the report: *Mary may$_{deontic}$ write the report and she will*. However, there is some conceptual indeterminacy here, for S may not be expecting with epistemic certainty that Mary will write the report if S lifts the prohibition (i.e. gives permission): *Mary may$_{deontic}$ write the report – possibly she will in fact do so, possibly she won't*. The DSM should then also show a vector in P at m'' = 0 = midpoint, as also shown in Figure 10.3. What is interesting about this is then that deontic *may* coincides with its epistemic meaning, as we need, given its polysemy.

What the diagram thus shows us is that conceptualisations in P have *two* 'background' conceptualisations, or two kinds of conceptual space. One is the O-space in which there is represented the prohibition *Mary must not write the report*. The other is the R-space, where S knows what is happening in his/her real world and this includes the idea that Mary will engage in writing the report at some point in the future relative to S. This configuration thus matches the linguistic intuition that understanding (8a) *Mary may write the*

10.2 Deontic reflections

report raises the notion (i) that there is a will on Mary's part to engage in the action and simultaneously raises (ii) the notion of an obligation to refrain from, that is a prohibition on engaging in the action in question, and (iii) the conceptual reversal of the obligation, where it is imagined that Mary will or might perform the action. In this kind of model, then, permission is a complex conceptual structure which rests on conceptualise obligation which in turn rests on a conceptualiser's assumptions about his reality.

There remains the question of the meaning of *Mary may not write the report* suggested in (8b(ii)) *Mary not-may write the report*, in which the permission-giving meaning of *may* is counterfactual (there is no permission to do x), which is tantamount to a prohibition.

If we compare (1b) *Mary must not write the report* and (8b) in the sense of (8b(ii)), both can be understood as prohibitions. This is an apparent problem for Figure 10.3, since *Mary may not write the report* might be expected, *qua* prohibition, to appear at the counterfactual end of m'', but because it seems conceptually similar to *Mary must not write the report* we might also expect it to coincide with the latter. One might argue that this is only a problem if we assume that the *may* form has to remain in *P*. In fact, of course, *may* is polysemous and can appear in *O* – which is the solution proposed here. Nonetheless, though they are logically and conceptually similar, there does seem to be some semantic difference between the two prohibition forms *must not* and *may not*.

There is perhaps another way of approaching this problem. Note first that *must*-prohibitions (Mary must not write the report) do not have the conceptual form *Mary not-must write the report*: it is the writing of the report that is entertained counterfactually not the deontic force that is negated. By contrast, the *may*-prohibitions have the conceptual form, as we have seen, *Mary not-may write the report*. That is to say, in the latter case, we can regard the entire *P*-space (the coordinate system R'') as being reversed (as undergoing yet another reflection transformation) producing a third-order space R'''. Now, complicated as it may seem, this does have interesting results, as the DSM in Figure 10.4 suggests.

What we have in Figure 10.4 is a complex configuration of conceptual spaces in which (i) S takes it as real in *R* that Mary is intending to write the report, (ii) conceptualises an obligation world R' (that is, *O*), in which Mary's intention is reversed (it is 'real' in this *O*-world), (iii) *O* is reversed to give R'' (i.e. the *P*-world) in which obligations are reversed and permission is granted, (iv) *P* is reversed to give R''' and it is here that permission is withdrawn. The difference between *must*-prohibitions and *may*-prohibitions then emerges: *may*-prohibitions withdraw permissions. *May*-prohibitions sit on top of permissions, which sit on top of obligations, which reverse the real-world expectations. Note also that we get an alignment between the *may*-prohibition

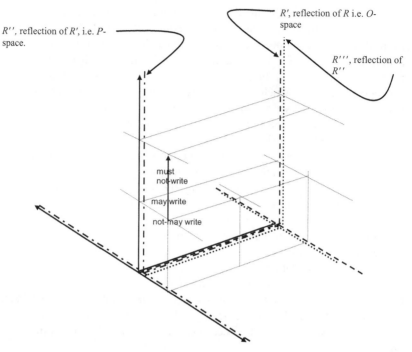

Figure 10.4 Conceptual structure of *may*-prohibition: removal of prohibition

in R''' and the *must*-prohibition of R'', which is desirable since it helps to capture the conceptual similarity between the two. In the process of stacking up these copied reflections the conceptual content of each new space (coordinate system), at least for English deontic modal conceptualisations, seems to become less dense and more rarefied, and this is perhaps not unexpected. Indeed it is possible that this is the reason why *may not* prohibitions, once having appeared in English, have tended to become rarer in use.

10.3 The deontic source

So far I have attempted to model deontic English models in terms of essentially epistemic contexts. What of the force-dynamics account of modal concepts? First, it is worth formally stating a few obvious points about conceptually represented force and actual force in the physical and social world of action.

It is important to distinguish the conceptual representation activated by *must* constructions from their pragmatic effect. In the conceptualisation, there is an ideal normative counterfactual R' relative to S in which S or some other source is represented as 'putting pressure' on some discourse referent d_j,

10.3 The deontic source

causing a represented action in R'. No such action has been physically performed at the point when a speaker utters a linguistic expression that conceptually represents it. In the conceptual representation such an action is conceptually counterfactual in R' and also counterfactual in the physical world in which S utters the utterance. But of course the intention of S in performing the utterance is to 'put pressure on' the hearer and in some sort of social pragmatic chain ultimately on Mary. This may (or may not) lead to her carrying out the desired action.

Let us now consider how force is modelled in the relevant DSM. The important point here is that we do not need to add any further formal apparatus. The DST framework already includes the geometrical concept of vectors, and vectors can be conventionally understood as 'force', in the metaphorical sense of Talmy, Sweetser and Johnson. Using the geometrical vector notation, rather than ad hoc pictures, is justified since metaphorical force has physical force as its source image schema, and physical force can be conventionally modelled in the sciences and applied sciences in terms of geometric vectors. In this sense, the adoption of the geometric framework of vectors in coordinate systems is arguably well motivated.

Now in the conceptual representation modelled in DSMs, and in the real sociophysical world, the source of the 'pressure' can be S or some other source in some other conceptual 'location' in the DSM. In the latter case the identity of the source is understood pragmatically among participants in the utterance event. If we wish, we can draw a DSM with a distal coordinate for, say, The Law, Tradition, God, etc. There is another kind of source that can be linked to constitutive rules and Wittgenstein's idea of games as well as ad hoc tasks-in-hand: 'in chess pawns are not allowed to move backwards', 'if you are writing a paper it must have a conclusion', etc. But it is not necessary to pursue this point here. All these kinds of deontic sources are complex cognitive frames and scripts that are not part of the semantics per se of *must*. However, such deontic sources belong to a cognitive frame invoked or presupposed in joint communicative action, though not always explicitly.[10]

In this analysis there are some significant differences in relation to Winter and Gärdenfors (1995), who reduce deontic meaning to 'social power'. The problem with this proposal is that, while deontic modals clearly have the potential to be used by powerful social actors, it is not difficult to imagine situations where neither the power of the speaker is evident nor can the

---

[10] In formal semantics, e.g. the classic account of modalilty in Kratzer (1977), what I am calling 'deontic source' roughly corresponds to 'conversational background' defined, approximately expressed, as the set of propositions known to a speaker in a context that is itself a subset of a possible world. In practice deontic expressions can be exploited by speakers by *not* specifying which among specifiable different deontic sources is being assumed.

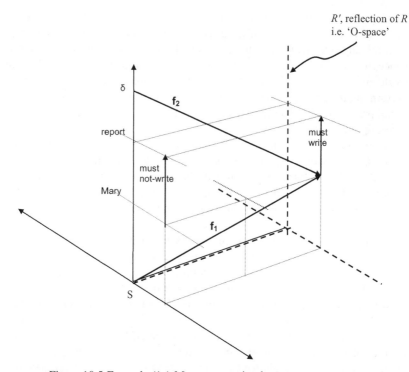

Figure 10.5 Example (1a) Mary must write the report

sanctioning source the speaker may be indicating be always be treated as 'power' – at least without a very general definition of power. For instance, moral principles do not have to be treated as power sources, at least not in the sense of 'social power', without long and substantial disputation. The range of possible normative sources that a speaker may invoke is indeterminate. Rather than seek to specify the source of deontic force, the DST model simply assumes a deontic source δ, for δέον, *deon*, 'that which is binding'.[11] Some δ is always presupposed in deontic conceptualisations. It is a very abstract concept and it is of some interest that such an abstract generalised 'moral imperative' is coded into presumably all languages.

But what concerns us here is a more technical matter: how to incorporate 'force' into deontic DSMs. Relating the deontic source δ to events in O is a relatively simple matter, given the use of vectors in DST, since vectors are readily interpretable as kinds of causal force and are in any case already

---

[11] The Greek word is the neuter form of the present participle of the Greek verb generally translated as 'bind'.

needed for modelling causative verbs (cf. Wolff and Zettergren 2002; Chilton 2005). Figure 10.5 gives an idea of how this can be done.

The proposal here is that a vector **f** representing moral force, a metaphorical concept, has its starting point at some deontic source. This may be S, as shown by force vector **f**, or some source distant from S, as shown by force vector **f**$_2$. The head of such vectors applies the modal force to a particular discourse entity, which is here the source (*Mary* in our example) of the action or event (*write*) in the DSM. What the DSM thus represents is a deontic force acting on Mary and causing her writing of the report. This configuration models the conceptualisation according to which δ or S is in some way the causal agent of the desired action. Figure 10.5 shows two deontic force vectors for illustrative purposes. Either is possible in understanding sentence (1a), but it seems unlikely in practice that two denotic sources would appear in the same DSM, perhaps because divided moral authority is not conceptualised.

In this model of deontic force in an epistemic universe, the modal force vector does not simply set up the reflection coordinates O, it represents a quasi-causal force resulting in Mary carrying out the action in the desired (but counterfactual) world O: the source is real, however, and so is Mary who has coordinates that are labelled in R. All coordinates in DST carry through from R into other embedded coordinate systems. As Lyons (1977: 791) puts it: 'We can also carry out the psychological process of trans-world identification across real and imaginary worlds of various kinds.' What we are modelling, then, in Figure 10.5 is a vector that carries force across to a *desired counterfactual* coordinate space, in which Mary still has a coordinate point on d'.

We might ask: how should we consider the deontic force with regard to the second-order reflection coordinate space P? Does it disappear with the reversing of obligation? The vector modelling does not lead to this conclusion. The force vector **f** remains, and this seems to be appropriate, for even when permission is granted and obligation is reversed, the prior existence is assumed of the initial imposing of obligation (the deontic force **f**).

It is important to note that the DSM in Figure 10.5 does not represent a real-life speaker applying illocutionary force on a real-life hearer. Rather it represents what the communication participants have to jointly represent mentally as a consequence of the form of the utterance, whether or not, for example, Mary does in reality write the report in response to what S utters.

10.4 Thoughts on *ought*

An implicit claim of this chapter has been that the epistemic aspect of conceptualisation is in some way prior to the deontic. What is being tested here is the idea that deontic meanings do in fact presuppose epistemic meanings of the kind we have modelled on the *m*-axis. As it happens, this idea follows from the way the three-dimensional space of DST is initially set up. But it was not set up for the purpose of arguing that epistemic meaning is prior to deontic; it is independently motivated and it explains a wide range of linguistic conceptualisations. What I have done is try to see what happens if we try to model deontic meanings in terms of DST assumptions, and the resulting analysis is, I believe, logically (and conceptually) consistent within itself. This is not to say that the force-dynamic account is absent from the story of deontics in DST, though it is not used to account for epistemic concepts in the way in which deontic 'force' is used by some cognitive linguists to account for epistemic meanings. It seems entirely plausible to think of deontic meaning in terms of some sort of force concept. This proposal is also already present in the DST framework in the form of vectors viewed as force. It is also worth noting that vectors are indeed also part of our account of epistemic meaning, vectors viewed as position markers relative to S, since that epistemic certainty/uncertainty may be related to spatial closeness or remoteness.

The DST model of deontic meaning is not simply a formalised redescription of what has already been said in the semantic literature, including the cognitive-linguistic literature, but carries some fundamental new claims. The most important aspect of the DST account of deontic expressions concerns the epistemic, specifically the counterfactual, dimension. The epistemic dimension, rather than the deontic dimension, is taken to be fundamental. In fact, it is taken to be fundamental to all conceptualisations associated with linguistic expressions. The three dimensions of the basic DST integrate the dimensions of language-linked cognition that approximate to Langacker's notion of 'grounding' (Langacker 1991: ch. 6). Any utterance presupposes that the conceptualiser takes a position with respect to perceived time, relative salience of entities attended to and degree of realness. Langacker refers to the latter as 'epistemic distance'. And this is precisely what DST attempts to model in a precisely spatial manner.

Where does this leave the Sweetser account which views epistemic meanings as metaphorical projections of prior 'root' meanings? There are a number of issues that need to be addressed and I shall only outline them for further reflection. The first point to consider is that the classic force-dynamic account is cast in diachronic terms and need not be a cognitively accurate picture of conceptualisation of deontic and epistemic modality in the adult human speaker. Even as a diachronic account the force-dynamic model makes best sense only in relation to lexical change, for we have no grounds to suppose that speakers in some early phase of a language-using culture first had only deontic concepts and expressions and no epistemic ones, only later acquiring epistemic conceptualisations and expressions. We might argue that such a culture did not much use epistemic expressions in its communication and only in a relatively late phase came to need epistemic expressions derived from its earlier deontic ones: but there is no real warrant for that assumption and little if any discussion of the point. Second, evidence is often adduced from first-language acquisition studies that seem to indicate that English-speaking children acquire deontic expressions before epistemic one. Doubts have been raised about the relevance of these findings (Papafragou 1998, Livnat 2002, Papafragou and Ozturk 2006). Moreover, it is not necessary to conclude that the order of acquisition confirms a supposed priority of deontic cognition over epistemic cognition in the adult cognitive system underlying the use of linguistic expressions.

In any event, the DST account that I have presented is not incompatible with the force-dynamic one, in particular because force itself is naturally incorporated in the vector geometry of DST and is required for the modelling of deontic source, which is inevitably invoked in deontic expressions. What is distinctive about the DST analysis, however, is the claim that it makes for the complex conceptual interplay between epistemic and deontic 'worlds'.

Premack and Premack (1994) argue that the form of moral beliefs can be distinguished from their contents and set out to establish this form on the basis of observation and experiment with young children. They propose '[t]hat just as there are primitives from which one builds sentences so are there primitives from which one builds moral beliefs'. These primitives do not affect the content of either a sentence or a moral belief, but the form. Thus, 'while the content will vary widely over cultures, the form will not' (Premack and Premack 1994: 161). Their view of the content is relativistic and the content seems to consist mainly of what one might call 'customs' rather than ethical principles. But it appears that they regard the 'form', consisting of 'primitives', as universal. 'The primitives are: intention; positive/negative (right/wrong); reciprocation; possession; power and group' (Premack and Premack 1994: 161). The point here is not to argue with these specific claims (though one might want to do so), but to take note that Premack and Premack themselves point out that the primitives do not contribute to the explication of the concept of *ought*, a concept that many philosophical treatments of morality treat as an irreducible primitive. They put forward the idea, however, that children's moral 'primitives' have the logical form of an 'expectancy'.

There are some similarities but also some important differences between this approach and the theoretical analysis I have offered in this chapter. It is not clear, for one thing, that studying children's moral concepts tells us much about adult moral concepts, although Premack and Premack seem to assume, in their concluding remarks, that this is the case. As I have already noted, the DST approach is concerned with the logical form of the adult conceptions that are associated with words like *ought*. In common with Premack and Premack, the DST analysis does indeed propose a universal *form* but not content. As to the content, my own view is that this is a matter for philosophical ethics and that universal ethical principles are not excluded. But so far as cognitive linguistics is concerned, what analysis of *must, ought, should*, etc. tells us is that *ought* concepts (i.e. deontic concepts) are dependent on epistemic ones, specifically counterfactuality, and that simultaneously they are dependent on a source of 'deontic force'. Such a source need not be linked with social power. That is, the content of morality need not be reduced to kinds of social power as appears to be done: it is possible for the mind to have a sense (concept) of *ought* as a special kind of impulsion whose source is nameless. Nor need *ought* be reduced to 'expectancy'. What is central is the link with the envisioning of an alternative world.

11 Concluding perspectives

> Ce qui fait donc que de certains esprits fins ne sont pas géomètres, c'est qu'ils ne peuvent du tout se tourner vers les principes de géométrie; mais ce qui fait que des géomètres ne sont pas fins, c'est qu'ils ne voient pas ce qui est devant eux ...
>
> Blaise Pascal, *Pensées*, 670 (Sellier)

This final chapter considers some of the questions that arise from the ideas and analyses that have been put forward in this book and will attempt some clarifications. There are certainly inherent limitations in the approach developed here. The geometrical approach, even with the abstract projections that I have argued for, cannot explain the whole of linguistic lexical and constructional meaning. Should that worry us? I believe not. There is no reason to suppose that any one theoretical framework is capable of shedding light on *all* aspects of language. There is probably no grand unified theory of language, for the human language ability, and the human practice of language, may not be one grand unified thing. We may need several different kinds of theory, given that there are many different dimensions to language structure and meaning, possibly because of its evolution over biological, and perhaps more importantly, over cultural time. If this is the case, DST is just one theory that may shed light on one aspect of language. That is the gamble underlying the DST approach. It is a gamble based on the idea – a well-motivated idea, I think – that spatial experience (and accompanying experience of directionality and force) is fundamental in cognition and language, but is projected metaphorically to more than conceptualisation of the perceptible physical world. How do we explore this possibility? A tool for modelling is needed. The obvious tool is geometry, which itself developed culturally as a conceptual tool for modelling physical space, was famously axiomatised by Euclid, and in more recent times was extended into the modelling of highly abstract 'spaces'.

This book is an exercise in theoretical linguistics. Both 'natural' and 'human' sciences have their theoretical components. Observation of 'facts' does not lead to explanation, unless one is an out-and-out empiricist. Experiment is always theory-driven, even if it subsequently modifies the theory.

282   Concluding perspectives

Sometimes it is necessary to re-theorise, especially if scientific orthodoxies have become rigid. But this book is not entirely theoretical; it does not come from nowhere. It comes from the many linguistic observations that have taken careful note of the unexpected extent of spatial cognition in the conceptualisations (meanings) connected with linguistic forms.

11.1   Questions

How far does a space-based approach go? The geometrical approach is not intended to explain all semantic phenomena. What it seeks to do is explore the outer limits of a geometrical description. This is not just a pictorial geometry, as some diagrams in cognitive linguistics have been, but an informal adaptation of what Euclid, and later Descartes, taught us. The geometrical ingredients are three orthogonal axes, given a conceptual interpretation, upon which a few geometrical transformations can be operated. In those axes we have relational elements: vectors that represent direction and distance. What I have proposed is that these initially spatial systems provide the conceptual scaffolding for more abstract conceptualisations, especially those encoded in grammatical constructions. The claim is that this approach is able to unify a significant number of linguistic phenomena in a descriptively insightful way. It is possible that the limits of what DST can describe corresponds to the contours of some significant 'component' of cognitive-linguistic meaning-making in humans – though the metaphor 'component' is likely to make us think in ways that are too mechanistic. In any case, this idea is purely speculative.

The geometric character of DST, and the diagrams used throughout this book, may give the impression that the theory is abstract, disembodied and quasi-mathematical. However, as I have repeated mathematics itself can (albeit controversially) be seen as an immensely subtle and rich system built up rooted in embodied experience (Lakoff and Núñez 2000). It can be argued that this is especially the case for geometry in general and for vectors and reference frames in particular. Vectors are an abstract form of pointing, the very meaning of deixis. Tomasello (1999, 2008) has argued that pointing is peculiarly human and likely to be linked with the emergence of human language. To function communicatively pointing depends on shared cooperative attention, and, it may be pointed out, on a sense of direction and distance, within shared coordinates.

Throughout the book I have put forward particular 'models' of grammatical constructions and predicates, or better, of the conceptualisations activated by such constructions and predicates. Each particular model has a particular configuration. But each particular model is derived from the general model (three axes, transformations and vectors). This general model does not

11.1 Questions

represent *everything* about language, just a part of language, a part that concerns relationships among conceptualised entities, events and the speaker–subject's 'point of view'. The term 'model' is itself much debated but what I mean by it in the book is, roughly, an explicit set of relationships, positions and relative distances which can be graphically represented and which by hypothesis structurally correspond to (some subpart of) the structure of conceptualisations induced by linguistic expressions.

In the philosophy of science a distinction is often made between epistemic explanation as opposed to realist (causal) explanation. According to the 'epistemic' view of explanation a model increases our understanding but may not be a direct description of the causation of a phenomenon under consideration. Looking at it in this light, the DST model hopes to help us grasp something about abstract linguistic structure(s). According to a realist view the model describes literally something that exists in objective reality. In this case, the geometrical models of DST would show that the linguistic conceptualisations actually have geometrical form (or its neural analogue) in brain processes. Further, it might appear that the abstract conceptualisations are derived from physical spatial conceptualisations. If this were being claimed to be the case, it might be close to a causal explanation, answering a question of the kind '*Why* do linguistically activated conceptualisations have the form they do?' To which the answer would be: because they are derived from spatial conceptualisations. However, it seems to me that to assert a realist account may be too bold. But this does not mean that the DST does not have explanatory power: it can still provide a unified understanding of a certain aspect of linguistically based conceptualisation. For some thinkers, the claim to unify descriptions of complex phenomena under a simple set of relations and operations would in itself constitute a kind of 'explanation'.

The shape of one's explanatory theory, or model, depends on the *explanandum*, the puzzle one sets oneself to start with. If one asks the question how it is that humans can get an infinite set of novel sentences from a finite set of symbols, one comes up with the combinatorial system of generative syntax in all its varieties. This was an exciting development that emerged in the 1950s and gave linguists a new way of looking not only at language structure but also at the human mind. Theories of syntax were expected to resolve apparent irregularities and unify diverse phenomena. And syntax was expected to be entirely arbitrary, interfacing with, but distinct from, the assignment of meaning. To cut a long story very short, this abstract symbol-manipulation approach to language structure has constantly run up against meaning. One of the most striking examples is the need to bring 'thematic roles' ('theta-roles'), such as *agent* and *patient* into formal combinatorial rules. In generative syntax theta-roles were treated as syntactic properties of verbs (Chomsky 1981). Fillmore's thematic roles, however, always had a semantic flavour (Fillmore

1968). In practice syntactic and semantic considerations are hard to separate. Successive studies of theta-roles have become semantically enriched and linked to embodied cognition. Intuitively, it is hard to maintain that the distinction between doing something to somebody and being on the receiving end of an action is not inherently meaningful to the human mind. As we shall see below, the relation between a particular verb and the noun(s) it 'takes' can be seen as a natural relation emergent from the action scenario that the verb denotes, which in turn is derived from motor cognition and its neural substrate.

Such considerations may lead us to focus on meaning itself and its intertwining with linguistic structure, as the main object of explanation.

11.2 Space, the brain and language

Cognitive linguistics is an enterprise that seeks to model the kinds of conceptualisation, otherwise known as 'meanings', that are associated with linguistic structures. It is normal in science for such models to start off being highly abstract, tentative and contestable. But the ultimate goal is to understand how these conceptualisations are related to human be-ing. This means the embodied mind: brains are part of the body, minds are somehow based in brains, and meanings in minds.

Deictic Space Theory is an attempt to explicitly model certain kinds of conceptualisations that are prompted by language. The initial hypothesis is that spatial cognition is in some way fundamental. This implies that at some level linguistic conceptualisation is dependent on the neurological structure of the human brain. But this should not imply reductionist physicalism. Rather than saying conceptualisation is determined by brain structure – I would not say that – I prefer to use a version of the notion of supervenience (see Chapter 2, Section 2.2 of this book). Though I have neither space nor competence to go into philosophical details, I want to say that the mind, and thus the mind's conceptualisations, whether linguistic or not, are supervenient on, not *determined* by, the physical brain. It is clear from neuroscientific research, however, that the brain does have specialised areas operative in the different aspects of language. It would be impossible to do justice here to a fraction of the neuroscientific data concerning language localisation that have emerged and continue to be amassed but it is also impossible to resist the temptation to summarise some of the results that have a bearing on DST.

11.2.1 Spatial orienting, frames of reference and two streams of attention

This subsection looks at the relevance to linguistics of the two-stream hypothesis of visual attention (see, for example, Ungerleider and Mishkin 1982, Goodale and Milner 1992, Milner and Goodale 1995). The original

11.2 Space, the brain and language

hypothesis claims, on the basis of lesion studies and imaging evidence, that optic input into the primary visual cortex is sent from there into two different regions of the brain, on each side, via two neural pathways. One, the ventral pathway, communicates to the inferior temporal lobe and is involved in the recognition of objects. It is often referred to as the 'what' pathway. The second pathway, the dorsal stream, often referred to as the 'where' pathway, runs to the parietal lobe and is concerned with object location relative to the perceiver. Subsequent studies have added detail and moved beyond a simple separation of functional pathways, bringing out the interaction between them and their reciprocal connections with other networks, as well as other functions. In line with the original hypothesis, dorsal and ventral attention systems are generally recognised and ongoing research has pursued lines that concern lateralisation and several connections with brain regions surrounding the sylvian fissure. These lines of research are of extreme interest for the evolution and structure of human language.

An important finding is that the attentional system is lateralised (Fox et al. 2006). The dorsal system is found in both hemispheres (intraprietal sulcus, junctions of the precentral and superior frontal sulcus); the ventral system is found predominantly in the right hemisphere (temporoparietal junction and ventral frontal cortex). Another finding of Fox et al. (2006) that is important for cognitive linguistics is that the neural pathways of human attention can be active even in the absence of external stimuli, which means that 'abstract' conceptualisation arising from spatial orientation and attention processes could be represented in these areas. In related neuroanatomical investigations spatial orientation and attention are discussed in terms of egocentric and allocentric reference frames and the regions of the brain that support them. There is some variation in the research reports, but imaging studies of brain lesions associated with hemispatial neglect indicate that the neural substrates of allocentric frames are in the posterior superior and middle temporal gyri, while egocentric frames are processed in the intraparietal sulcus and the temporoparietal junction (Gramann et al. 2006, Hillis 2006, Iachini et al. 2009, Chechlacz et al. 2010).

Interesting language-relevant lateralisation for spatial attention has also been observed. Iachini et al. (2009) hypothesise that lateralisation in egocentric and allocentric reference frames cross-cuts with the distinction between near and far space as well as with the distinction between metric and categorical spatial representations – distinctions that are fundamental in DST. The right hemisphere is more involved in processing metric coordinate systems and the left in categorical spatial relations. And it appears that the right hemisphere network is more specialised in egocentric frames, linked with reaching and grasping actions in near (peripersonal) space, for which fine distance measures are needed, and that the left hemisphere system is weakly lateralised for allocentric frames with categorical spatial relations in distal (extrapersonal, i.e. out-of-reach) space (Kosslyn 1987, Kemmerer 1999, 2010, Trojano et al. 2006).

The hippocampal formation, which is crucial for place memory, navigation and event memories (as noted in Chapter 2), is neurally linked with the regions processing reference frames, as outlined already in the work of O'Keefe, Burgess and colleagues (O'Keefe and Nadel 1978, Burgess et al. 2003). More recent work indicates that the hippocampus plays a role in disambiguation and grammatical structure (McKay 2011). Duff and Brown-Schmidt (2012) argue that the declarative memory system of the hippocampus plays a crucial role in the main features of language, in particular in the arbitrary binding of sign and meaning, and in flexible on-line processing of the inherently underspecified and multiperspectival meanings found in language use. It seems likely that the hippocampus, parahippocampal and entorhinal regions are important for any space-based model of language because of their role in place-related event memories (event schemas decoupled from stimuli) and the formation of allocentric coordinates.

Behrendt's model of spatial attention includes links between the adjacent and connected parietal, temporal, entorhinal, parahippocampal and hippocampal regions (Behrendt 2011: 540–3 and 306–28). The dorsal visual stream feeds a bilateral attention system involving frontal eye fields and posterior parietal cortex (intraparietal sulcus, superior parietal lobule) in the posterior parietal cortex, which processes spatial information in egocentric reference frames, with the medial parietal lobe, especially the precuneus (which, incidentally, is associated with self-representation) encoding head-centred coordinates. The intraparietal sulcus encodes coordinates centred on different body locations and calculates 'intention vectors' within such coordinates for limb actions (Behrendt 2011: 309; see also Colby and Goldberg 1999). The ventral visual stream links to the forward part of the temporal cortex and to a right ventral frontoparietal attention network (middle and inferior frontal gyri, temporoparietal junction) parietal. Importantly, the right ventral attention network is homologous with the language areas classically associated with Broca's and Wernicke's areas in the left hemisphere (Behrendt 2011: 327). Both dorsal and ventral streams connect to the parahippocampal region and the hippocampus itself, and thence to different parts of the hippocampus. We have here a description of neural architecture broadly corresponding with the theoretical architecture of DST.

Information from both visual streams converges on the CA3-dentate gyrus region, where place and event memories in allocentric reference frames are represented. Allocentric representations connect entities in space independently of the self but interact with egocentric coordinates: neuroanatomical research generally speaks of the transformation of egocentric into allocentric

11.2 Space, the brain and language

reference frames and vice versa. According to Behrendt's (2011) summary of research such transformations are effected in the hippocampus (regions CA1 and subiculum) and posterior cingulate cortex. The parietal lobe may also be implicated in transformations (Colby and Goldberg 1999). In geometric terms, such neuronally encoded transformations would be translations: the shift of the origin of a set of axes from a distal location to a location centred on the head, for example, or an effector limb. As in DST, distal objects are the potential site for the origin of a further reference frame, and such a reference frame is always embedded within an overarching self-centred reference frame: without this structure neither transitive nor intransitive action is possible, nor perhaps is sense of self.

The DST reference-frame structure integrates deictic attentional space with deictic time. The role of the hippocampus in the neural attentional structure is interesting in this connection, since the hippocampus is involved not only in spatial orienting but in memory – linking places to events and orienting self to personal events in personal time. Left frontoparietal regions, especially the left parietal sulcus, have been found to be involved in temporal orienting (directing attention on focusing on moments in time and time intervals when events are anticipated), while it is the right parietal sulcus that is active during egocentric spatial orienting (Coull and Nobre 1998, Coull *et al.* 2000). When spatial orienting and temporal orienting are combined it was found that the right parietal areas are active (Coull 1998).

Behrendt's comment on the interaction of spatial and temporal attention is revealing. Speaking of the evolution of consciousness, the attentional system and the role of the hippocampus, he says:

All conscious experience is 'situated': conscious experience is always allocentric, in that event memory formation places landmarks or objects within a framework of space and time. Phenomenal space and time are not dissimilar: the abstract domain of time gets its relational structure from the concrete domain of space (Boroditsky 2000). By continuously encoding allocentric representations of landmarks, that is, encoding event memories, the hippocampus creates phenomenal space and time themselves, consistent with philosophical insights that we do not live in an objective framework of space and time but that the spatially and temporally extending world that appears to surround us is a fundamentally subjective and private creation. (Behrendt 2011: 540)

This phenomenal space and time created by the hippocampus is also consistent with the fundamental structure of DST and the account of linguistically mediated time conceptions outlined in Chapter 5. It should be noted once again, however, that the d-axis in DST does not stand for concrete space directly, but for spatialised attention.

The neuroscience research appears to show that spatial orientation and attention are processed in a dense network of connections in the periyslyvian region of the brain. What is intriguing about this is the fact that in the right hemisphere the spatial attention systems are homologous with the left-lateralised language systems, and that in the left hemisphere itself the two systems – spatial and linguistic – overlap one another (Fox et al. 2006, Behrendt 2011). This observation has led some neuroscientists to speculations that happen to be of major importance for the theoretical claims of DST. In this line of speculation particularly important are studies of hemispatial neglect – a condition suffered by people with lesions found mainly in the right hemisphere. In this condition people pay no or only limited attention to the contralesional side of their visual field. For example, if a right-handed person suffers right brain damage, they suffer visual neglect on their left (Hillis 2006; see McGilchrist 2010 for the wider significance of this condition). For the vast majority of right-handers this is the expected pattern but cases are found where it is *left* brain damage that is associated with the kind of left visual field neglect that is typically associated with *right* brain damage – in all cases in regions involved in the spatial orienting and attention systems outlined above.[1]

Suchan and Karnath (2011) investigate some sufferers of the atypical pattern, i.e. people with spatial neglect of the left visual field but whose lesions are not in the right spatial orienting regions but in the left. These patients had lesions in parts of the left temporal and parietal cortex that are part of the language-processing area. In other words, unlike in the usual pattern, these patients have *both* language and spatial systems in the left perisylvian cortex. They suffered both varying degrees of spatial neglect and varying types of aphasia, so these subjects have both language system and the spatial orienting system in the left hemisphere. The areas affected are homologous with areas affected in the typical case of spatial neglect after right brain damage.

Suchan and Karnath's study corroborates what is already indicated by the cumulative results from other studies which show that predominantly right-lateralised spatial orienting areas are homologous with predominantly left-hemisphere language areas. Further, Suchan and Karnath strongly suggest that in the course of evolution spatial orienting structures have been exapted for language:

A representation of spatial orienting in left hemisphere language areas thus might be a phylogenetic relict in humans, though this representation stays subdominant in the vast majority of individuals. (Suchan and Karnath 2011: 3067)

In non-human primates spatial orienting and attention appear to be processed in both hemispheres. In the transition from monkeys to humans, the

---

[1] The patients studied by Suchan and Karnath (2011) were not atypically right-lateralised for language. There is apparently no obvious explanation for the co-occurrence of left-lateralised language and left-lateralised spatial orienting in these rare cases.

development of language in the left hemisphere caused a partial shift of spatial orienting to the right hemisphere (Karnath 2001, Karnath *et al.* 2001, Cai *et al.* 2013).

Rather than thinking of this phenomenon as a 'shift', it is perhaps better to think in terms of an evolutionary restriction in left spatial orienting functions, or better still a modification. It may be that the spatial orienting structures in the left hemisphere became 'abstracted' and are manifest in the fundamental structures of modern human language, in particular in the fundamental conceptual organisation that is found not only in overtly spatial expressions (such as prepositions and demonstratives) but also, as proposed in DST, in verb schemas, conceptual distancing and grammatical constructions. This is a complex area and I have doubtless oversimplified. Nonetheless the findings, taken cumulatively, point to a tight connection between spatial orienting and language. DST is a partial theory of language that has good consistency with this strand of neuroscience.

11.2.2 Mirror neurons and action schemas

The egocentric and allocentric frames of reference that are part of the spatial orienting and attention system are integral with action schemas, both for intransitive locomotion, intransitive actions without locomotion and transitive action on the environment. Navigating the environment requires allocentric and egocentric frames, action on the environment such as reaching and grasping requires egocentric frames. It is the reaching and grasping actions that have been linked with mirror neurons, which have in turn been linked with language.

Mirror neurons are neurons that fire when an individual observes another individual executing a reaching and grasping action but not itself executing that action. They were first studied in the brains of macaque monkeys, in area F5 of the frontal cortex, which is said to be homologous to Broca's area in humans. Brain imaging experiments on humans show activation in the area close to or within Broca's area, when humans simply observe another person's arm or hand action (e.g. Di Pellegrino *et al.* 1992, Rizzolatti *et al.* 1996, Rizzolatti *et al.* 1997, Gallese *et al.* 1996, Rizzolatti *et al.* 2001, Stamenov and Gallese 2002, Behrendt 2011). The superior temporal sulcus, inferior parietal lobule and posterior parietal lobe have also been reported as active in the same experiments (Rizzolatti *et al.* 2001, Gallese *et al.* 2002). It is not surprising that these areas are also reported as the sites of egocentric and allocentric coordinate representations. Neuroimaging studies have found evidence of mirror neurons also in other areas: the supplementary motor area, hippocampus, parahippocampal gyrus and entorhinal cortex – which, as noted above, are also important for spatial orienting (Keysers *et al.* 2010). Although these reports are still to some extent controversial, they seem strong enough to warrant further hypotheses. Certainly, the presence of mirror neurons both in the language-related areas and in the spatial orienting areas seems to be significant for the DST framework.

Mirror neurons fire automatically in response to a real-world event such as an individual grasping an object, and not only as part of the observer's own plan for action. This firing has been taken by commentators to correspond with an internal mental representation of the observed event. However, it is not clear exactly what the nature of this representation is, nor how it might come to be associated with arbitrary sounds (or gestures) in human language. There has been a considerable amount of speculation around such questions (Rizzolatti and Arbib 1998, Hurford 2004, Arbib 2005, 2006, 2012 Fogassi and Ferrari 2007, 2012). It is frequently said that mirror neurons enable individuals to *understand* the actions of others – a formulation that implies a claim that 'understanding' is reducible to the firing of mirror neurons. This is an implication that deserves further philosophical discussion. It can only be noted here that a general claim has emerged that the phenomenon of understanding in linguistic communication is 'grounded' in neural motor schemas for action, and perhaps consists in 'simulation'.

This is a view that makes most sense in relation to the conceptual schemas associated with those grammatical categories found in human languages (verbs such as *hold*, *hit*, *see*) that are prototypically associated with various types of human action. However, it seems to make less sense as an account of how nouns (such as *tree* and *computer*) are represented and understood, though it is conceivable that certain types of object are so tightly connected with types of action that they might be part and parcel of a representational frame associated with mirror neurons for action. It is likely that the representation of noun referents (and also the wider understanding of verbs) is represented neurally in a widely distributed form that includes the right hemisphere.

What exactly might the process have been that took us, over evolutionary time, from mirror neurons in the monkey observing a conspecific's actions to the hypothesised linguistic function of mirror neurons in the human left frontal and parietal areas? Arbib (2005, 2006) argues that the language of modern humans involved several evolutionary stages, all built up on physical grasping, mirror neurons for grasping, and the emergence of simple imitation in pre-hominids. In the hominid stages, complex imitation for grasping develops, providing the ability to recognise, modify and repeat schematic mental representations of physical actions. At the next stage, protosign, a hand-based system of communication, emerges, gradually detaching itself from the mirror neuron system and moving towards protospeech, in which there develops the possibility of non-iconic conventionalised association between vocalised signs and concepts. Each stage provides 'scaffolding' for the next. Arbib suggests that the eventual development of human language is cultural rather than biological, and leads from action-object schemas with their mirror neuron underpinning to syntactic verb-argument structures. Whatever the details, the central point here is the striking co-occurrence in Broca's area of mirror neurons for action and the neural networks for language.

Arbib's theory seems to be consistent with DST, though there are two points where the emphasis is different. First, it is important to recall that the 'scaffolding' for language must consist not only of the action schemas whose substrate is the mirror neuron system but also of the spatial reference frames within which actions are embedded. Second, the extent to which action and space schemas carry through into apparently abstract conceptualisation remains an open question. Arbib rejects mirror neurons as a basis for more abstract concepts, while cognitive linguistics hypothesises the grounding of the latter in the former, as does DST.

11.2.3 Lateralisation and abstraction

When an individual observes the action of another and the mirror neurons fire, they are somehow decoupled from action (Borg 2007, Keysers 2010). To understand how mirror neurons might be involved in the emergence of language one has to surmise that mirror neurons, as internal representations of actions, became not only detached from motor activity but ceased to be automatic responses to observed actions; rather, they have in some way become coupled with linguistic items (which could have been originally gestural, then later sounds or visual marks). Mirror neurons would also have to be seen as providing the substrate for generalised schemas (see Gallese 2003, 2005, Gallese and Lakoff 2005), rather than fine-grained perceptions of particular events. In other words we have to posit a biological (or perhaps cultural) transition from mirror neurons to the kinds of cognitive generalisation that are available for lexical items and grammatical constructions.

Throughout this book I have invoked the notion of 'abstractness' and presupposed some kind of evolution from perception and cognition of what is spatial and concrete to what is schematic, categorical and generalised. This is where lateralisation enters the picture. Many of the processes that seem to be dominant in various regions of the left side of the brain appear to involve, precisely, the schematic, abstract and categorical, as argued extensively by McGilchrist (2010). This is consistent with the left-lateralisation of the human brain for language.

The notion of schema, or more specifically of 'image schema', has been central in the development of cognitive semantics – for example in Lakoff and

Johnson (1980), Lakoff (1987) and in the particularly clear account in Johnson (1987), who acknowledges the Kantian origin of the idea. In cognitive grammar, Langacker's diagrams all presuppose conceptual schematisation. In developmental psychology the idea of schema was elaborated by Piaget (e.g. Piaget 1954, Piaget and Inhelder 1971) and has been further developed by Mandler (2004). There is a convergence with the theorisations surrounding the discovery of mirror neurons. In Chilton (2009) I proposed a geometric model based on a prehension schema that could be easily linked with action schemas derived from mirror neurons. In that theoretical study I was concerned with probing the semantics of the verb *get*, suggesting that its various meanings can be naturally understood in terms of a reaching and grasping schema, modelled in an abstract vector-and-coordinates geometry.

More recently, empirical studies in neuroscience and neurolinguistics have turned their attention to abstracted (that is, schematic or categorical) representations in the brain in connection with the conceptualisation of space. Around the time that cognitive linguists were becoming interested in image schemas, Kosslyn (1987) distinguished theoretically between categorical and coordinate representations of space, a distinction by corroborated lesion and neuroimaging studies and linked with studies of lateralisation. The right hemisphere predominantly represents space in metrical viewer-centred coordinates. The origins of such neurally instantiated coordinates are located at different parts of the body (retinas, limbs, head, for example). These representations have to be precise because they are linked to motor actions such as navigating to goals and manipulating diversely configured objects. In general this fits with the tendency of right-hemisphere functions to be fine-grained and engaged, as well as with the idea that the right hemisphere is attuned to 'depth', that is fine degrees of distance from self needed for the manipulation of objects. What is involved here is visually guided action but also attention systems independent of executed actions.

The left hemisphere on the other hand deals in schematic categories. Spatial representations, in particular spatial representations of the kind encoded in languages (prepositions and demonstratives are the salient examples), are abstractly relational. English prepositions for example communicate relations between a figure and ground, as expressed in pairs such as *above/below*, *in front of/behind*, etc. Similarly, deictic expressions such as *here/there* do not denote precise places but relative positions or regions, with fuzzy boundaries, in egocentric coordinates. And demonstrative pairs like *this/that* do not denote precise places but degrees of relative distance egocentrically, relationally and with great variability (Kemmerer 1999; cf. Enfield 2003). There is now evidence from neuroimaging and lesion studies (Kemmerer 1999, 2006a, Damasio *et al.* 2001) strongly indicating that processing of these abstract linguistically coded spatial concepts takes place in the left inferior parietal lobe (supramarginal and angular gyri), the area receiving input by way of the dorsal 'where' pathway. Near–far representations prompted by language – those linked with the demonstratives, for example – are likely to be linked with separate circuits processing peripersonal space (the reaching–grasping space) and extrapersonal space (Kemmerer 1999, 2010; cf. also Coventry et al. 2008). We might hypothesise that the left dorsal pathway and the left inferior parietal lobe provide abstract versions of the homologous right hemisphere areas. Kemmerer (1999) suggests that it is the left parietal–occipital–temporal (POT) junction (thought to have expanded relatively late in hominid evolution) that is specifically involved in abstraction processes that subserve language:

Perhaps one of its functions is to restructure spatial representations in such a way that their content becomes more abstract and hence more available for cognitive purposes, as opposed to being limited to sensorimotor control. (Kemmerer 1999: 57)

Kemmerer extends this line of speculation to offer a possible neuroanatomical account of 'language-specific semantic structure', in particular the binary near–far distinction, locative prepositions, aspectual properties of verb meaning and thematic roles.

The finding that specific brain locations support spatial abstractions fits well with the theoretical proposals made throughout this book. For DST, categorical distance, that is relative as opposed to metric distance, is fundamental. The basic axis structure of DST consists of abstracted proximal–distal (peripersonal–extrapersonal relations) (see Chapter 2). This permits figure–ground (foreground–background) structure that is required for locative prepositions. It also permits the modelling of agent–patient relations in the same terms, treating agents as prototypically proximal (see Chapter 3 for this and transformations of the prototypical order). The conceptual difference between simple and progressive aspects of verb meaning can also be modelled in terms of closeness as opposed to remoteness (Chapter 4). Kemmerer further speculates that it is the POT region that contributes to the general tendency of human cognition, much discussed in cognitive linguistics, to project spatial representation into even more abstract domains such as time, causation and possession – projections that are built into the basic architecture of DST.

Subsequent work in lesion studies and neuroimaging also bears out this approach. Amorapanth et al. (2012), making explicit connections with cognitive linguistics, define a schema as 'any kind of representation (external or cognitive) where perceptual detail been abstracted away from a complex scene or event while preserving critical aspects of its analog qualities', specifically those that are 'spatial-relational'(Amorapanth et al. 2012: 226). These researchers were interested in whether different brain regions processed spatial relations between objects expressed in three different ways:

(i) in rich perceptual detail, (ii) in iconic schematic format and (iii) in the arbitrary signs of a human language. Accordingly, photographs of objects, geometric line-diagrams and English prepositions were presented to subjects. The subjects had frontal and/or parietal lesions in either the left hemisphere or the right hemisphere. It was found that patients with left-hemisphere damage (frontal, temporal gyri and premotor and primary motor cortex) showed significant impairment in an experimental task involving words or schemas, prompting the inference that these left-hemisphere areas are involved in the linguistic representation of schematic spatial relations. Interestingly, the general results indicated that the right hemisphere played a role in *non*-verbal processing of schemas and pictures: the right hemisphere (possibly the supramarginal gyrus) distinguishes between rich pictorial representations and schematic representations and may play a role in abstracting schematic from rich representations, making them available as categorical non-verbal structures available for association with linguistic expressions.[2] This kind of research supports a spatial abstraction (or schema) theory of linguistic meaning.

11.2.4 Theories of language and the brain

Linguists have linked specific aspects of language structure with the two-streams hypothesis of visual attention. Landau and Jackendoff (1993) link each stream with a particular grammatical category. They propose that the ventral (or 'what') stream underlies common nouns denoting physical objects, since the neurological descriptions present the ventral stream as instrumental in the qualitative identification of objects. The dorsal ('where') stream, since it is concerned with the location of objects relative to the perceiver, is linked by Landau and Jackendoff with prepositions, since the dorsal stream is involved in the location by orientation and distance within egocentric reference-frame coordinates. However, as Hurford (2003) points out, prepositions are also allocentric (they can relate one object to another), and allocentric relations are associated in non-linguistic perception with brain regions other than those traversed by the dorsal stream. Hurford (2003) himself, starting from the formal logic-based account of linguistic meaning, argues that the ventral 'what' and the dorsal 'where' streams correspond respectively to the predicate and the arguments of simple logical propositions. It is important to note that predicates may be of two kinds at least: they may be qualities of arguments or relations (spatial or actional) between them. It is the qualities that the ventral stream may deliver. The DST approach is to model

---

[2] Kemmerer suggests the left POT is involved in linguistic abstraction while it is *non*-linguistic pictorial abstraction that Amorapanth *et al.* link with *right* supramarginal gyrus, a structure that lies in the POT.

11.2 Space, the brain and language

entities (arguments) explicitly as locations (corresponding with dorsal stream functions) and relations between entities, thus unifying the account in terms of spatial representation and leaving out the posited function (predicating qualities) of the ventral stream.

In seeking to correlate precise brain regions with different components of language, researchers are inevitably influenced by the particular theory of language they happen to favour. For example, Hurford's (2003) use of the two-stream hypothesis is dependent on his assumption of Fregean argument–predicate formulae in formal semantics. In neuroanatomical work, where in the brain you look for a particular language component depends on what components your theory of language contains – for instance, whether your theory takes language to be autonomous or connected to other brain systems, and how (if at all) it relates syntax to meaning. Working in the opposite direction, the prevailing neural models influence the theorisation of language. The two-stream model of visual attention is widely accepted, though controversies over details continue. But even if there were consensus among neuroscientists on the two streams, the way the model influences language theory is highly variable. This is of course as one should expect; it is the way science advances. A brief and limited review is in order at this point.

Hickok and Poeppel (2004), followed by Saur et al. (2008), propose a ventral system and a dorsal system around the sylvian fissure that subserve not only visual attention and object identification but also certain aspects of language processing. The 'dual stream model' of Hickok (2009) claims that a bilateral ventral stream (superior and middle temporal lobe) processes speech for comprehension and that a left-dominant dorsal stream (involving temporoparietal junction linked with premotor cortex) turns speech signals into articulatory representations. As regards syntax, Hickok and Poeppel (2004), who are not explicit about what they assume 'syntax' or 'grammar' to be, note that lesion and imaging studies challenge the long-held view that syntactic processing is primarily supported by the inferior frontal gyrus (Broca's area). Syntactic processing can in fact can survive lesions to Broca's area. Other brain areas are clearly involved: for example, the anterior temporal lobe, which appears also to process semantic content, is active during what is referred to as syntactic activity. How such data are interpreted depends on how the relation between meaning and grammar is theorised.

Significantly, Hickok and Poeppel note that the involvement of multiple brain areas is not surprising, given that on-line language processing must require the integration of several linguistic dimensions, including grammatical relations, lexical meaning, intonation and prosody.

Grodzinsky and Friederici (2006) and Friederici (2009) use a particular theory of language (generative syntax in Chomsky's minimalist version) that models syntax in terms of arbitrary symbol manipulation and is segregated from semantics. They argue for the location of specific syntactic knowledge components (as defined by the theory of syntax they adopt) in the left frontal gyrus, temporal gyrus and temporal sulcus. Phrase structure building, for example, is said on the basis of lesion and fMRI observations to take place in the left frontal operculum, ventral premotor cortex (roughly Broca's area) and anterior superior temporal gyrus (next to Wernicke's area), with syntactic integration taking place in the posterior section of the superior temporal gyrus (Wernicke's area). The fibres connecting these areas are said to be crucial for the processing of complex sentences and for certain operations right-hemisphere areas are claimed to come into play. While these observations seem precise, the fact remains that the areas identified are also noted by other researchers as being involved in semantic processes, and indeed spatial processes, and do not seem therefore to unequivocally point to autonomous non-semantic syntax.

Consistently with these findings Griffiths *et al.* (2013) establish the importance for language processing of the two pathways that were already known to be implicated: the arcuate fasciculus (a tract connecting Broca's and Wernicke's areas dorsally) and the extreme capsule (ventrally connecting Broca's and Wernicke's areas along the superior temporal gyrus). Like the researchers mentioned in the preceding paragraphs, Griffiths and colleagues also view syntax as a set of combinatorial operations, 'underpinning language comprehension'. The nature of this underpinning is not made clear and, though comprehension must involve meaning, semantics seems to be seen as something separate.

Using neuroimaging methods with left-hemisphere-damaged patients, these researchers administered test sentences – sentences that were syntactically anomalous, sentences that were syntactically normal but had some lexical oddity, and sentences in active or passive voice. This approach of course reflects the theoretical paradigm of syntax-centred generative linguistics. Deficits in response times on the tests correlated with anatomical disruption of the left arcuate fasciculus and extreme capsule, the connecting pathways between Brocas's and Wernicke's areas. The authors report that in the patients whose syntactic performance was impaired semantic processes were intact and the conclusion was drawn that the two pathways are involved in specifically syntactic processing. However, I suggest that the assumptions behind the measuring of semantic comprehension are not clear and the syntactic–semantic separation itself may conceivably be influenced by the theory of syntax that was assumed. A particular issue relates to the fact that verbs determine the syntactic frames they license, but this relationship can be seen to derive from the semantics of the verbs and is not arbitrary. To this extent syntax and semantics would not be expected to be neurally segregated. The situation is likely to be more complex, however. Meaningful but schematic representations linked to verbs might be processed in the left-hemisphere language structures referred to by Griffiths and colleagues, while being additionally linked to distributed areas of the brain, including the right hemisphere, for richer and more detailed semantic representations and pragmatic processing. In general, while dual-stream models remain in the neuroscience of language, the evidence seems increasingly to be that language is dependent on wide connectivity between the classic language areas (for a review see Dick and Tremblay 2012), as well as other areas.

Work in neuroscience has shown signs of moving towards a cognitive-linguistics theory of language. Pulvermüller (2010), for instance, despite beginning with what appears to be formal generative theory, provides evidence for an 'embodied' view of semantics and a 'syntax embedded in semantics'. Pulvermüller and Fadiga's (2010) review of neuroimaging studies of aphasic patients supports the general claim of embodied theories of language – that the mirror neuron system and left perisylvian sensorimotor areas are crucially involved in semantics and grammatical constructions. On this issue there is increasing convergence of neuroscience, experimental psycholinguistics and cognitive linguistics. While it is accepted that the abstract symbol manipulation view of language may be needed for some aspects of sentence processing, the phenomenon of meaning itself, including the meanings of particular grammatical constructions, comes from embodied grounding (for endorsement of this approach see e.g. Feldman and Narayanan 2004 and the review by Fischer and Zwaan 2008).

The striking overlap of language functions and spatial–motor functions in certain left cortical areas is striking. It has been shown experimentally that these overlapping functions are connected. What would happen if the brain were dealing with an action schema for pushing something in one direction while simultaneously uttering a sentence about pulling in the opposite direction? If the action-based conjecture concerning linguistic meaning is plausible, one would expect some kind of disruption, since spatial action schemas and linguistic action expressions should be in concert. Glenberg and Kaschak (2002), expanding a series of experiments reported in Kaschak and Glenberg (2000), tested this idea by eliciting subjects' responses to *sentences* implying action directed toward self or action directed away from self, while performing an *action* in the same or opposite direction. Importantly, the test sentences included simple physical actions ('close the drawer/open the drawer'), concrete ditransitive transfer verbs ('Courtney handed you the notebook/you handed Courtney the notebook'), dative construction transfer sentences, and 'abstract' versions of such transfer verbs ('Liz told you the story/you told Liz the story'). The results, put in very summary form, showed that 'merely understanding a sentence can facilitate or interfere with a physical response' (Glenberg and Kaschak 2002: 560), and that it is deictic *action* not simply near/far *location* that is involved. Results of this kind are consistent with embodied meaning theories, such as cognitive grammar and construction grammar, but not consistent with abstract operation theories of grammar.

Generalising to a theory of grammatical meaning, which they call the 'Indexicality Hypothesis', Glenberg and Kaschak propose that words and phrases are linked to perceptions of their real-world referents. They also connect the processing of semantically acceptable sentences to experiential relations between action and objects (a process they call 'meshing' of 'affordances', the latter being, for example, the perceptual–cognitive adjustment of grasp to a particular object). Furthermore, they argue, constructions such as the double object construction, which basically express the physical transfer of objects, are extended to 'abstract' transfer meanings. Similarly abstract language about causation is derived from forceful bodily action on the environment, as argued by Talmy (2000 [1988]). But the precise underlying neural mechanism for such 'projections' from concrete to abstract has not, to my knowledge, been conclusively accounted for, although the process may be similar to the notion of 'abstraction' discussed earlier, which has been linked by some researchers to categorisation and to the mirror neuron system. The model of grammar that matches this approach well is found in frame semantics (as developed in Lakoff 1987, Fillmore *et al.* 1988, Goldberg 1995, 2003, Croft 2001, Bergen and Chang 2005; see also Hoffmann and Trousdale 2013).

As it happens, the Glenberg and Kaschak experiment works with what are the essential ingredients of DST – deictic concepts expressed by words like *near* and *far* (and thus with egocentric reference frames) combined with deictic directions toward or away from self. This kind of study tends to bear out the assumption that I have made throughout this book and in particular the assumption made in Chapter 3, Section 3.2, that a concrete directed-force schema (modelled as a force vector) underlies abstract transfer verb meanings. A further inference would be that it is activity in the left parietal and inferior frontal regions, linked to mirror neurons and frames of reference, that constitutes such verb meanings.

Construction grammar, unlike generative grammar, does not take sentences to be the product of non-semantic combinatorial rules that rearrange grammatical symbols (NP, VP, etc.). Rather it takes sentences to consist of unitary constructions that are inherently meaningful and modelled as strings of grammatical symbols that are themselves inherently meaningful. The meaning comes from learned association with event schemas, in which the semantics of verbs plays a crucial role. Different verbs are associated with subtly different constructions and event types. Construction grammar shares many of the theoretical principles of Langacker's cognitive grammar, though the latter, like DST, makes far more use of schematic diagrams to model constructions.

Kemmerer (2000, 2006b) explicitly adopts a cognitive grammar framework. In this approach grammaticality judgements are not seen as the application of abstract rules but as the natural acceptability of particular verbs (specifically, action verbs) in particular construction schemas that have their own idiosyncratic meaning. Kemmerer has also found a dissociation between the ability to understand specific verb semantics on the one hand and, on the other hand, the ability to understand semantic differences among grammatical constructions, suggesting that verb meaning and constructional meaning are processed in different areas of the left hemisphere (Kemmerer 2000, 2003). The brain regions implicated for impairment in constructional understanding were the left inferior prefrontal region and the left anterior supramarginal gyrus. Kemmerer conjectures that neurons for constructional meanings will be found to be close to neurons for specific verb meanings, in the frontal–temporal–parietal networks and perhaps close to the networks handling syntactic processes for clause structure.

The picture that is emerging, then, is one in which neuroanatomical studies indicate an organisation of language that resides in the left perisylvian structures (including classic Broca's and Wernicke's areas). Some of these structures are assumed to involve systems for analysing formal morphosyntactic relations within clauses. Note, however, that even sentence organisation and the processing of it could turn out to be abstract but in a sense embodied, as in the hypothesis that syntactic relations derive from motion detectors that respond to sequence (Pulvermüller 2010). There remain unanswered questions both for linguistic theory and for neuroanatomical topography. However, the progress and direction of neurolinguistic research provides a degree of corroboration for the type of abstract modelling that DST outlines.

Chapter 3 of the present book is consistent with the construction grammar approach, distinguishing different types of verb meaning and relating them to the kinds of conceptual structure they are associated with by virtue of their meaning and as a function of alternating perspectives on objectively identical events. However, construction grammar still describes the phenomena it seeks to model in arbitrary symbols or in English words (e.g. the caused motion construction Subject–Verb–Object means 'X causes Y to move along path Z' describes the sentence 'Bert knocked the vase onto the floor'). DST explores a metalanguage – natural vectors, coordinate systems and referent points in those systems – that is cognitively and physically motivated and can thus mediate between abstract modelling and embodied meaning.

11.3 Deictic Space Theory and the brain

Deictic Space Theory proposes a motivated formalism for modelling fundamental conceptual structures that are prompted by linguistic input and which, in the opposite direction, shape the linguistic output in language production. The most general hypothesis is that linguistically prompted conceptual representations are consistent with, and in evolution exapted from, more basic cognitive functions – in particular, visual attention, frames of reference, locomotion and action schemas, all of which arise from an organism's interaction with physical space. Below I summarise some entirely speculative hypotheses concerning the ways in which the fundamental architecture of language that DST seeks to model may be linked with brain structures.

11.3.1 Abstraction

The ability to derive abstractions and categories from spatial structure, locomotion and action on the environment underlies the conceptual space postulated in DST. This conceptual space makes it possible for attention to be focused on highly abstract relations between self and the world and relations between independent entities in space and time. The mirror neurons found around the left perisylvian fissure may be crucial in the formation of schematic representations of action events. It seems plausible to propose that the spatial reference frames within which actions take place are abstracted into the schematic dimensions of the conceptual space hypothesised in DST. Mirror neurons are not the only systems accounting for abstraction and categorisation, since nouns, for example, prototypically stand for abstracted entity categories as distinct from action categories.

11.3.2 Frames of reference

Deictic Space Theory proposes three fundamental reference frames defined on three axes. The d-axis is a scale of attentional distance relative to the self. It may correspond with the processing of near (peripersonal) and far (extrapersonal) space in the dorsal stream. The research reports point to the involvement of the intraparietal sulcus and temporoparietal junction, with involvement also of the hippocampal formation. The d-axis is the egocentric frame in which the self focuses attention on referents (linguistically, noun phrases) and acts on them. The actions, encoded in language as verbs and prepositions attached to noun phrases, are modelled as vector relations between points in DST, which may correspond with schematic effects produced by mirror neurons.

Within the egocentric space are embedded allocentric frames of reference. Each entity with a location in the base egocentric frame has its own potential allocentric frame of reference, which can also be conceptualised as egocentric relative to that entity. Schematic allocentric frames have their neural substrate in the posterior superior, middle temporal gyri and again the hippocampal structures. In linguistic conceptualising, egocentric and allocentric frames

11.3 Deictic Space Theory and the brain

would be co-active according to DST, which embeds allocentric representations in the basic egocentric frame of the speaker.

The t-axis models egocentric experience of time as conceptualised via linguistic structures. Personal experience of time depends on the hippocampus and the left frontoparietal regions. The temporal scale may be seen as a metaphorical abstraction from space-based egocentric reference frames in the parietal cortex; it may also involve abstraction from the hippocampus derived from episodic memory process.

A third scale (the m-axis) is posited in DST in order to account for a significant set of human linguistic abilities, viz. those related to the linguistic encoding of reality assessments. What the neural correlates of this dimension might be is unclear but the hypothesis is that they may be connected with the spatial scale of proximity and distance (the peripersonal–extrapersonal gradient). It may be relevant to look for the neural correlates of risk and probability assessment in action planning as the evolutionary basis for higher-level epistemic judgements about the environment. It is also plausible to think that the epistemic m-axis is a product of the development (evolutionary or cultural) arrived at by abstraction from the d-scale. However, in actual language use, the way a speaker assigns epistemic distance depends on judgements about what is real and such judgements are the product of sensory engagement with the world linked to cultural knowledge and inferential processes – a set of connections that is not likely to be the product solely of the left hemisphere's language areas.

11.3.3 Spatial relations and prepositions, action schemas and verbs

Categorical spatial relations between a locandum and a location are modelled in a highly abstract form in DST, since DST is itself not a model of physical space: on the d-axis either the locandum or the location is foregrounded (i.e. is relatively close to S), and the location is at the tail of the vector. DST does not directly model the categorical spatial relations denoted by prepositions; these are processed in the left temporoparietal junction, the region also for egocentric reference frames. It is worth adding that DST claims (see Chapter 8) that the clause 'complementisers' *that* and *to* are high-abstraction spatial concepts linked with an embedded reference frame.

Two other types of relation are modelled by vectors in DST and labelled for specific meanings: locomotion from one point to another and transitive/intransitive actions by an agent on an object. Locomotion schemas may involve the hippocampus, which is active in navigating the environment, as well as motor cortex governing the organism's self-propulsion, and the perception of movement of other bodies in the environment. The supplementary motor cortex, responsible for planning movement, could conceivably be the area where schematic representations of locomotions are generated. With regard to transitive action schemas, the extensive literature on mirror neurons suggests that the directed action vectors in DST can be related to activation of mirror neurons in the perisylvian region. Subregions of the left frontal–temporal–parietal network appear to be activated for specific types of action associated with specific verbs (reflected in the 'labelling' of vectors). The basic meaning of material-action verbs is equated with simulation in distinct regions. The neural sunbstrate of perception and cognition verbs remains an open question but may involve simulations in the visual-attention pathways. Extensions of basic locomotion and action schemas are postulated for seemingly 'abstract' meanings. The reach–grasp action schema in peripersonal space is the most fundamental relation and underlies the natural unit vector in DST. More speculatively, embedded in egocentric reference frames it is the core of the sense of self and other.[3]

11.3.4 Transformations

The term *transformation* in DST refers to a natural spatial operation, corresponding to a geometrical transformation of a coordinate system. Such natural transformations arise in orienting and navigating in physical space. It is possible to imagine oneself at a distal point in one's egocentric reference frame and to imagine oneself using that point for a new set of coordinates; one can also imagine another person doing the same. It appears that people readily adopt the spatial perspective of others (Samson *et al.* 2010). Similarly, the abstract, non-spatial transformations in DST can copy the base egocentric system to a distal point in the base system. There is thus a correspondence between spatial transformations used in spatial orienting and certain kinds of grammatical construction that embed one set of coordinates inside another. For example, a belief sentence such as *(John believes that Mary wrote the report* (see Chapter 7) embeds new coordinates at the distal point *that*.

In the neuroscientific literature, the term 'transformation' refers to an organism's switching between egocentric coordinates centred at different locations on the body and between these and allocentric frames with coordinates located at landmarks. Such mental translations are reported to involve the hippocampus, the parietal lobe and posterior cingulate cortex.

---

[3] The right temporoparietal junction, whose left homologue processes egocentric reference frames for grasping, is linked with sense of agency, reorienting of attention (a transformation of coordinates), the ability to adopt the perspective of others, and theory of mind (Decety and Lamm 2007). The role of the temporoparietal junction in both hemispheres means that it is of particular interest for DST because of the centrality of S (self and subject as speaker) in that theory.

11.3 Deictic Space Theory and the brain

In DST a further type of transformation – the reflection transformations – is used in order to model highly abstract linguistic conceptualisations involving counterfactuality (Chapters 6 and 10). I know of no specific studies of transformation in studies of spatial cognition, though it is found in linguistic analyses of certain spatial prepositions (Levinson 1996, 2003). However, there are some indirect empirical indications. Humans do sometimes seem to perform a mirror reflection of their own coordinates when they project their own left hand and right hand on to a counterpart in face-to-face encounters ('canonical encounters': Clark 1973). However, it may be questioned (Levinson 2003: 85–8) whether a true reflection is involved here, or rather a mixture of speaker's egocentric axes and a reflection of their front–back axis only (or translation and rotation of the front–back axis). Levinson is disinclined to accept the role of a true reflection transformation in English spatial expressions that might be thought to use them (e.g. *the cat is in front of the tree*). Nonetheless there is some cogent albeit indirect evidence for the neural reality of reflection transformations that comes from lesion studies and concerns reflections in mirrors. Normally people do not confuse real space with space reflected in a mirror – which implies that they recognise, and thus their brains unconsciously represent, a reflection transformation. This suggests that people normally have a dual representation, one of the real space and one of the reflected space, as also suggested in the DST diagrams of Chapters 6 and 10.

Ramachandran *et al.* (1997) describe a condition that they call 'mirror agnosia', in which patients reach for objects behind a mirror or literally 'in' it, and may even verbally assert that objects are literally 'in' it. This phenomenon occurs in patients with hemispatial neglect in the left visual field resulting from lesions in the right parietal–occipital–temporal area. It can be inferred that normally it is the right parietal lobe that is involved in cognitive alternations (that is, geometric transformations) between a real-world reference frame and a reversed non-real world as reflected in a mirror, though Ramachandran and colleagues do not describe what may be going on in quite these terms.[4] DST postulates mirror reflection (combined with a translation) in order to account for some complex conceptualisations that are effects of particular grammatical constructions – counterfactual conditionals and deontically modalised expressions. Though there are many questions to resolve, the point is that mirror transformations seem to be neurally instantiated and in principle available for linguistic abstraction.

---

[4] They refer, interestingly, to 'a rather peculiar form of dual representation or "mental diplopia"' (Ramachandran *et al.* 1997: 647).

11.3.5 Recursion and the Deictic Space Theory

The ability to create indefinitely many complex sentence structures is generally acknowledged to be the core property of human language, at least since Chomsky (1957, 1963, 1965, 1995). DST accounts for this ability in terms of embedding rather than recursive rules, which are over-powerful and have not yet been reliably located in brain structures. Moreover, evidence from a number of languages strongly suggests that in practice recursive embedding is limited to a depth of three for central embedding (Karlsson 2007). By contrast, neural correlates of frames of reference appear to be well established and provide the potential for abstraction and for nesting.

Allocentric frames must always be in some way embedded in egocentric frames at some level of cognition. At the simplest spatial level, if I perceive a tree, it is in my egocentric frame, and within it I can attend to a rock that is located in an allocentric reference frame relative to that tree, or I may attend to another person acting on another tree within their own egocentric frame...and so on. In DST an *embodied* recursive mechanism is proposed in terms of the nesting of abstracted reference frames within one another; allocentric frames are always nested within the base egocentric frame and further egocentric and allocentric frames are potentially embedded within one another. Thus embedding is present in the fundamental domain of spatial cognition and action and is available for exaptation to language in the course of evolution. This approach to the recursive property of human language differs considerably from that of Chomskian generative grammar, as does the account of recursion developed in the work of Arbib (2005, 2006, 2012) and Jackendoff and Pinker (Jackendoff 2002a, 2011; Jackendoff and Pinker 2005).

In the wider context of discussion, recursion should be distinguished from embedding (nesting). Recursion in linguistics has generally been defined in terms of symbol manipulation algorithms, rules that can apply endlessly to their own output to produce constituent-within-constituent structures. The claim of Hauser, Chomsky and Fitch (2002) is that the human language faculty 'in a narrow sense' is uniquely defined by the possession of such a symbol manipulating ability, acquired by the human species alone during evolution. However, it can be argued that what language displays is an abstraction from spatial cognition and action in the spatial environment. Recursion algorithms, of the kind formulated in Chomskian syntax generate embedding structures, but embedding does not have to be defined by algorithms. The embedding we see in linguistic structure may derive, for example, from the embodied experience of hierarchical nesting of parts within whole objects, of whole objects within an environment, and of action sequences within action sequences. Such part-whole hierarchies occur across behavioural and cognitive domains, for example, action, vision and music (Arbib

2005; Jackendoff 2011; Jackendoff and Pinker 2005). It may be noted that vision and action are fundamentally dependent on reference frames, and that in general DST's claim about the natural nesting of reference frames support the arguments of Arbib, Jackendoff and Pinker concerning recursivity In any event, it seems that DST has the capacity to handle embedding, including the problematic phenomenon of central embedding (see Appendix).

11.3.6 Deictic space and the dorsal stream

Is DST compatible with the two-streams theory of attention? Yes and no. Certainly, much of DST can be said to correspond in key respects with the dorsal 'where' stream in so far as it models 'locations' in an (abstract) egocentric space that is characterised by near–far relations in peripersonal and extrapersonal space and incorporates relations between entities corresponding either to cognised spatial relations or to motor action schemas represented in mirror neurons. However, allocentric reference frames are underpinned by temporal and hippocampal areas which are 'ventral'. Despite this, DST corresponds predominantly with the dorsal stream 'where' functions, in that it does not 'see' the semantic details of objects which the two-streams model attributes to ventral processes, but has access only to schematic 'labels'. The rich knowledge we have about objects from experience is stored in conceptual frames in ventral and widely distributed neural networks. This is not to say that there is no interplay between the dorsal and the ventral streams (*mutatis mutandis*, semantic and grammatical information): interplay between the two streams is now part of the modified two-stream theory.

An important part of the present account is that the fundamental scaffolding DST describes is likely to correspond with dorsal-stream features that are part of the classic left-hemisphere language areas, broadly Broca's and Wernicke's areas. The left hemisphere, and the basic structure of language itself, deal in abstractions and categorisations. This does not mean that in language *use* the right hemisphere is not involved – indeed it is crucial for contextualisation and implied meaning, without which language would not work as a means of communication. It is possible, however, that the left hemisphere can become in some sense decoupled from the contribution of the right hemisphere and that this is essential to creative uses of language while running the risk of dysfunctional or pathological disconnection.

11.4 Deictic Space Theory and the mind

While DST seeks to ground itself in the brain's representation of external reality, this does not mean reducing mental activity to physical brain states. The complexities of consciousness and the mind–body problem are matters that are beyond the scope of this book, but such questions won't go away, so something, however crude, needs to be said about them and about what it is that any linguistic theory, including DST, is modelling. The notion of supervenience is useful to a degree, but is not entirely satisfying. We can say that mind is supervenient on brain states, in the sense that without a brain there is no mind and that if two mind states differ then the underlying brain states also differ. The unsatisfying part here is the deterministic implication: what changes the brain state? Can it be the mind itself? This is not all. Phenomenologically, one's experience of mental states gives them the 'feel' of being independent of any brain activity determining them, especially when ideas are circulated by way of language among minds in a community. Explanation can be attempted in terms of layers of supervenience but the supervenience relation itself is territory for exploration. The problem is that if the relation between the underlying and the supervenient levels is too tight, we get a materialist reduction, if it is too loose the relation of mind to brain becomes inscrutable. There is a kind of double vision here, an oscillation between the mechanistic urge to find structured components on the one hand and intuitive holistic experience on the other.

What we may be seeing here is that is in all probability a product of the way the human mind–brain works – two different modes of knowledge and knowledge construction. McGilchrist (2010) argues at length, and on the basis of wide range of neuroscientific data, that the complex tensions in human thinking arise from the different modes of the left and right hemispheres of the human brain. It is important to avoid oversimplified dichotomies but given the predominant left-lateralisation of language the evidence that the left hemisphere deals in the reductive and the abstract cannot be left out of the picture, particularly in cognitive linguistics, which is precisely concerned with finding the neurally embodied correlates of abstract conceptualisation. Moreover, if McGilchrist is right, the 'double vision' of the cognising mind is bound to inhere in our theorising. Not surprisingly, linguistic theories have searched for components, rules and structure, while seeking to reconcile these with the ungraspable nature of meaning and the seemingly messy complexities of context. The tendency to split our theoretical understanding of language into 'competence' and 'performance', morphosyntax on the one hand and semantics–pragmatics on the other, is a natural outcome, perhaps reflecting something essential to the theorising process, but also reflecting the left-hemisphere bias of language itself. This is not to say, however, that in practice and experience language is integrated and seamless, the two hemispheres working together in the real world; theory will need, ultimately, to reflect this fact.

The theory put forward in this book does indeed risk looking machine-like. However, two points are important. The first is that the theory is a theory of

11.4 Deictic Space Theory and the mind

embodied language and the geometric diagramming reflects the intuitive experience of being in space and time – since natural geometry itself may be one of the embodied brain's activities. Second, the language faculty itself does have a left-hemisphere core, albeit one that is interconnected with right-hemisphere processes of understanding the world. So, it might be argued, one should not be all that surprised to find that a model of the basic architecture of the language ability has an abstract and structured character.

A particular way of putting some of the questions raised above is this: how can a theory that rests on the neuronal representation of space, movement and action illuminate the creative abstractions that language use can convey? The concept of 'creativity' is not new in linguistic theory. Chomsky claimed (e.g. Chomsky 1965) that 'creativity' is a defining feature of language and seeks to account for it in terms of formal recursion. While it might reasonably be thought that the concept of creativity involved here is rather limited, some form of constrained recursion is clearly part of the ability to produce, 'generate', new sentences and new conceptualisations. But it is not sufficient for any wider understanding of the creative potential of language to produce new meaning. Taking a fuller view of creativity, cognitive linguists have concentrated on conceptual metaphor and blending as crucial to linguistic creativity (Lakoff and Johnson, 1980, 1999, Lakoff 1987, Lakoff and Núñez 2000, Fauconnier and Turner 2002). The process of abstraction itself is perhaps the most fundamental means of creatively transcending physical space and action. While DST implies conceptual creativity in its claim that abstraction is derived from spatial cognition and motor activity, it is not about metaphor as such but about the potential for language to provide conceptual scaffolding that provides a means for the human mind to move away from physical referentiality not only for the purposes of planning but also for what we commonly call creative thinking in a multiplicity of domains.

The theoretical approach of the present book (especially Chapters 6 and 10) extended the basic geometrical idea (overextended some might say), in order to model some of the abstraction processes that may derive from spatial cognition. Essentially these have to do with processes that I model in terms of geometrical transformations, which I take to be derived from natural perceptual–cognitive processes of humans.

Approaches that concentrate on computational models and neural grounding confront an obvious property of human language – the ability to refer to and think about what is beyond the here and now. Arbib, for example, speaks of verb tenses and other devices that 'express the ability to recall past events or imagine future ones' and 'communicate about other "possible worlds" ', looking to the 'cognitive machinery' of episodic memory in the hippocampus and for action planning in the frontal cortex (Arbib 2005: 109). It is doubtless true that these regions are involved, but in the form of the categorised abstractions that are typical of the left-hemisphere language areas. In language it seems likely that there is not a direct linkage from verb tense forms and their schematic meanings and epistemic–modal expressions to the basic memory and orienting functions of the hippocampus and frontal cortex, but rather one mediated by abstracted and categorical representations. For the meaningful use in real life of tense forms, however, reference to times and to possibilities (and impossibilities) must be connected to the neural substrates of episodic memory, communally shared history representations, and to reality-checking mechanisms in other, especially right-hemisphere, brain regions.

The t-axis of the deictic space in DST is not directly referential with respect to time, but to representations of time, and not directly or simply to right-hemisphere experience of the flux of time, but to abstracted representation of time in terms of the relation to one's current 'position' in time. This language-relevant representation of time is likely to be partly at least based on metaphorical abstraction from deictic space to a bidirectional linear conception of time – at least in English and similar languages – in which times are regions and sometimes points relative to one another and to the subject S.

There seem to be overlaps and analogies between reference to time beyond the now and the reference to 'possible worlds'. Future worlds can also be thought of as possible worlds; both are non actual. Intuitively this does not seem quite right, since future worlds are connected to intention, volition and expectation. Moreover, in language expressions of time and expressions of modality are able to combine, i.e. they largely do not exclude one another. While the epistemic m-axis of DST may be conceptually connected with, perhaps derived from, deictic distance in spatial experience, it seems to me that in the deictic space model it should be represented as orthogonal, both to the t-axis and the d-axis (*pace* Jaszczolt 2009; see also Chilton 2013).

How does the basic language scaffolding modelled in DST relate to the real? The d-axis postulates things that exist from S's point of cognitive viewpoint, whether or not they do from anyone else's. In real-world language use there is normally connection with intuitions, perhaps dominantly made in the right hemisphere, concerning what is experientially real. Setting aside metaphysical questions, there are various possible ways in which the deictic space relates to independent reality. One possibility is that the referent entities do exist or are communally believed to exist. Another is that S is simply mistaken on one or both of these counts. Another is that S is mistaken because of a pathological decoupling of the left-hemisphere language system from right-hemisphere engagement with the world (i.e. the first two possibilities, including mistakes about reality), as could conceivably happen in psychopathological conditions in which we might think of the left-language system as going into independent overdrive. A further possibility, not a pathological

11.4 Deictic Space Theory and the mind

one, is that human minds can set up a mental space that is *knowingly* decoupled ('suspension of disbelief'), as in creative mental states that involve language (not all do, of course).

The m-axis reflects the mechanisms languages all have for consciously indicating some degree of cognitive uncertainty on the part to S as to what is actual– and for proposing positive possibilities. In normal language use one has therefore to assume that m-axis locations depend on epistemic judgements as to what exists in time and space and probably links to the kind of cognitive engagement characteristic of the right hemisphere's grasp of reality, though also perhaps from conscious or unconscious inferencing from evidence. Again, however, similarly to what may be said about the d-axis, the m-axis, if to some degree spinning independently of reality-checks, can be creative (the 'what if...' thought experiment) or pathological ('what if aliens are controlling my thoughts...?'). In the latter condition there would be also interaction with d-axis presuppositions of entities that are not reality-checked by non-linguistic parts of the brain.

The ability to engage in counterfactual conceptualisation, not only in the sense of envisaging possible worlds, but in the sense of worlds that the subject knows to be counter to fact in the sense of contrary to (i.e. S's) reality, is one of the most intriguing manifestation of the capacity of language structure to take us beyond the here and now. It may be the case that without language it is difficult to conceptualise, or at least to exercise sustained reasoning about, entities and events that are known not to be real. The utility of counterfactuals as one element in shared planning for future action is fairly self-evident ('if we do/don't do x, then y and z would happen/not happen...'). What is less evident is how such conceptualisations might be neurally instantiated. By using simple geometrical ideas, I have rather speculatively suggested that mirror-transformation of axes allows us to see a way in which even such highly abstract conceptualisations might be grounded in embodied spatial cognition, neurally active in brain regions responsible for reference frames. The actual uses to which counterfactual conditional thinking is put is, however, not easily reducible to determination by the brain.

Another way of going beyond the here and now is to go beyond *my* here and now into the mind of another person. Somehow individual humans, and perhaps some other animals, know that other individuals have minds that are similar to but not identical to their own; they have a theory of mind. It seems very likely that a number of verbs and their grammatical constructions in all languages support the ability to fix and communicate representations of one's representations of the states of another person's mind: x knows y, x wants y, wants to y, believes that y, etc. Some counterfactual conditionals, usually without their *if*-clauses, combined with such theory-of-mind constructions, enable us to speak and reason about absent people's minds, including those of the dead: *this is what she would want, she would have wanted this*. Such complex conceptualisations are essential to human consciousness, continuity and culture. They involve the *t*-axis, bringing the past into the present; pragmatically they can influence present decisions and future actions.

Whether language is necessary for the acquisition of theory of mind, even if the potential is innate, is a question that requires investigation. What is involved in theory-of-mind sentences is a transposition of a schematic copy of S's egocentric axes to that of another person, as modelled in DST. This picture is in line with the idea of stepping into another's place, as expressed in different terms by Kant amongst others. This corresponds with versions of theory of mind that understand it in terms of empathy and perspective taking. The accounts of theory of mind that see its neural basis as grounded in 'mirror' neurons make sense in this context, except that mirror neurons do not, so far as I can see, constitute strict geometrical reflections of an observed action but schematic simulations that place a simulacrum self in the place of an other. It may of course be the case that the linguistic conceptualisations of other minds via particular grammatical conceptualisations are additional to spontaneous everyday 'reading' of the intentions of others that do not require language. Linguistically mediated conceptualisations may be a creative cultural extension in the life of individuals in communities (Gallese and Goldman 1998, Gallese *et al.* 2004) but they are rooted in the spatial.

The ability to simulate other minds and take their perspective has been held to be part of moral judgement. I will not enter here into the far-reaching question as to whether moral values should be seen as embedded in a subjective perspective or in a neutral and objective 'view from nowhere' (Nagel 1986) – an intriguing question for DST. I have, however, broached the question (Chapter 10) of deontic conceptualisations associated with modal verbs, tense forms and other expressions. These forms encode highly abstract concepts of compulsion or binding force but not the background value systems they operate on. What 'should be done', 'ought to be done', 'must be done', etc. is some action that an individual speaker believes to be universally right or is right in or for a particular situation. Value systems themselves that underpin such beliefs are bound up with social customs, legal institutions, philosophical and religious systems of thought, and possibly with innate empathy. Languages develop vocabulary and concepts for values historically but these are distinct from concepts of moral force and are developed in complex discourse communities over time. The role of deontic expressions is to provide the element of moral force.

The main reason for the complexity of the conceptual structures associated with deontic modals is that categories of deontic meanings (broadly, obligation and permission) require conceptualisation of actions and states of affairs that are either not actual or are actually counterfactual. Chapter 10 of the present book is not concerned with principles of right and wrong as such. It is concerned merely with the ways in which linguistic expressions of deontic 'force' appear to work conceptually. The account of deontic force I have put forward depends on epistemic counterfactuality modelled as a mirror transformation. Typically, a deontic sentence such as 'you ought to do so-and-so' presupposes that what you ought to do is counter to fact, counter to what you actually are doing or are understood to be intending. If there is anything in this approach, then deontic constructions in language are perhaps the most striking of the linguistically enabled abstractions that can derive from concrete embodied experience, specifically from spatial experience and the experience of oriented physical force. The geometry that was proposed in Chapter 10 is not, of course, what Spinoza had in mind in the title of his *Ethics*, which I invoked at the head of that chapter. He was thinking of methods of rational proof by axiom and inference. But it is important to note that Spinoza saw the rational intellect as united with the body. His contemporary Pascal was equally clear that 'geometric' method must be complemented with the intuitive and the physical. The thinking of both philosophers is perhaps compatible with the notion of an embodied geometry.[5]

11.5 In conclusion: Deictic Space Theory and metaphor

Metaphor has not been the central focus of this book. Yet there is a sense in which any linguistic theory depends on it, including the present theory. Metaphorical mappings in fact ground the whole idea of deictic space and the relations between its three axes – mappings from the bodily experience of direction and distance in physical space to abstract concepts of time, reality and irreality. Apparently abstract meanings of verbs are themselves metaphorically connected to physical actions and spatial relationships – we may *grasp* an idea, *see* a solution and *move toward* a conclusion. The even more abstract meanings of grammatical constructions – for example, constructions coding mental representations of caused motion or events that never happened or could never happen – these are also rooted in the embodied spatial

---

[5] Interestingly, the psychologist John Macnamara also coupled geometry and morality. Like Spinoza, he was interested in how ethics can be axiomatised on the basis of simple initial principles (Macnamara 1991). Macnamara's argument is essentially an analogy between moral intuitions as the ground of moral systems and spatial intuitions as the ground of axiomatised geometries; he is not making a direct connection between morality and geometry. My own use of geometry in the context of morality is concerned not so much with an innate sense of right and wrong but with the sense of 'oughtness'. To repeat, Chapter 10 suggests that this sense may be derived from, not simply analogous to, a fundamental spatial ability that includes (i) the ability to 'reflect' – conceptualise the world as radically other than it is – and (ii) the bodily experience of a force directed to the self and to which a source may or may not be attributed.

reference frames, reference frames that we need in order to manipulate objects and navigate the environment.

It is tempting to think that metaphor is a just a way of doing a theory, a handy way of thinking about something that is not in itself metaphorical. But it may be rather the other way around. It seems likely that the metaphorical ability itself, however it evolved or was learned, actually made possible the scaffolding of language by abstracting schemas from physiological engagement with physical space. The move from concrete representation to abstract representation is a mapping across domains of experience and thus metaphorical. This is not the end of the process, since metaphors become entrenched and conventionalised. What I am trying to describe in the abstract deictic space and schematic relations within it may be the result of a metaphorical process that has roots in spatial orientation and action. But what remains is an abstract framework in the spirit of geometry – only part of the great complexity of real-world linguistic conceptualisation.

Appendix

Embedding of relative clauses and natural recursion at discourse entity points

Deictic Space Theory is not a model of processing but an attempt to outline the products of processing. It makes no claim to explain how the brain moves from a phonemic (or graphemic) sequence to the conceptualisations proposed in DSMs. Nor does it model the semantic richness of lexemes, which is treated as a separate though intrinsically connected part of the mind–brain's language system. However, recursion can be the site of expanded semantic richness for NPs evoked by NPs. The embedding of relative clauses in NPs is a case in point. DST can incorporate this in the way outlined below. It should be noted that there are practical limits on the depth of embedding that language-users produce: see Karlsson (2007) and cf. Pulvermüller (2010) on the need for a neural account of these constraints.

Consider a centrally embedded relative clause structure of the classic kind:

(1)   The car hit the post

(2)   The car the thief drove hit the post

(3)   The car the thief the cops chased drove hit the post

(4)   ?The car the thief the cops the chief sent chased drove hit the post.

The diagram is the DSM for sentence (3), which has two degrees of embedding. An unexpected and curious consequence of the coordinate–vector framework is that it leads to a congruent account of the embedding of relative clauses (or rather the basic structure of their conceptual effects). The English word order of sentence (3) gives relative attentional distance (foregrounding) for the NPs, i.e. labelled entities, on the d-axis. The vectors for *drove* and *chased* have the appropriate direction for the agent–patient relations that a reader of the sentences mentally constructs. The DSM does not of course seek to model this process, merely its conceptual outcome. The *hit* vector on the

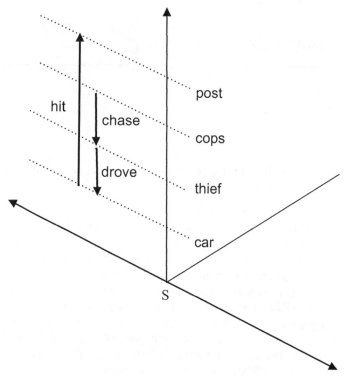

Appendix Figure Two degrees of central embedding

left models the relative direction and distance for the relation between car and post in sentence (3).

The interesting implication of this type of modelling is the following. If the *drove* and *chased* vectors are added together and then added to the *hit* vector, the result by vector addition is the 'shortening' of the *hit* vector, reducing it to the matrix sentence *the car hit the post*.

In regard to (4) most readers find this very awkward. Even sentence (3) requires extra effort. From a processing point of perspective the usual explanation is based on limitations of working memory. It is also worth noting that, drawn onto a DSM, the result would be increased deictic distance. Demonstratives in natural languages rarely exceed three positions of deictic distance. This is perhaps consistent with the sense that (2) is the most 'natural'.

References

Amorapanth, P., Kranjec, A., Bromberger, B., Matthew Lehet, M., Widick, P., Woods, A.J., Kimberg, D. Y. and Anjan Chatterjee, A. 2012. Language, perception, and the schematic representation of spatial relations. *Brain and Language*, **120**(3): 226–36.

Andersen, R. A. 1995. Coordinate transformations and motor planning in posterior parietal cortex. In M. S. Gazzaniga (ed.), *The Cognitive Neurosciences*, pp. 519–32. Cambridge, MA: MIT Press.

Anderson, J. 1971. *The Grammar of Case: Towards a Localist Theory*. Cambridge University Press.

Anton, H. and Rorres, C. 1991. *Elementary Linear Algebra*. New York: Wiley.

Arbib, M. A. 2005. From monkey-like action recognition to human language: an evolutionary framework for neurolinguistics. *Behavioral and Brain Sciences*, **28**(2): 105–67.

(ed.) 2006. *From Action to Language via the Mirror Neuron System*. Cambridge University Press.

2012. *How the Brain Got Language: The Mirror System Hypothesis*. Oxford University Press.

Asher, N. 1987. A typology for attitude verbs and their anaphoric properties. *Linguistics and Philosophy*, **10**(2): 125–97.

Asher, N. and Lascarides, A. 2003. *Logics of Conversation*. Cambridge University Press.

Bach, E. 1986. The algebra of events. *Linguistics and Philosophy*, **9**(1): 5–16.

Bahr, G. S. 2003. Psychologists' belief in visual emission. *American Psychologist*, **58**(6–7): 494–6.

Baron-Cohen, S. 1995. *Mindblindness: An Essay on Autism and Theory of Mind*. Cambridge, MA: MIT Press.

Baron-Cohen, S., Leslie, A. M. and Frith, U. 1985. Does the autistic child have a theory of mind? *Cognition*, **21**(1): 37–46.

Baron-Cohen, S., Ring, H., Moriarty, J., Schmitz, B. and Ell, P. 1994. Recognition of mental state terms: clinical findings in children with autism and functional neuroimaging study of normal adults. *British Journal of Psychiatry*, **165**(5): 640–9.

Behrendt, R.-P. 2011. *Neuroanatomy of Social Behaviour: An Evolutionary and Psychoanalytic Perspective*. London: Karnac Books.

Bergen, B. and Chang, N. 2005. Embodied construction grammar in simulation-based language understanding. In J.-O. Östman and M. Fried (eds.), *Construction*

Grammar(s): Cognitive and Cross-Language Dimensions, pp.147–89. Amsterdam: Benjamins.
Berti, A. and Rizzolatti, G. 2002. Coding near and far space. In H.-O. Karnath, D. Milner and G. Vallar (eds.), *The Cognitive and Neural Bases of Spatial Neglect*, pp. 119–30. Oxford University Press.
Bloom, P., Peterson, M., Nadel, L. and Garrett, M. (eds.) 1996. *Language and Space*. Cambridge, MA: MIT Press.
Bolinger, D. 1977. *Meaning and Form*. London: Longman.
Borg, E. 2007. If mirror neurons are the answer, what was the question? *Journal of Consciousness Studies*, **14**(8): 5–19.
Boroditsky, L. 2000. Metaphoric structuring: understanding time through spatial metaphors. *Cognition*, **75**(1): 1–28.
Borsley, R. 1996. *Modern Phrase Structure Grammar*. Oxford: Blackwell.
Brisard, F. 2002. The English present. In F. Brisard (ed.), *Grounding: The Epistemic Footing of Deixis and Reference*, pp. 251–98. Berlin: Mouton de Gruyter.
Brugmann, K. 1885. *Griechische Grammatik*. Munich: Beck.
Bühler, K. 1990 [1934]. *The Theory of Language: The Representational Function of Language*, translated by D. F. Goodwin. Amsterdam: Benjamins. [Translation of Bühler, K. 1934. *Sprachtehorie: Die Darstellungsfunktion der Sprache*. Stuttgart: Fischer.]
Burgess, N. 2002. The hippocampus, space, and viewpoints in episodic memory. *Quarterly Journal of Experimental Psychology, Section A*, **55**(4): 1057–80.
Burgess, N., Jeffery, K. J. and O'Keefe, J. 2003. *The Hippocampal and Parietal Foundations of Spatial Cognition*. Oxford University Press.
Byrne, R. 2007. *The Rational Imagination: How People Create Alternatives to Reality*. Cambridge MA: MIT Press.
Cai, Q., Van Der Haegen, L. and Brysbaert, M. 2013. Complementary hemispheric specialization for language production and visuospatial attention. *Proceedings of the National Academy of Sciences of the USA*, **110**(4): E322–30.
Carlson, L. 1981. Aspect and quantification. In P. Tedeschi and A. Zaenen (eds.), *Tense and Aspect*, pp. 31–64. New York: Academic Press.
Carlson, L. A. 2010. Parsing space around objects. In Evans and Chilton (eds.), pp. 115–37.
Carlson-Radvansky, L. A. and Irwin, D. E. 1993. Frames of reference in vision and language: where is above? *Cognition*, **46**(3): 223–44.
Carlson-Radvansky, L. A., Covey, E. and Lattanzi, K., 1999. 'What' effects on 'Where': functional influences on spatial relations. *Psychological Science*, **10**(6): 516–21.
Carlson-Radvansky, L. A., West, R., Taylor, H.A. and Herdon, R. W. 2002. Neural correlates of spatial term use. *Journal of Experimental Psychology: Human Perception and Performance*, **28**(6): 1391–407.
Carreiras, M., Carr, L., Barber, H. A. and Hernandez, A. 2010. Where syntax meets math: right intraparietal sulcus activation in response to grammatical number agreement violations. *NeuroImage*, **49**(2): 1741–9.
Chechlacz, M., Rotshtein, P., Bickerton, W.-L., Hansen, P.C., Deb, S. and Humphreys, G.W. 2010. Separating neural correlates of allocentric and egocentric neglect: distinct cortical sites and common white matter disconnections. *Cognitive Neuropsychology*, **27**(3): 277–303.

Chilton, P. 2005. Vectors, viewpoint and viewpoint shift: toward a discourse space theory. *Annual Review of Cognitive Linguistics*, **3**: 78–116.
  2007. Geometrical concepts at the interface of formal and cognitive models: *Aktionsart*, aspect and the English progressive. *Pragmatics and Cognition*, **15**(1): 91–114.
  2009. Get and the grasp schema: a new approach to conceptual modelling in image schema semantics. In V. Evans and S. Pourcel (eds.), *New Directions in Cognitive Linguistics*, pp. 331–70. Amsterdam: Benjamins.
  2010a. From mind to grammar: coordinate systems, prepositions, constructions. In Evans and Chilton (eds.), *Language, Cognition and Space*, pp. 499–514.
  2010b. The conceptual structure of deontic meaning: a model based on geometrical principles. *Language and Cognition*, **2**(2): 191–220.
  2013. Frames of reference and the linguistic conceptualization of time: present and future. In Jaszczolt and Saussure (eds.), pp. 236–58.
Chomsky, N. 1957. *Syntactic Structures*. The Hague: Mouton.
  1963. Formal properties of grammars. In R. D. Luce, R. Bush and E. Galanter (eds.), *Handbook of Mathematical Psychology*, vol. 2, pp. 323–418. New York: Wiley.
  1965. *Aspects of the Theory of Syntax*. Cambridge, MA: MIT Press.
  1981. *Lectures on Government and Binding: The Pisa Lectures*. Dordrecht: Foris.
  1995. *The Minimalist Program*. Cambridge, MA: MIT Press.
Chung, S. and Timberlake, A. 1985. Tense, aspect and mood. In T. Shopen (ed.), *Language Typology and Syntactic Description*, vol. 3, *Grammatical Categories and the Lexicon*, pp. 202–58. Cambridge University Press.
Clark, H. H. 1973. Space, time, semantics and the child. In T. E. Moore (ed.), *Cognitive Development and the Acquisition of Language*, pp. 28–64. New York: Academic Press.
Colby, C. L. and Goldberg, M. E. 1999. Space and attention in parietal cortex. *Annual Review of Neuroscience*, **22**(3): 19–49.
Comrie, B. 1976. *Aspect: An Introduction to the Study of Verbal Aspect and Related Problems*. Cambridge University Press.
Cosmides, L. and Tooby, J. 2000. Consider the source: the evolution of adaptations for decoupling and metarepresentations. In D. Sperber (ed.), *Metarepresentation: A Multidisciplinary Perspective*, pp. 53–115. Oxford University Press.
Coull, J. T., and Nobre, A. C. 1998. Where and when to pay attention: the neural systems for directing attention to spatial locations and to time intervals as revealed by both PET and fMRI. *Journal of Neuroscience*, **18**(18): 7426–35.
Coull, J. T., Frith, C. D., Büchel, C. and Nobre, A. C. 2000. Orienting attention in time: behavioural and neuroanatomical distinction between exogenous and endogenous shifts. *Neuropsychologia*, **38**(6): 808–19.
Coventry, K. R., and Garrod, S. 2004. *Saying, Seeing and Acting: The Psychological Semantics of Spatial Prepositions*. London: Taylor & Francis.
Coventry, K. R., Prat-Sala, M. and Richards, L. 2001. The interplay between geometry and function in the comprehension of *over, under, above*, and *below*. *Journal of Memory and Language*, **44**(3): 376–98.
Coventry, K. R., Valdés, B., Castillo, A. and Guijarro-Fuentes, P. 2008. Language within your reach: near–far perceptual space and spatial demonstratives. *Cognition*, **108**(3): 889–95.

Coventry, K. R., Lynott, D., Cangelosi, A., Monroux, L., Joyce, D. and Richardson, D. C. 2010. Spatial language, visual attention, and perceptual simulation. *Brain and Language*, **112**(3): 202–13.

Croft, W. 2001. *Radical Construction Grammar: Syntactic Theory in Typological Perspective*. Oxford University Press.
  2012. *Verbs: Aspect and Causal Structure*. Oxford University Press.

Croft, W. and Cruse, D. A. 2004. *Cognitive Linguistics*. Cambridge University Press.

Dalimier, C. 2001. Introduction. In C. Dalimier (ed. and transl.), *Apollonius Dyscolus: Traité des conjonctions*. Paris: Vrin.

Damasio, H., Grabowski, T. J., Tranel, D., Ponto, L. L., Hichwa, R. D. and Damasio, A. R. 2001. Neural correlates of naming actions and of naming spatial relations, *NeuroImage*, **13**(6): 1053–64.

Damourette, J. and Pichon, E. 1911–36. *Des mots à la pensée: Essai de grammaire de la langue française*. Paris: d'Artrey.

Dancygier, B. 1998. *Conditionals and Prediction: Time, Knowledge and Causation in Conditional Constructions*. Cambridge University Press.
  2002. Mental space embeddings, counterfactuality, and the use of *unless*. *English Language and Linguistics*, **6**(2): 347–77.

Dancygier, B. and Sweetser, E. 2005. *Mental Spaces in Grammar: Conditional Constructions*. Cambridge University Press.

Deane, P. D. 1996. On Jackendoff's conceptual semantics. *Cognitive Linguistics* **7**(1): 35–91.

Decety, J. and Lamm, C. 2007. The role of the right temporoparietal junction in social interaction: how low-level computational processes contribute to meta-cognition. *Neuroscientist*, **13**(6): 580–93.

Di Pellegrino, G., Fadiga, L., Fogassi, L., Gallese, V. and Rizzolatti, G. 1992. Understanding motor events: a neurophysiological study. *Experimental Brain Research*, **91**(1): 176–80.

Dick, A. S. and Tremblay, P. 2013. Beyond the arcuate fasciculus: consensus and controversy in the connectional anatomy of language. *Brain*, **135**(12): 3529–50.

Diessel, H. 1999. *Demonstratives: Form, Function and Grammaticalization*. Amsterdam: Benjamins.
  2006. Demonstratives, joint attention, and the emergence of grammar. *Cognitive Linguistics*, **17**(4): 463–89.

Dixon, R. M. W. 1982. *Where Have All the Adjectives Gone? And Other Essays in Semantics and Syntax*. The Hague: Mouton.
  1984. The semantic basis of syntactic properties. *Proceedings of the 10th Annual Meeting of the Berkeley Linguistics Society*, pp. 583–95.
  1991. *A New Approach to English Grammar, on Semantic Principles*. Oxford: Clarendon Press.

Dowty, D. R. 1977. Toward a semantic analysis of verb aspect and the English 'imperfective' progressive. *Linguistics and Philosophy*, **1**(1): 45–77.
  1979. *Word Meaning and Montague Grammar*. Dordrecht: Reidel.
  1986. The effects of aspectual class on the temporal structure of discourse: semantics or pragmatics? *Linguistics and Philosophy*, **9**(1): 37–61.

Duff, M. C., and Brown-Schmidt, S. 2012. The hippocampus and the flexible use and processing of language. *Frontiers in Human Neuroscience*, **6**: 1–9.

Dummett, M. 2006. *Thought and Reality*. Oxford: Clarendon Press.
Eckardt, R. 2006. *Meaning Change in Grammaticalization: An Enquiry into Semantic Reanalysis*. Oxford University Press.
Enfield, N. J. 2003. Demonstratives in space and interaction: data from Lao speakers and implications for semantic analysis. *Language*, **79**(1): 82–117.
Evans, V. 2004. *The Structure of Time: Language, Meaning and Temporal Cognition*. Amsterdam: Benjamins.
  2009. *How Words Mean: Lexical Concepts, Cognitive Models and Meaning Construction*. Oxford University Press.
Evans, V. and Chilton, P. (eds.) 2010. *Language, Cognition and Space: The State of the Art and New Directions*. London: Equinox.
Evans, V. and Tyler, A. 2004a. Rethinking English 'prepositions of movement': the case of *to* and *through*. *Belgian Journal of Linguistics*, **18**(1): 247–70.
  2004b. Spatial experience, lexical structure and motivation: the case of *in*. In G. Radden and K. Panther (eds.), *Studies in Linguistic Motivation*, pp. 157–92. Berlin: Mouton de Gruyter.
Fauconnier, G. 1994. *Mental Spaces: Aspects of Meaning Construction in Natural Language*. Cambridge University Press.
  1996. Analogical counterfactuals. In G. Fauconnier and E. Sweetser (eds.), *Spaces, Worlds and Grammar*, pp. 57–90. Chicago, IL: University of Chicago Press.
  1997. *Mappings in Thought and Language*. Cambridge University Press.
Fauconnier, G. and Turner, M. 2002. *The Way We Think*. New York: Basic Books.
Feldman, J. and Narayanan, S. 2004. Embodiment in a neural theory of language. *Brain and Language*, **89**(2): 385–92.
Fillmore, C. J. 1968. The case for case. In E. W. Bach and R. T. Harms (eds.), *Universals in Linguistic Theory*, pp.1–88. New York: Holt, Rinehart & Winston.
  1997 [1975]. *Lectures on Deixis*. Stanford, CA: CSLI Publications. [Fillmore, C. J. 1975. *Santa Cruz Lectures on Deixis, 1971*. Bloomington, IN: Indiana University Linguistics Club.]
  1982a. Towards a descriptive framework for spatial deixis. In R. J. Jarvella and W. Klein (eds.), *Speech, Place and Action*, pp. 31–59. Chichester: Wiley.
  1982b. Frame semantics. In Linguistic Society of Korea (ed.), *Linguistics in the Morning Calm*, pp.111–37. Seoul: Hanshin Publishing Company.
  1985. Frames and the semantics of understanding. *Quaderni dei Semantica*, **6**(2): 222–53.
Fillmore, C. J., Kay, P. and O'Connor, C. 1988. Regularity and idiomaticity in grammatical constructions: the case of *let alone*. *Language*, **64**(3): 501–38.
Fischer, M. H. and Zwaan, R. 2008. Embodied language: a review of the motor system in language comprehension. *Quarterly Journal of Experimental Psychology*, **61**(6): 825–50.
Fleischmann, S. 1989. Temporal distance: a basic linguistic metaphor. *Studies in Language*, **13**(1): 1–50.
Fodor, J. D. 1983. *The Modularity of Mind*. Cambridge, MA: MIT Press.
Fogassi, L. and Ferrari, P. F. 2007. Mirror neurons and the evolution of embodied language. *Current Directions in Psychological Science*, **16**(3): 136–41.
  2012. Cortical motor organization, mirror neurons, and embodied language: an evolutionary perspective. *Biolinguistics*, **6**(3–4): 308–37.

Fox, M. D., Corbetta, M., Snyder, A. Z., Vincent, J. L. and Raichle, M. E. 2006. Spontaneous neuronal activity distinguishes human dorsal and ventral attention systems. *Proceedings of the National Academy of Sciences of the USA*, **103**(26): 10 046–51.
Francez, I. and Koontz-Garboden, A. 2012. Semantic variation and the grammar of property concepts. University of Chicago and University of Manchester ms. http://ling.auf.net/lingbuzz/
Frawley, W. 1992. *Linguistic Semantics*. Hillsdale, NJ: Erlbaum.
Frege, G. 1956 [1915]. The thought: a logical enquiry. *Mind, New Series*, **65**(259): 289–311. [First published in the *Beiträge zur Philosophie des Deutschen Idealismus*, 1915–19.]
Friederici, A. D. 2009. Pathways to language: fiber tracts in the human brain. *Trends in Cognitive Sciences*, **13**(4): 175–81.
Gallese, V. 2003. A neuroscientific grasp of concepts: from control to representation. *Philosophical Transactions of the Royal Society of London, Series B*, **358**: 1231–40.
  2005. Embodied simulation: from neurons to phenomenal experience. *Phenomenology and the Cognitive Sciences*, **4**(1): 23–48.
Gallese, V. and Goldman, A. 1998. Mirror neurons and the simulation theory of mindreading. *Trends in Cognitive Sciences*, **2**(12): 493–501.
Gallese, V. and Lakoff, G. 2005. The brain's concepts: the role of the sensorymotor system in reason and language. *Cognitive Neuropsychology*, **22**(3–4): 455–79.
Gallese, V., Fadiga, L., Fogassi, L. and Rizzolatti, G. 1996. Action recognition in the premotor cortex. *Brain*, **119**(2): 593–609.
Gallese, V., Fogassi, L. Fadiga, L. and Rizzolatti, G. 2002. Action representation and the inferior parietal lobule. In W. Prinz and B. Hommel (eds.), *Attention and Performance, vol. 19, Common Mechanisms in Perception and Action*, pp. 334–55. Oxford University Press.
Gallese, V., Keysers, C. and Rizzolatti, G. 2004. A unifying view of the basis of social cognition. *Trends in Cognitive Sciences*, **8**(9): 396–403.
Gallistel, C. R. 1990. *The Organization of Learning*. Cambridge, MA: MIT Press.
  1999. Coordinate transformations in the genesis of directed action. In B. O. M. Bly and D. E. Rummelhart (eds.), *Cognitive Science*, pp. 1–42. New York: Academic Press.
  2002. Language and spatial frames of reference in mind and brain. *Trends in Cognitive Sciences*, **6**(8): 321–2.
Gärdenfors, P. 2000. *Conceptual Spaces: The Geometry of Thought*. Cambridge, MA: MIT Press.
Geach, P. 1967. Intentional identity. *Journal of Philosophy*, **64**(20): 627–32.
Geertz, C. 1973. *The Interpretation of Cultures*. New York: Basic Books.
Givón, T. 1982. Evidentiality and epistemic space. *Studies in Language*, **6**(1): 23–49.
  2005. *Context as Other Minds: The Pragmatics of Sociality, Cognition and Communication*. Amsterdam: Benjamins.
Glenberg, A. M. and Kaschak, M. P. 2002. Grounding language in action. *Psychonomic Bulletin and Review*, **9**(3): 558–65.
Goldberg, A. E. 1995. *A Construction Grammar Approach to Argument Structure*. Chicago, IL: University of Chicago Press.
  2003. Constructions: a new theoretical approach to language. *Trends in Cognitive Science*, **7**(5): 219–224.

Goldberg, A. E., and Jackendoff, R. 2004. The English resultative as a family of constructions. *Language* **80**(3): 532–68.
Goldsmith, J. and Woisetschlaeger, E. 1982. The logic of the English progressive. *Linguistic Inquiry* **13**: 79–89.
Goodale, M. A., and Milner, A. D. 1992. Separate visual pathways for perception and action. *Trends in Neuroscience*, **15**(1): 20–5.
Goodman, N. 1947. The problem of counterfactual conditionals. *Journal of Philosophy*, **44**(5): 113–28.
Gramann, K., Müller, H. J., Schönebeck, B. and Debus, G. 2006. The neural basis of ego- and allocentric reference frames in spatial navigation: evidence from spatio-temporal coupled current density reconstruction. *Brain Research*, **1118**: 116–29.
Griffiths, J. D., Marslen-Wilson, W. D., Stamatakis, E. A. and Tyler, L. K. 2013. Functional organization of the neural language system: dorsal and ventral pathways are critical for syntax. *Cerebral Cortex*, **23**(1): 139–47.
Grodzinsky, Y. and Friederici, A. D. 2006. Neuroimaging of syntax and syntactic processing. *Current Opinion in Neurobiology*, **16**(2): 240–6.
Gruber, J. S. 1976. *Lexical Structures in Syntax and Semantics*. Amsterdam: North Holland.
Haiman, J. 1983. Iconic and economic motivation. *Language*, **59**(4): 781–819.
  1985. *Natural Syntax: Iconicity and Erosion*. Cambridge University Press.
Haspelmath, M. 1997. *From Space to Time: Temporal Adverbials in the World's Languages*, Munich: Lincom Europa.
Hauser, M. D., Chomsky, N. and Fitch, W. T. 2002. The faculty of language: what is it, who has it, and how did it evolve? *Science*, **298**: 1565–6.
Heine, B. 1993. *Auxiliaries: Cognitive Forces and Grammaticalization*. Oxford University Press.
Herskovits, A. 1986. *Language and Spatial Cognition: An Interdisciplinary Study of Prepositions in English*. Cambridge University Press.
Hickok, G. 2009. The functional neuroanatomy of language. *Physics of Life Reviews*, **6**(3): 121–43.
Hickok, G. and Poeppel, D. 2004. Dorsal and ventral streams: a framework for understanding aspects of the functional anatomy of language. *Cognition*, **92**(1–2): 67–99.
Hill, C. 1982. Up/down, front/back, left/right: a contrastive study of Hausa and English. In J. Weissenborn and W. Klein (eds.), *Here and There: Cross-Linguistic Studies on Deixis and Demonstration*, pp. 11–42. Amsterdam: Benjamins.
Hillis, A. E. 2006. Neurobiology of unilateral spatial neglect. *Neuroscientist*, **12**(2): 153–63.
Hoffmann, T. and Trousdale, G. (eds.) 2013. *Oxford Handbook of Construction Grammar*. Oxford University Press.
Hopper, P. and Traugott, E. C. 2003. *Grammaticalization*. Cambridge University Press.
Horn, L. R. 1989. *A Natural History of Negation*. Chicago, IL: University of Chicago Press.
Hurford, J. 2003. The neural basis of predicate–argument structure. *Behavioral and Brain Sciences*, **23**(6): 261–83.
  2004. Language beyond our grasp: what mirror neurons can, and cannot, do for the evolution of language. In O. Kimbrough and U. Griebel (eds.), *Evolution of*

Communication Systems: A Comparative Approach, pp. 297–313. Cambridge, MA: MIT Press.
  2007. *The Origins of Meaning*. Oxford University Press.
Iachini, T., Ruggiero, G., Conson, M. and Trojano, L. 2009. Lateralization of egocentric and allocentric spatial processing after parietal brain lesions. *Brain and Cognition*, **69**(3): 514–20.
Israel, M. 2004. The pragmatics of polarity. In L. R. Horn and G. Ward (eds.), *The Handbook of Pragmatics*, pp. 701–23. Oxford: Blackwell.
Jackendoff, R. 1975. On belief contexts. *Linguistic Inquiry*, **6**(1): 53–93.
  1976. Toward an explanatory semantic representation. *Linguistic Inquiry*, **7**(1): 89–150.
  1990. *Semantic Structures*. Cambridge, MA: MIT Press.
  2002a. *Foundations of Language*. Oxford University Press.
  2002b. The natural logic of rights and obligations. In R. Jackendoff, P. Bloom and K. Wynn (eds.), *Language*, pp. 67–95. *Logic, and Concepts*. Cambridge, MA: MIT Press.
  2011. What is the human language faculty? Two views. *Language*, **87**(3): 586–624.
Jackendoff, R. and Pinker, S. 2005. The nature of the language faculty and its implications for evolution of language (Reply to Fitch, Hauser, and Chomsky). *Cognition*, **97**(2): 211–25.
Jarvella, R. J. and Klein, W. (eds.) 1982. *Speech, Place and Action*. Chichester: Wiley.
Jaszczolt, K. M. 2009. *Representing Time: An Essay on Temporality as Modality*. Oxford University Press.
Jaszczolt, K. and Saussure, L. de (eds.) 2013. *Time: Language, Cognition and Reality*. Oxford University Press.
Jeannerod, M. 1997. *The Cognitive Neuroscience of Action*. Oxford: Blackwell.
Johnson, M. 1987. *The Body in the Mind: The Bodily Basis of Meaning, Imagination, and Reason*. Chicago, IL: Chicago University Press.
Johnson-Laird, P. 1983. *Mental Models*. Cambridge University Press.
Kamp, H. 1984. A theory of truth and semantic representation. In J. Groenendijk, T. M. V. Janssen and M. Stokhof (eds.), *Truth, Interpretation and Information*, pp. 115–43. Dordrecht: Foris.
Kamp, H. and Reyle, U. 1993. *From Discourse to Logic*. Dordrecht: Kluwer.
Kamp, H., van Genabith, J. and Reyle, U. 2011. Discourse representation theory. In D. M. Gabbay and F. Guenthner (eds.), *Handbook of Philosophical Logic*, vol. 15, pp. 125–394. Dordrecht: Springer.
Karlsson, F. 2007. Constraints on multiple central embedding of clauses. *Journal of Linguistics*, **43**(2): 365–92.
Karnath, H.-O. 2001. New insights into the functions of the superior temporal cortex. *Nature Reviews Neuroscience*, **2**(8): 568–76.
Karnath, H.-O., Ferber, S. and Himmelbach, M. 2001. Spatial awareness is a function of the temporal not the posterior parietal lobe. *Nature*, **411**: 950–53.
Kaschak, M. P. and Glenberg, A. M. 2000. Constructing meaning: the role of affordances and grammatical constructions in sentence comprehension. *Journal of Memory and Language*, **43**(3): 508–29.
Kaufmann, S., Condoravadi, C. and Harizanov, V. 2005. Formal approaches to modality. In W. Frawley (ed.), *The Expression of Modality*, pp. 71–105. Berlin: Mouton de Gruyter.

Kemmerer, D. 1999. 'Near' and 'far' in language and perception. *Cognition*, **73**(1): 35–63.
  2000. Grammatically relevant and grammatically irrelevant features of verb meaning can be independently impaired. *Aphasiology*, **14**(10): 997–1020.
  2003. Neuropsychological evidence for the distinction between grammatically relevant and irrelevant components of meaning. *Behavioral and Brain Sciences*, **26**(6): 684–5.
  2006a. The semantics of space: integrating linguistic typology and cognitive neuroscience, *Neuropsychologia*, **44**(9): 1607–21.
  2006b. Action verbs, argument structure constructions, and the mirror neuron system. In Arbib (ed.), pp. 347–73.
  2010. A neuroscientific perspective on the linguistic encoding of categorical spatial relations. In Evans and Chilton (eds.), *Language, Cognition and Space*, pp. 139–68.
Kenny, A. 1963. *Action, Emotion and Will*. London: Routledge.
Keysers, C. and Gazzola, V. 2010. Social neuroscience: mirror neurons recorded in humans. *Current Biology*, **20**(8): R353–R354.
King, J. A., Burgess, N. and Hartley, T. 2002, Human hippocampus and viewpoint dependence in spatial memory. *Hippocampus*, **12**(6): 811–20.
Kosslyn, S. M. 1987. Seeing and imagining in the cerebral hemispheres: a computational approach. *Psychological Review*, **94**(2): 148–75.
Kövecses, Z. 2000. *Metaphor and Emotion: Language, Culture, and Body in Human Feeling*. Cambridge University Press.
Kratzer, A. 1977. What 'must' and 'can' must and can mean. *Linguistics and Philosophy*, **1**(1): 337–55.
  1981. The notional category of modality. In H.-J. Eikmeyer and H. Rieser (eds.), *Words, Worlds and Contexts*, pp. 38–74. Berlin: Mouton de Gruyter.
Lakoff, G. 1987. *Women, Fire and Dangerous Things: What Categories Reveal about the Mind*. Chicago, IL: University of Chicago Press.
Lakoff, G. and Johnson, M. 1980. *Metaphors We Live By*. Chicago, IL: University of Chicago Press.
  1999. *Philosophy in the Flesh: The Embodied Mind and its Challenge to Western Thought*. New York: Basic Books.
Lakoff, G. and Núñez, R. E. 2000. *Where Mathematics Comes From: How the Embodied Mind Brings Mathematics into Being*. New York: Basic Books.
Landau, B. 2003. Axes and direction in spatial language and cognition. In van der Zee and Slack (eds.), pp. 18–38.
Landau, B. and Jackendoff, R. 1993. 'What' and 'where' in spatial language and spatial cognition. *Behavioral and Brain Sciences*, **16**(2): 217–38.
Langacker, R. W. 1987. *Foundations of Cognitive Grammar*, vol. 1. Stanford, CA: Stanford University Press.
  1991. *Foundations of Cognitive Grammar*, vol. 2. Stanford, CA: Stanford University Press.
  1995. Viewing in grammar and cognition. In P. W. Davis (ed.), *Alternative Linguistics: Descriptive and Theoretical Models*, pp. 153–212. Amsterdam: Benjamins.
  1999. Virtual reality. *Studies in the Linguistic Sciences*, **29**(2): 77–103.
  2001. The English present tense. *English Language and Linguistics*, **5**(2): 251–72.

2011. The English present: temporal coincidence vs. epistemic immediacy. In A. Patard and F. Brisard (eds.), *Cognitive Approaches to Tense, Aspect, and Epistemic Modality*, pp. 45–86. Amsterdam: Benjamins.
Leclercq, M. and Zimmermann, P. 2002. *Applied Neuropsychology of Attention: Theory, Diagnosis, and Rehabilitation*. London: Psychology Press.
Leslie, A. M. and Frith, U. 1988. Autistic children's understanding of seeing, knowing and believing. *British Journal of Developmental Psychology*, 6(4): 315–324.
Levin, B. 1993. *English Verb Classes and Alternations: A Preliminary Investigation*. Chicago, IL: University of Chicago Press.
Levinson, S. 1996. Frames of reference and Molyneux's questions: cross-linguistic evidence. In Bloom *et al.* (eds.), pp. 109–69.
  2003. *Space in Language and Cognition: Explorations in Cognitive Diversity*. Cambridge University Press.
  2004. Deixis. In L. R. Horn and G. Ward (eds.), *The Handbook of Pragmatics*, pp. 97–121. Oxford: Blackwell.
Levinson, S. C., and Meira, S. 2003. 'Natural concepts' in the spatial topological domain: adpositional meanings in crosslinguistic perspective – an exercise in semantic typology. *Language*, 79(3): 485–516.
Lewis, D. 1973. *Counterfactuals*. Oxford: Blackwell.
Livnat, Z. 2002. From epistemic to deontic modality: evidence from Hebrew. *Folia Linguistica Historica*, 23(1–2): 107–14.
Lonergan, B. 1957. *Insight: A Study of Human Understanding*. London: Longman.
Lucy, J. 1992. *Language Diversity and Thought: A Reformulation of the Linguistic Relativity Hypothesis*. Cambridge University Press.
Lyons, J. 1975. Deixis as the source of reference. In E. L. Keenan (ed.), *Formal Semantics of Natural Language*, pp. 61–83. Cambridge University Press.
  1977. *Semantics*. Cambridge University Press.
MacKay, D. G. 2011. Hippocampus. In P. C. Hogan (ed.), *The Cambridge Encyclopedia of the Language Sciences*, pp. 357–59. Cambridge University Press.
Macnamara, J. 1991. The development of moral reasoning and the foundations of geometry. *Journal for the Theory of Social Behaviour*, 21(2): 121–50.
Malchukov, A., Haspelmath, M. and Comrie, B. 2010. Ditransitive constructions: a typological overview. In A. Malchukov, M. Haspelmath and B. Comrie (eds.), *Studies in Ditransitive Constructions: A Comparative Handbook*, pp. 1–64. Berlin: De Gruyter Mouton.
Mandler, J. M. 2004. *The Foundations of Mind: The Origins of Conceptual Thought*. Oxford University Press.
McGilchrist, I. 2010. *The Master and his Emissary: The Divided Brain and the Making of the Western World*. New Haven, CT: Yale University Press.
Michaelis, L. 1998. *Aspectual Grammar and Past Time Reference*. London: Routledge.
  2004. Type shifting in construction grammar: an integrated approach to aspectual coercion. *Cognitive Linguistics*, 15(1): 1–67.
  2006. Time and tense. In B. Aarts and A. McMahon (eds.), *The Handbook of English Linguistics*, pp. 220–34. Oxford: Blackwell.
Milner, A. D. and Goodale, M. A 1995. *The Visual Brain in Action*. Oxford University Press.

Nagel, T. 1986. *The View from Nowhere*. Oxford University Press.
O'Keefe, J. 1996. The spatial prepositions in English, vector grammar, and the cognitive map theory. In P. Bloom *et al.* (eds.), pp. 277–316.
  2003. Vector grammar, places, and the functional role of the spatial prepositions in English. In van der Zee and Slack (eds.), pp. 69–85.
O'Keefe, J. and Burgess, N. 1996. Geometric determinants of the place fields of hippocampal neurons. *Nature*, **381**: 425–8.
O'Keefe, J. and Nadel, L. 1978. *The Hippocampus as a Cognitive Map*. Oxford: Clarendon Press.
Paillard, J. 1991. Motor and representational framing of space. In J. Paillard (ed.), *Brain and Space*, pp. 163–82. Oxford University Press.
Palmer, F. 1986. *Mood and Modality*. Cambridge University Press.
Papafragou, A. 1998. The acquisition of modality: implications for theories of semantic representation. *Mind and Language*, **13**(3): 370–99.
  2000. *Modality: Issues in the Semantic–Pragmatic Interface*. Amsterdam: Elsevier.
Papafragou, A. and Ozturk, O. 2006. The acquisition of epistemic modality. In A. Botinis (ed.), *Proceedings of ITRW on Experimental Linguistics in ExLing-2006*, pp. 201–4. Online archive: International Speech Communication Association.
Petitot, J. 1995. Morphodynamics and attractor syntax: constituency in visual perception and cognitive grammar. In R. F. Port and T. van Gelder (eds.), *Mind as Motion: Explorations in the Dynamics of Cognition*, pp. 227–81. Cambridge, MA: MIT Press.
Piaget, J. 1954 [1937]. *The Construction of Reality in the Child*. New York: Basic Books. [Piaget, J. 1937. *La construction du réel chez l'enfant*. Paris : Delachaux et Niestlé.]
  1967 [1926]. *The Child's Conception of the World*. Totowa, NJ: Littlefield, Adams & Co. [Piaget, J. 1926. *La représentation du monde chez l'enfant*. Paris: Presses universitaires de France.]
Piaget, J. and Inhelder, B. 1971 [1966]. *Mental Imagery in the Child*. London: Routledge & Kegan Paul. [Piaget, J. 1966. *L'image mentale chez l'enfant: Études sur le développement des représentations imaginées*. Paris: Presses universitaires de France.]
Pinker, S. 1997. *How the Mind Works*. New York: Norton.
Piwek, P., Beun, R-J. and Cremers, A. 2007. 'Proximal' and 'distal' in language and cognition: evidence from deictic demonstratives in Dutch. *Journal of Pragmatics*, **40**(4): 694–718.
Portner, P. 2009. *Modality*. Oxford University Press.
Postal, P. 1974. *On Raising: One Rule of English Grammar and its Theoretical Implications*. Cambridge, MA: MIT Press.
Premack, D. and Premack, A. J. 1994. Moral belief: form versus content. In L. A. Hirschfeld and S. A. Gelman (eds.), *Mapping the Mind: Domain Specificity in Cognition and Culture*, pp. 149–68. Cambridge University Press.
Premack, D. and Woodruff, G. 1978. Does the chimpanzee have a 'theory of mind'? *Behavioral and Brain Sciences*, **4**: 515–26.
Previc, F. H. 1990. Functional specialization in the lower and upper visual fields in humans: its ecological origins and neurophysiological implications. *Behavioral and Brain Sciences*, **13**(3): 519–75.

Pulvermüller, F. 2010. Brain embodiment of syntax and grammar: discrete combinatorial mechanisms spelt out in neuronal circuits. *Brain and Language*, **112**(3): 167–79.
Pulvermüller, F. and Fadiga, L. 2010. Active perception: sensorimotor circuits as a cortical basis for language. *Nature Reviews Neuroscience*, **11**(5): 351–60.
Quine, W. V. O. 1960. *Word and Object*. Cambridge, MA: MIT Press.
Radden, G. 2004. The metaphor TIME AS SPACE across languages. In N. Baumgarten, C. Böttger, M. Motz and J. Probst (eds.), *Übersetzen, Interkulturelle Kommunikation, Spracherwerb und Sprachvermittlung – das Leben mit mehreren Sprachen: Festschrift für Juliane House zum 60. Geburtstag*, pp. 225–38. Bochum: AKS-Verlag.
Ramachandran, V. S., Altschuler, E. L. and Hillier, S. 1997. Mirror agnosia. *Proceedings of the Royal Society of London, Series B*, **264**: 645–7.
Recanati, F. 2007. *Perspectival Thought: A Plea for (Moderate) Relativism*. Oxford University Press.
Regier, T. and Carlson, L. 2001. Grounding spatial language in perception: an empirical and computational investigation. *Journal of Experimental Psychology: General*, **130**(2): 273–98.
Reichenbach, H. 1947. *Elements of Symbolic Logic*. New York: Macmillan.
Rizzolatti, G. and Arbib, M. A. 1998. Language within our grasp. *Trends in Neurosciences*, **21**(5): 188–94.
Rizzolatti, G., Fadiga, L., Gallese, V., Fogassi, L. 1996. Premotor cortex and the recognition of motor actions. *Cognitive Brain Research*, **3**(2): 131–41.
Rizzolatti, G., Fadiga, L., Fogassi, L. and Gallese, V. 1997. The space around us. *Science*, **277**: 190–1.
Rizzolatti, G., Fogassi, L. and Gallese, V. 2001. Neurophysiological mechanisms underlying the understanding and imitation of action. *Nature Reviews Neuroscience*, **2**(9): 661–70.
Rosenbaum, P. 1967. *The Grammar of English Predicate Complement Constructions*. Cambridge, MA: MIT Press.
Rothstein, S. 2004. *Structuring Events*. Oxford: Blackwell.
Russell, B. 1905. On denoting. *Mind, New Series*, **14**(56): 479–93.
Ryle, G. 1949. *The Concept of Mind*. London: Hutchinson.
Samson, D., Apperly, I. A., Braithwaite, J. J., Andrews, B. J. and Bodley Scott, S. E. 2010. Seeing it their way: evidence for rapid and involuntary computation of what other people see. *Journal of Experimental Psychology: Human Perception and Performance*, **36**(5): 1255–66.
Saur, D., Kreher, B. W., Schnell, S., Kümmerer, D., Kellmeyer, P., Vrya, M.-S., Roza, U., Musso, M., Glauche, V., Abel, S., Huber, W., Rijntjes, M., Hennig, J. and Weiller, C. 2008. Ventral and dorsal pathways for language. *Proceedings of the National Academy of Sciences of the USA*, **105**: 18 035–40.
Saussure, L. de. 2003. *Temps et pertinence: Élements de pragmatique cognitive du temps*. Brussels: de Boeck & Larcier.
  2013. Perspectival interpretations of tenses. In Jaszczolt and de Saussure (eds.), pp. 46–69.
Saussure, L. de and Morency, P. 2012. A cognitive–pragmatic view of the French epistemic future. *Journal of French Language Studies*, **22**(2): 207–23.

Siewierska, A. 1998. Languages with and without objects. *Languages in Contrast*, **1**(2): 173–90.
Sinha, C. and Jensen de Lopez, K. 2000. Language, culture and the embodiment of spatial cognition. *Cognitive Linguistics*, **11**(1–2): 17–41.
Sperber, D. (ed.) 2000. *Metarepresentation: A Multidisciplinary Perspective*. Oxford University Press.
Stalnaker, R. 1968. A theory of conditionals. In N. Rescher (ed.), *Studies in Logical Theory*, pp. 98–112. Oxford: Blackwell.
Stamenov, M. I., and Gallese, V. 2002. *Mirror Neurons and the Evolution of Brain and Language*. Amsterdam: Benjamins.
Sthioul, B. 1998. Temps verbaux et points de vue. In J. Moeschler, J. Jayez, M. Kozlowska, J.-M. Luscher, B. Sthioul and L. de Saussure (eds.), *Le temps des événements*, pp. 197–220. Paris: Kimé.
Strawson, P. F. 1950. On referring. *Mind New Series*, **59**(235): 320–44.
Suchan, J. and Karnath, H.-O. 2011. Spatial orienting by left hemisphere language areas: a relict from the past? *Brain*, **134**(10): 3059–70.
Sweetser, E. 1982. Root and epistemic modals: causality in two worlds. *Proceedings of the 8th Annual Meeting of the Berkeley Linguistics Society*, pp. 484–507.
  1990. *From Etymology to Pragmatics: Metaphorical and Cultural Aspects of Semantic Structure*. Cambridge University Press.
  1996. Mental spaces and the grammar of conditional constructions. In G. Fauconnier and E. Sweetser (eds.), *Space, Worlds and Grammar*, pp. 318–33. Chicago, IL: University of Chicago Press.
Talmy, L. 2000 [1983]. How language structures space. In L. Talmy, 2000, vol. 1, pp. 177–254. [Earlier version in H. L. Pick and L. Acredolo (eds.), *Spatial Orientation: Theory, Research and Application*, pp. 225–82. New York, Plenum Press, 1983.]
  2000 [1985]. Lexicalization patterns. In L. Talmy, 2000, vol. 2, pp. 21–135. [Earlier version in T. Shopen (ed.), *Language Typology and Syntactic Description*, vol. 3, *Grammatical Categories and the Lexicon*, pp. 66–168. Cambridge University Press, 1985.]
  2000 [1988]. Force dynamics in language and cognition. In L. Talmy, 2000, vol. 1, pp. 409–70. [Earlier version in *Cognitive Science*, **2**(1): 49–100.]
  2000. *Toward a Cognitive Semantics*, 2 vols. Cambridge, MA: MIT Press.
Taniwaki, Y. 2005. Achievement verbs and progressive meanings. *Journal of Osaka Sangyo University. Humanities*, **117**: 41–54.
Thom, R. 1970. Topologie et linguistique. In A. Haeflinger and R. Narasimham (eds.), *Essays on Topology and Related Topics*, pp. 226–48. New York: Springer.
Tomasello, M. 1999. *The Cultural Origins of Human Cognition*. Cambridge, MA: Harvard University Press.
  (ed.) 2003. *The New Psychology of Language: Cognitive and Functional Approaches to Language Structure*, vol. 2. Mahwah, NJ: Erlbaum.
  2008. *Origins of Human Communication*. Cambridge, MA: MIT Press.
Traugott, E. C. 1989. On the rise of epistemic meanings in English: an example of subjectification in semantic change. *Language*, **65**(1): 31–55.
Trojano, L., Conson, M., Maffei, R. and Grossi, D. 2006. Categorical and coordinate spatial processing in the imagery domain investigated by rTMS. *Neuropsychologia*, **44**(9): 1567–9.

Tyler, A. and Evans, V. 2001. Reconsidering prepositional polysemy networks: the case of *over*. *Language*, **77**(4): 724–65.
  2003. *The Semantics of English Prepositions: Spatial Scenes, Embodied Meaning and Cognition*. Cambridge University Press.
Ungerleider, L. G. and Mishkin, M. 1982. Two cortical systems. In D. J. Ingle and R. J. W. Mansfield, *Analysis of Visual Behavior*, pp. 549–86. Cambridge, MA: MIT Press.
Vandeloise, C. 1991 [1986]. *Spatial Prepositions: A Case Study from French*. Chicago, IL: University of Chicago Press. [Vandeloise, C. 1986. *L'espace en français: sémantique des prépositions spatiales*. Paris: Seuil (Travaux Linguistiques).]
van der Zee, E. and Slack, J. (eds.) 2003. *Representing Direction in Language and Space*. Oxford University Press.
van Hoek, K. 2003. Pronouns and point of view: cognitive principles of coreference. In M. Tomasello (ed.), *The New Psychology of Language: Cognitive and Functional Approaches to Language Structure*, vol. 2, pp. 169–94. Mahwah, NJ: Erlbaum.
Van Valin, R. D. 2001. *An Introduction to Syntax*. Cambridge University Press.
van Zomeren, A. H., and Brouwer, W. H. 1994. *Clinical Neuropsychology of Attention*. Oxford University Press.
Vendler, Z. 1957. Verbs and times. *Philosophical Review*, **66**(2): 143–60.
Weiss, P. H., Marshall, J. C., Wunderlich, G., Tellmann, L., Halligan, P. W., Freund, H. J., Zilles, K. and Fink, G. R. 2000. Neural consequences of acting in near versus far space: a physiological basis for clinical dissociations. *Brain*, **123**(12): 2531–41.
Werth, P. 1997a. Conditionality as cognitive distance. In A. Athanasiadou and R. Dirven (eds.), *On Conditionals Again*, pp. 243–71. Amsterdam: Benjamins.
  1997b. Remote worlds: the conceptual representation of linguistic *would*. In J. Nuyts and E. Pederson (eds.), *Language and Conceptualisation*, pp. 84–115. Cambridge University Press.
  1999. *Text Worlds: Representing Conceptual Space in Discourse*. London: Longman.
Whiten, A. (ed.). 1991. *Natural Theories of Mind: Evolution, Development and Simulation of Everyday Mindreading*. Oxford: Blackwell.
Widdows, D. 2004. *Geometry and Meaning*. Stanford, CA: CSLI Publications.
Wierzbicka, A. 1988. *The Semantics of Grammar*. Amsterdam: Benjamins.
Wimmer, H. and Perner, J. 1983. Beliefs about beliefs: representation and constraining function of wrong beliefs in young children's understanding of deception. *Cognition*, **13**(1): 103–28.
Winer, G. A., and Cottrell, J. E. 2002. Fundamentally misunderstanding visual perception: adults' belief in visual emissions. *American Psychologist*, **57**(6–7): 417–24.
Winter, S. and Gärdenfors, P. 1995. Linguistic modality in expressions of social power. *Nordic Journal of Linguistics*, **18**(2): 137–65.
Wolff, P. 2007. Representing causation. *Journal of Experimental Psychology*, **136**(1): 82–111.
Wolff, P. and Zettergren, M. 2002. A vector model of causal meaning. In W. D. Gray and C. D. Schunn (eds.), *Proceedings of the 24th Annual Conference of the Cognitive Science Society*, pp. 944–9. Mahwah, NJ: Erlbaum.

Yu, N. 1998. *The Contemporary Theory of Metaphor: A Perspective from Chinese.* Amsterdam: Benjamins.

Zwarts, J. 1997. Vectors as relative positions: a compositional semantics of modified PPs. *Journal of Semantics*, **14**(1): 57–86.

  2003. Vectors across spatial domains: from place to size, orientation, shape, and parts. In van der Zee and Slack (eds.), pp. 39–68.

  2005. Prepositional aspect and the algebra of paths. *Linguistics and Philosophy*, **28**(6): 739–79.

  2010. Forceful prepositions. In Evans and Chilton (eds.), pp. 193–213.

Zwarts, J. and Winter, Y. 2000. Vector space semantics: a model-theoretic account of locative prepositions. *Journal of Logic, Language and Information*, **9**(2): 171–213.

Index

Aktionsart, 6, 105–6, 113; *see also* event types
Amorapanth, P., 293–4
anaphora, 5, 44, 49, 131, 157, 168, 200–1, 205
Arbib, M., 290–1, 304, 307
Asher, N., 5, 200–1
attention, 28, 32–3, 42–3, 50–1, 66, 73, 85–8, 90, 92–3, 95–7, 118, 164, 181, 184, 188, 217–19, 227, 232, 239, 247, 284–5, 292, 295, 300, 313
  joint, 11, 32, 34, 141, 159, 201, 218, 275, 282

Bach, E., 107
Bahr, G. S., 86
Baron-Cohen, S., 205–6
Behrendt, R.-P., 286–7
Bergson, H., 36
blending Theory, 5, 159, 163, 176
Bolinger, D., 214, 223
brain, 9, 72, 206, 283–303, 308–9, 313
  arcuate fasciculus, 296
  Broca's area, 286, 289, 291, 295–6, 299, 305
  extreme capsule, 296
  hippocampus, 27, 286–7, 289, 300–2, 305, 308
  lateralisation, 285–6, 291–2
  left hemisphere, 285–6, 288–9, 292, 296, 299, 301, 305–6
  mirror neurons, 289–91, 297–8, 300, 302, 305, 310
  right hemisphere, 184, 206, 285, 288–90, 293–4, 296–7, 305, 308–9
  Wernicke's area, 286, 296, 299, 305
Brisard, F., 145
Brouwer, W. H., 86
Brown-Schmidt, S, 286
Bühler, K., 8, 48, 141
Burgess, N., 286
Byrne, R., 160

Carlson, L. A., 28
Chomsky, N., 3, 19, 237, 295, 304, 307

Chung, S., 37
cognitive frame, 44–5, 146–7, 201, 275
cognitive grammar, 117–18, 136, 292, 298
cognitive operators
  and as, 204
  if as, 163, 167, 169
cognitive operators, tense forms as, 107, 117, 245
  instancing, 117–24, 126–7, 129, 131–2, 134–42, 149, 219–21, 232, 245–7
  presencing, 116–17, 124–32, 134–5, 137, 140, 142, 148–50, 152, 155, 166, 219–21, 240–3, 245–6, 249, 251, 253
complement clauses, 210–15
  ing constructions, 219, 239–43, 250–5
  it constructions, 222–5
  that constructions, 180–3, 186–200, 215, 254
  to constructions, 225–7, 229–39, 251, 255
  zero constructions, 219–21, 244–9
Comrie, B., 106
conditional sentences, 133, 159–72, 176–7, 262
construction grammar, 298–9
Cottrell, J. E., 85–6
counterfactual conditionals, 158–63, 165, 171, 173–7, 262, 303, 310
Croft, W., 6–7, 51, 73, 106

Dancygier, B., 38, 160–1, 171, 176
dative shift, *see* verbs, ditransitive
de dicto and de re, 45, 168, 201
Deane, P. D., 72
Default Semantics, 144–5
deixis, 5–6, 8–9, 11–12, 17, 19, 27, 29, 32–8, 40–1, 43–4, 47–8, 50, 53, 62, 70, 75, 82, 93–4, 103–5, 107–8, 118–19, 121, 126, 129, 131–5, 137, 139–42, 144–5, 147–8, 150, 152, 154, 156, 159, 163, 165, 168, 176–7, 180–2, 184, 186, 188–9, 194, 201, 217–18, 222–3, 225, 230, 260, 282, 287, 292, 297–8, 308, 311–12, 314

330

Index

demonstratives, 11, 32, 41, 181–2, 289, 293, 314
deontic source, 260, 268, 274–7, 279
Descartes, R., 282
diagrams, 2
Discourse Representation Theory (DRT), 4–5, 200
displacement, *see* vectors, translation
Dixon, R. M. W., 214
Dowty, D. R., 106–7
Duff, M. C., 286
Dummett, M., 119

Euclid, 85, 281, 282
Evans, V., 28, 216
event types; *see also Aktionsart*
  accomplishment, 110, 113–15, 124, 250
  achievement, 110, 115–16, 124, 250
  activity, 112–13, 123–4, 150
  process, 107, 110–16, 124, 136, 138–9, 150
  semelfactive, 111–12, 124
  state, 108–10, 112–14, 119–21, 126–9, 136, 139, 179, 253
eventuality *see Aktionsart*; event type
evidentiality, 227, 245

Fadiga, L., 297
Fauconnier, G., 46, 49, 158–60, 167, 179, 200, 205, 260
figure–ground (foreground, background), 3, 14, 31–3, 40, 51–5, 62, 65–7, 74, 79, 93–4, 96–7, 100, 131, 225, 227, 293, 301, 313
Fillmore, C. J., 19, 146, 211, 283
force dynamics, 13, 72–3, 256–7, 260, 262, 274, 278–9
frames of reference; *see* reference frames
Francez, I., 59
Frawley, W., 12, 37–8, 65, 260–1, 265, 269
Frege, G., 119, 136–7, 295
Friederici, A. D., 295

Gallistel, C. R., 8, 47
Gärdenfors, P., 8, 259–60, 275
Geach, P., 201, 217
Geertz, C., 146
generative grammar, 19, 87, 99, 211, 222, 228, 237, 283, 291, 295–8
geometry, 1, 6–9, 12–13, 16–17, 19–20, 28–9, 40, 48, 76, 104, 117, 119, 121, 146, 149, 156–7, 172, 176, 200, 222, 231, 237, 262, 279, 281, 292, 307, 311
Gestalt psychology, 51
Glenberg, A. M., 297–8
Goldsmith, J., 139

Goodman, N., 158
Griffiths, J. D., 296–7
Grodzinsky, 295
Gruber, J., 87, 89

Heidegger, M., 36, 127
Hickok, G., 295
Hopper, P., 150
Hume, D., 158
Hurford, J., 294–5
Husserl, E., 36

Iachini, T., 285
idealized cognitive model, 146
Inhelder, B., 292

Jackendoff, R., 294
Jaszczolt, K., 35, 42, 144, 152
Jeannerod, M., 8
Johnson, M., 262, 275, 292

Kamp, H., 5, 200
Kant, I., 292, 310
Karnath, H.-O., 288–9
Kaschak, M. P., 297–8
Kemmerer, D., 293
Kenny, A., 107
Koontz-Garboden, A., 59
Kosslyn, S. M., 292

Lakoff, G., 146, 292
Landau, B., 294
Langacker, R. W., 6–7, 17, 19, 37–8, 46, 55, 73, 83, 87, 111, 115, 117, 135, 139–41, 145, 182–3, 187, 214, 216, 219, 221, 223, 225, 227, 237, 245, 249, 257–8, 262, 278, 292, 298
Lascarides, A., 4
Leclercq, M., 86
Levin, B., 211
Levinson, S. C., 9, 16, 19–20, 23, 53, 303
Lewis, D., 158, 161, 177
lexical aspect, 6, 105–7; *see also Aktionsart*; event type
Lonergan, B., 145–6
Lyons, J., 11, 157, 277

Mandler, J. M., 292
McGilchrist, I., 127, 206, 291, 306
McTaggart, J. M. E., 35
mental space theory, 5, 49, 157, 162, 167, 176–7, 179, 200, 260
mental state, 38, 56, 173, 176, 183, 186, 195, 201, 206–7, 209, 215, 230–8, 239, 243, 246, 254–5, 306, 309

Index

metaphor, 3, 5, 11, 13, 15–17, 20, 33, 35, 37, 48, 50, 55–6, 69–70, 80, 82–3, 86–7, 89, 93, 95, 110, 125, 132, 150, 161–2, 210, 218, 238, 242, 248, 257–60, 262, 275, 277–8, 281–2, 301, 307–8, 311–12
Mill, J. S., 158
minimalist syntax, 295
mirror agnosia, 303
mirror transformation, *see* transformations, reflection transformation
modality, 5, 12, 131, 134, 143–5, 147–9, 152, 155–6, 163, 165, 215, 220
   deontic, 38–9, 176, 238, 255–80, 303, 310–11
   epistemic, 12, 14, 16, 36–44, 47, 72, 118–19, 133–77, 179–83, 186–99, 207–9, 215, 218, 221, 223, 225–7, 231–2, 235–6, 239–40, 241, 245–7, 250, 254, 256–63, 265–6, 268–9, 271–2, 277–80, 301, 308–9, 311
Morency, P., 154–5

Nagel, T., 310
navigation, 13, 26, 286, 289, 292, 301–2, 312

O'Keefe, J., 9, 27–8, 286

Paillard, J., 8
Pascal, B., 311
passive construction, 44, 50, 55, 77, 94, 99–103, 296
peripersonal space, 31–2, 34, 40–2, 75–6, 127, 150, 217, 219, 223, 293, 302
peripersonal time, 42, 118–19, 126, 129, 140, 166, 219–20
Petitot, J., 8
Piaget, J., 85, 292
Pinker, S., 3
Plato, 85
Poeppel, D., 295
point of view (perspective, viewpoint), 10, 15, 17–20, 25, 33, 35, 44, 48, 50, 53, 77, 87, 93, 107, 116, 130, 135, 150, 152, 156–7, 172–3, 182, 186, 189, 191, 198, 209–11, 226, 231–2, 239, 241–3, 245, 249, 251–4, 265, 283, 299, 302, 308, 310
pointing gesture, 11, 218, 282
prepositions, 8, 10, 13, 16–26, 29, 47–9, 51, 54–7, 62–5, 73, 75–6, 80, 87–9, 97, 99, 102, 104, 111, 134, 150, 214–19, 230, 237, 289, 292–4, 300–1, 303
presupposition, 38, 46, 62, 65, 76, 115, 130, 152, 161, 165, 173–4, 179, 183, 187, 189–93, 199–201, 209, 233, 245, 252, 258, 261, 268, 272, 275, 277, 309, 311

property predication, 58–60
public reference frames, 145–8
Pulvermüller, F., 297

Quine, W. V. O., 181, 201

Ramachandran, V. S., 303
reaching and grasping, 9, 15, 40–2, 93, 217, 285, 289–90, 292, 298, 302, 305, 311
recursion, 210, 291, 305, 307, 313–14
reference frames, 15–16, 18, 20–3, 28, 30, 35, 44–51, 53, 56, 60–1, 72, 104, 117, 126, 131–3, 135–6, 139, 141–2, 145–9, 151–7, 160, 178–9, 181, 183–7, 189–95, 198–9, 201, 209, 211, 217–18, 225–7, 231–9, 241–3, 245–8, 250, 254–5, 284–7, 289, 291, 294, 298, 300–3, 309, 312
Reichenbach, H., 135, 150
Reyle, U., 4
Ryle, G., 107

Saur, D., 295
Saussure, L. de, 154–5
semantic roles, 85, 90–1, 233, 235, 237, 239, 242–3, 248, 283–4
space
   physical space, 3–4, 9–12, 14, 20, 26, 29, 32, 35, 40, 43, 50–2, 56, 60–1, 63, 73, 87, 106, 145, 157, 217–18, 233, 281, 300–2, 307, 311–12
   spatial concepts and conceptualization, 1, 16, 29, 37, 56, 109, 257, 292–3, 301
   spatial orientation, 4, 10, 14, 17, 27, 55, 284, 287–90, 294, 302, 308, 312
Spinoza, B., 311
Stalnaker, R., 161
state of mind, *see* mental state
Sthioul, B., 154–5
Suchan, J., 288–9
Sweetser, E., 38, 160–1, 257–9, 262, 275, 278

Talmy, L., 6, 13, 19, 38, 65, 72–3, 256–9, 262, 275, 298
telic and atelic, 61–3, 107, 110–16, 123, 217
tense forms, 5, 15, 35, 42–3, 82, 105, 107, 110, 116–17, 132–5, 156, 160–1, 163–5, 168, 170, 172, 243, 268, 308, 310
   conditional, 161
   future, 43, 133, 144, 148–56, 165–7, 171
   past, 82, 114, 123–4, 131–2, 134, 160–1, 165–7, 170, 173, 247
   perfect, 133–4, 154, 160–1, 165, 176, 268–9, 271
   progressive, 36, 43, 88, 107–8, 113–14, 116–17, 120–1, 123–32, 134–7, 140, 142,

148–50, 154–5, 166, 219, 245, 266, 268–9, 271, 293
simple present, 36, 107, 117–29, 134, 136–40, 142, 145–9, 152, 156, 166–7, 169, 220–1, 232, 245–7, 293
theory of mind, 178, 184, 198, 205–9, 309–10
theta criterion, 237
Thom, R., 8
Timberlake, A., 37
time, 33–6, 42, 301, 308
Tomasello, M., 205, 282
transformations, 7–8, 17, 19–20, 23–6, 29, 48–50, 117, 123, 126, 132, 135–6, 156, 159, 179, 209, 225, 269, 282, 286–7, 293, 302–3, 307
  glide reflection, 174
  reflection transformation, xvi, 17, 141, 157, 174–6, 255, 269, 271, 273, 303, 309, 311
  translation transformation, 141, 156, 172, 255
transformations (syntactic), 19, 87, 99, 211, 222, 228, 237
  control (equi NP), 211–15, 221, 233, 235–7, 239–40, 242, 248–51
  raising, 55, 211–14, 221–2, 225, 233, 236–9, 242, 248, 251
Traugott, E. C., 150, 257
Turner, M., 158–9, 205
Tyler, A., 28, 216

van Hoek, K., 17
van Zomeren, A. H., 86
Vandeloise, C., 28
vectors, 3, 7, 10, 12–13, 26–9, 34, 40, 43, 47–8, 50–6, 61–3, 72–3, 77, 85–6, 89–91, 93, 97, 104, 109–27, 135, 137–40, 142, 148–50, 152, 154–5, 168–72, 175, 182, 186, 189–90, 195–6, 199–201, 205, 207, 216–18, 225–7, 232, 235, 237–9, 240, 243, 245–7, 260, 262–5, 268, 270, 272, 279, 282, 286, 292, 299–302, 314
  displacement vector, 60–71, 75, 81, 87, 89, 97, 218
  force vector, 7, 13, 52, 56, 71–3, 75–6, 78, 80–2, 87–9, 91–4, 96–7, 102, 237–9, 247, 262, 275–6, 298
  movement vector, 77, 79, 83, 91–2, 96, 102; see also displacement vector; translation vector
  position vector, 13, 37, 52–60, 67, 70, 75–6, 81, 83, 89–94, 97, 102, 187, 217, 232, 278
  translation vector, 13, 52, 68; see also displacement vector; movement vector
  zero vector, 47, 60, 94, 115, 186, 188, 190, 209
Vendler, Z., 106–7
verbs, 51
  and causation, 71–6, 79–83, 90–1, 96, 99–102, 220–1, 236, 249, 277
  change of state, 68–71, 80–3
  cognising, 91–3
  create, 103–4
  ditransitive, 94–7, 297
  emotion, 89–91
  inchoative, 82–3
  intransitive, 73, 80–3
  motion, 60–8
  perception, 83–9
  possession, 57–60
  resultative, 80–2
  spatial, 55–6
  transfer, 75–9
  transitive, 31, 55–6, 71–104, 112
visual emission, 85–7
visual streams, 284–6, 294–5, 297, 305
  and DST, 300, 305

Whorf hypothesis, 2, 86
Wierzbicka, A., 214, 216, 219, 237, 253
Winer, G. A., 85–6
Winter, S., 259–60
Wittgenstein, L., 15, 86, 275
Woisetschlaeger, E., 139

Zimmermann, P., 86
Zwarts, J., 28, 56